MR Imaging and Spectroscopy of Central Nervous System Infection

MR Imaging and Spectroscopy of Central Nervous System Infection

Edited by

Rakesh K. Gupta

Department of Radiodiagnosis
Sanjay Gandhi Postgraduate Institute of Medical Sciences
Lucknow, India

and

Robert B. Lufkin

Department of Radiological Sciences
University of California, Los Angeles Medical Center
Los Angeles, California

Kluwer Academic / Plenum Publishers
New York, Boston, Dordrecht, London, Moscow

Library of Congress Cataloging-in-Publication Data

MR imaging and spectroscopy of central nervous system infection/edited by Rakesh K. Gupta and Robert B. Lufkin.

p. cm.

Includes bibliographical references and index.

ISBN 0-306-46551-5

I. Central nervous system—Infections—Magnetic resonance imaging. 2. Nuclear magnetic resonance spectroscopy. I. Gupta, Rakesh K., 1955– II. Lufkin, Robert B.

RC359.5 .M7 2001

616.8'047548—dc21

2001023697

The cover illustration (taken by Anil Kumar) is of multiple tuberculous brain abscesses. T2, magnetization transfer MT-T1 and postcontrast MT-T1 MR imaging (central column, in descending order) through the third ventricle show abscesses in the right caudate nucleus, right frontal, and left periventricular regions. The lesions show hyperintense core and peripheral hypointense rim along with perifocal edema. On MT-T1 image these lesions show hypointense core with hyperintense rim. Postcontrast MT-T1 imaging shows enhancement of the peripheral rim. MT ratio from the wall of these abscesses varied form 18.3 to 20.4%. In vivo MR spectroscopy from the large right caudate nucleus lesion (right top) show only lipid and lactate (Lip/L) at 1.3 ppm. Ex vivo high resolution MR spectroscopy confirms the presence of Lip/L with no evidence of choline (right bottom). Histopathology of the wall (left bottom) shows dense mixed inflammatory infiltrate with no epitheloid granuloma. Ziehl–Neelsen stain of the wall shows numerous acid-fast bacilli present in cluster and individually, both in the intracellular and extracellular spaces.

ISBN: 0-306-46551-5

233 Spring Street, New York, N.Y. 10013

http://www.wkap.nl/

10 9 8 7 6 5 4 3 2 1

Printed in the United States of America

To our families

PREFACE

Central nervous system (CNS) infections continue to pose a serious problem in health care even with improved knowledge and treatment. Despite the introduction of newer antimicrobial agents and diagnostic techniques, the morbidity and mortality associated with CNS infections remain high. The morbidity associated with CNS infections may be even more important than the death rate especially in developing countries as neurological sequelae may deprive the survivors of intellect and physical ability, demeaning the quality of life and burdening health resources and social services. MR imaging is an important integral part of the protocol for the management of CNS infections and MR spectroscopy is increasingly being utilized in its management.

This work is an attempt to provide a comprehensive review of imaging and spectroscopy of the commonly encountered CNS infections in the clinical practice in developing and developed countries. The first chapter deals with basic physical principles of MR imaging and spectroscopy that will help beginners to understand the technical terms used in subsequent chapters. The remaining 10 chapters deal with clinical, pathological, MR imaging and spectroscopy features and their applications in CNS infections. This will help in giving a comprehensive understanding to readers with a background in clinical, radiological, basic MRI, and neurological sciences. The T2 hypointense lesions are a real diagnostic dilemma especially in developing countries for which an algorithm has been suggested in the concluding chapter.

Rakesh K. Gupta
Robert B. Lufkin

ACKNOWLEDGMENTS

I wish to thank all of the contributors of this book for giving so much of their time and effort to complete this book. I am very grateful to our students, Rajesh Verma Sanjeev Chawla, and SB Rao for the help they extended toward the preparation of this book, and Mr. Anil Malhotra for preparing the illustrations. I wish to also thank Dr. M. Husain, professor and head, Department of Neurosurgery, K.G. Medical College, and Dr. Lily Paul, Department of Pathology, Sanjay Gandhi Post-Graduate Institute of Medical Sciences, Lucknow, India, for their cooperation in providing the desired imaging and pathology material.

Rakesh K. Gupta

CONTENTS

Chapter 1

MAGNETIC RESONANCE IMAGING AND SPECTROSCOPIC TECHNIQUES

Shantanu Sinha , M. Albert Thomas , Amir S. Huda , and Robert B. Lufkin

The main strength of MR derives from the relatively large dynamic range of gray scale or intensity values between different types of tissues that the contrast mechanisms of this imaging modality generate. The microscopic properties of tissues that produce these differences in intensity values in MR images include proton density (ρ), spin–lattice relaxation time (T1), spin–spin relaxation time (T2), flow, susceptibility, magnetization transfer, and diffusion, among others. Of these, the first three mechanisms were the first to be utilized for clinical MR imaging. All three arise from microscopic magnetic properties of the nuclei, i.e., the protons of the water molecules that abound in human tissue. The MR imaging experiment consists of translating the differences in the value of these properties into differences in intensity values and mapping their spatial dependence, i.e., establishing a one-to-one correspondence between different points in the anatomy with different pixels in the image. Since there are a number of textbooks (1–3) devoted entirely to the physics and/or clinical aspects of MR imaging and spectroscopy, we will not consider the details in

---

Shantanu Sinha, M. Albert Thomas, Amir S. Huda, and Robert B. Lufkin • Department of Radiological Sciences, UCLA School of Medicine, University of California, Los Angeles 90095.

MR Imaging and Spectroscopy of Central Nervous System Infection, edited by Gupta and Lufkin. Kluwer Academic / Plenum Publishers, New York, 2001.

this chapter. Instead, we will describe aspects of MR imaging and spectroscopy in sufficient detail so as to enable the reader to both understand the techniques and tailor their use to the needs of the specific clinical questions pertaining to CNS infections.

The MR imaging experiment on the human anatomy utilizes the magnetic moment of the billions of protons that are part of the water and organic molecules making up a sizable portion of the human body and other biological tissues. In the absence of a magnetic field, these small magnets all have the same energy, very similar to a bunch of billiard balls scattered on the surface of a billiard table. The most noticeable part of an MR scanner is the huge magnet into which the patient is placed. With the patient inside the magnet, this very large number of protons or small magnets in the soft tissue of the body lineup parallel to the magnetic field, since this corresponds to a lower energy state of the spins, and creates a macroscopic magnetization. Stated alternatively, the single energy level that the magnetic spins occupied in free space is now split up into two energy levels, the lower one corresponding to the spins pointing along the main magnetic field B_0, and the upper one to the spins pointing in the opposite direction. At body temperatures, most of the spins occupy the lower energy level, with very few in the upper level. This results in the creation of a net equilibrium magnetic moment M_0, corresponding to the net algebraic sum of all the individual spins of the body and pointing along B_0. Next, a radio-frequency (RF) generator is used to produce electromagnetic (EM) energy at exactly the right frequency, such that the energy of the photons associated with this electromagnetic radiation corresponds exactly to the energy difference between these levels. An RF transmitting coil, integrated within the bore of the magnet, is used to transmit this RF energy and irradiate the body lying inside the bore of the magnet. When the energy of the RF photon and the energy level separation are equal, the resonance condition is achieved; the spins in the lower energy level (aligned along B_0) are excited up to the higher energy level (aligned opposite to B_0). This is very similar to the fact that one will need energy to twist a magnet pointing along a magnetic field to make it point in the opposite direction—except that in the MR experiment we are dealing with the microscopic sub-atomic world of electrons and protons. Conventionally, B_0 is presumed to be along the *Z*-axis and the RF magnetic field component along the *Y*-axis. The macroscopic implication of this excitation of spins is tilting the direction of M_0, the net magnetic moment of the spins, away from the exact direction of B_0. Following the same laws of physics that govern the gyration or precession of a spinning top about the vertical axis, M_0 precesses (i.e., goes round and round) about B_0 at an angular rate, called the Larmor precessional frequency, ω, given by

$$\omega = \gamma * B_0 \quad (1)$$

here γ is the gyromagnetic ratio of the proton, which indicates its magnetic strength, and B_0 is the magnetic field.

Before the RF pulse interacts with the ensemble of spins in the body, the spins, even though they align with the main magnetic field B_0, are said to be out of phase with each other. This situation is similar to the billiard balls lying scattered—even though they have the same potential energy, they are said to be in a disordered state. When the RF pulse, along the *Y*-axis, interacts with the net magnetization aligned along the *Z*-axis, it not only tilts M_0 away from the longitudinal direction but also brings the spins in phase with each other. In terms of the billiard balls analogy, this situation is similar to the balls being bunched together in one place on the table surface. Because the main field is still on, these spins in the *X–Y* plane rotate about the *Z*-axis, producing an alternating electric voltage across a detector coil if the axis of the detector coil is anywhere in the *X–Y* plane. The voltage detected by this coil is called the nuclear magnetic resonance signal and its magnitude is decided by M_0, which in turn is determined by the number of protons per unit volume or proton density ρ, and it fluctuates at the Larmor precessional frequency, ω_o.

After M_0 or a fraction thereof is tilted away from the *Z*-axis, a component, M_{0x}, is developed along the *X*-axis. As noted above, it rotates about the B_0 or *Z*-axis at frequency ω_o. Starting with a state in which the spins are all bunched along the *X*-axis, they start to fan out or dephase. As a consequence of this fanning out, the detected MR signal shows an exponential decay, called the free induction decay (FID), at a rate governed by a factor called T2, the spin–spin or transverse relaxation rate. The time taken for this decay to occur is on the order of 10 to 100 ms. Concurrently, another process takes place in which the excited spins start relaxing back to the *Z* direction. The rate at which M_{0x} relaxes to the B_0 direction is exponential and has a characteristic time, T1, the spin–lattice or longitudinal relaxation rate. This time, like T2, is tissue dependent and ranges from hundreds to thousands of milliseconds in biological systems. The gray scale values on an MR image reflect the magnitude of this detected voltage determined by the proton density and the spin–spin (T2) and spin–lattice (T1) relaxation processes. These factors, the proton density and the different rates of decay, T1 and T2, are tissue dependent and, more importantly, vary between cancerous and normal tissue, thus providing a mechanism for the detection of abnormal tissue even in the absence of contrast agents.

1. PULSE SEQUENCE TECHNIQUES FOR IMAGING

1.1. Conventional Spin-Echo Imaging

Through the history of the development of MR imaging, several methods of spatially encoding the MR signal have been developed. At present, two-dimensional Fourier transform (2D-FT) is the standard method for routine clinical imaging. The most widely used pulse sequence is called a spin-echo sequence (Fig.1). Since this sequence is treated in detail in several textbooks on MRI, only a brief description is given here. With the subject placed within the magnetic field B_0, the first 90° RF pulse (shown as $\pi/2$ on the first line in the Fig.1) is applied to tip the proton spins into the transverse (perpendicular) direction. The RF pulse is switched on in the presence of the slice-select gradient. Magnetic field gradients, which are linear variations of the magnetic field along the three orthogonal axes *X*, *Y* and *Z*, are used to encode the spatial information in the MR signal. They are shown on the third, fourth, and fifth lines of Fig.1 as the slice select, readout, and phase-encoding gradients, respectively.

After being tipped into the *X–Y* plane, as explained in the previous section, the transverse magnetization decays as the T2-decay curve shown on the second line in Fig.1. At a time TE/2, after the first 90° pulse (during this time spins dephase in the *X–Y* plane), a second RF pulse termed π or 180° pulse is applied to flip the spins to the opposite direction in the *X–Y* plane. As a result of this flipping, at an exact time TE/2 after the 180° pulse, the spins coalesce together in the *X–Y* plane to form what is known as the spin echo and is the MR signal from the spins irradiated by the RF. The slice select gradient is switched on during both the 90° and the 180° pulses and is instrumental in selecting a slab of protons in that particular direction. Similarly, the phase encoding gradient is switched along one of the two remaining axes of the MR image and encodes along that direction by means of phase. Finally, the last remaining axis is encoded by means of the readout gradient that is switched on during the collection of the spin-echo MR signal. Once the MR signal is collected, it is made to undergo two successive mathematical operations termed Fourier transformation, which transforms the signal from the time domain to a frequency domain. This mathematical operation is widely used whenever it is convenient to collect data in one form but easier to visualize and analyze in another. To cite other examples: Our visual perception is unable to Fourier transform or decompose white light into its various component frequencies; one requires either a prism or the mist in the atmosphere after rain to break up (i.e., Fourier transform) the white light into its rainbow (frequency) components. In contrast, the auditory sensory function of, in particular, trained musicians is able to perform this operation of Fourier transformation when it distinguishes the separate notes in the complex sounds emanating from a musical instrument. In MR imaging, the Fourier transformation decomposes the free induction decay or the decay of the MR signal with time detected by the receiver coil into its separate frequency components, so that a unique assignment with respect to spatial coordinates is made possible.

The raw data collected at one level of phase encoding result in a single line in *k*-space, along the K_x axis. After the collection of one line of data, the spin system is allowed to relax for a repetition time TR, before being subjected to the next cycle of RF and gradient pulses shown in Fig.1. This cycle is repeated 128 or 256 times. In each time TR, the phase-encoding level is incremented (see dashed lines within the phase-encoding gradient lobe in Fig.1) and the *k*-space is incremented by another line. Hence, the total scan time is 256 * TR. After all 256 lines of data are collected, 2D-FT is performed to yield the final MR image.

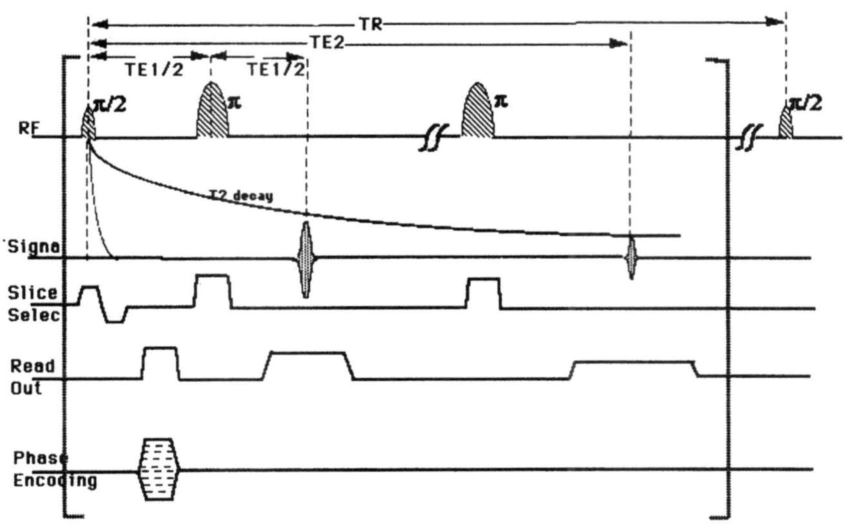

Figure 1. Pulse sequence diagram of a conventional spin-echo. The first line shows the radio-frequency pulses, with the first 90° and 180° separated by TE/2. It also shows an extra 180° associated with double spin-echo sequences, as well as the 90° pulse of the next TR. The next line shows the MR spin-echo signals, with a separation of TE/2 of the first from the 180°, and a second echo formed by the second 180° at TE2. The next three lines show the slice select, readout, and phase-encoding gradients. It should be noted that in this conventional spin-echo, one phase-encoding line is used for each TR, with the PE level being incremented for each of the next 256 levels.

Extrinsic parameters of the sequence, TE and TR, and intrinsic parameters such as ρ, T1, and T2 values of the protons in that pixel determine the pixel intensity in the spin-echo image. As explained below, each of these different intrinsic tissue contrast properties can be emphasized in the image by choosing different combinations of TE and TR. Other aspects of the image such as slice thickness, field of view, angle of orientation, resolution of the image, and others can be varied by changing different aspects of this pulse sequence such as the strength of the respective gradients, their timing, the type of RF pulse, and so on. In fact, one of the major strengths of MR imaging is the flexibility afforded in varying the geometric and contrast parameters of the images according to the clinical needs. In the following, we describe various modifications of this basic imaging process that increase the power of this imaging technique.

1.2. Gradient Echo Imaging

While conventional spin echo is still considered a reference standard for image quality, its main drawback is the relatively long total acquisition time. This is particularly true for the diagnostically importantT2-weighted scans, whose scan time can be as long as ~ 8 min [2000 ms (TR) * 256 (phase-encoding levels)]. Within this time, one can image (i.e., excite and acquire data from) a large number of [= TR /pulse sequence length = 2000 ms/100 ms ~ 20] slices necessary to cover almost the entire volume of interest, by the technique of interleaving slices. There is therefore a considerable impetus to attempt to reduce this total scan time, based on increasing throughput, decreasing patient comfort, and decreasing artifacts from gross and/or physiologic motion. The considerable degree of motion arising from physiologic processes of the body, such as respiration, cardiac, CSF pulsation, and peristalsis, gives rise to artifacts within the image, emphasized in sequences with long acquisition times.

One of the first results in this direction was the development of the field echo or the gradient echo sequence. The main reason for the long acquisition time of spin echoes is the length of the repetition time. TR needs to be that long (order of seconds) because the spins are tipped all the way to the transverse plane by the 90° RF pulse. It takes them a time on the order of the spin–lattice relaxation time T1 (i.e., ~2000 ms) to relax back to the longitudinal direction before they can be subjected to the next 90° pulse for the next phase-encoding level. The strategy in gradient echo imaging is not to tip the spins all the way to the transverse plane by 90°, but rather only by a fraction thereof, e.g., flip angle, θ ~ 30° (Fig.2). This still allows the projection of the net magnetization (i.e., M_0 * cos 30°), to be detected as the MR signal by the detector coil and processed to form the image. It takes only a fraction of the original TR for this relatively small transverse projection of the magnetization to relax back to the longitudinal direction before M_0 can be flipped by the next θ° pulse for the next phase-encoding level. A direct consequence of this strategy is that the total acquisition time is significantly reduced, since TR can be reduced from ~ 2000 ms to ~20 ms. Very importantly, the echo in these gradient echo pulse sequences is formed not by a 180° RF pulse which is responsible for the formation of the spin-echo. Rather, the projection of M_0 on the transverse plane is deliberately dephased by extra gradients in the readout direction, and refocused back to form a "gradient echo" by another gradient lobe of an equal but opposite magnitude. There are several important facets of these gradient echo images that are clinically relevant:

1. The height of the echo reflects T2* decay rather than T2 effects as in a spin echo. This new contrast mechanism differs from the conventional T2 in that it has contributions from extrinsic field inhomogeneity (ΔB_0) effects that arise from a variety of sources.

2. These field inhomogeneity effects in a spin-echo sequence are reversed by the 180° pulse, but persist in the gradient echo sequence because the echo in this sequence is created by the field gradients. This sensitivity to T2* rather than T2 in gradient echoes results in a faster signal intensity decay which results in a decrease in image quality compared to spin-echo images. To this day, in spite of great strides in hardware and magnet technology, this deterioration in image quality of gradient-echo images makes the spin echo the sequence of choice for routine clinical work. The absence of the second 180° pulse in gradient echoes allows a reduction in the pulse length of the sequence. One should remember, however, that it is not this reduction of the pulse sequence length but the drastic reduction in TR that is the primary reason for the reduction of the acquisition time.

3. Gradient-echo imaging technique nevertheless is extremely important at present, not so much for routine morphological imaging, as much as in niche applications in a variety of imaging techniques. Novel techniques such as 3D, cardiac gated, blood oxygen level dependent (BOLD) imaging, MR angiography, and others rely on gradient-echo imaging extensively.

4. The contrast reflected in gradient-echo images depends not only on the echo time TE and the repetition time TR as in spin echo, but also on a third parameter, the angle by which the magnetization vector is tipped or the flip angle θ. The resulting contrast has a relatively complicated dependence on these three parameters, dealt with more extensively in the section on contrast mechanisms.

5. The contrast dependence is made even more complicated by the advent of a whole host of modifications to the original gradient echo, such RF spoiling, gradient spoiler pulses, steady-state free precession, and others.

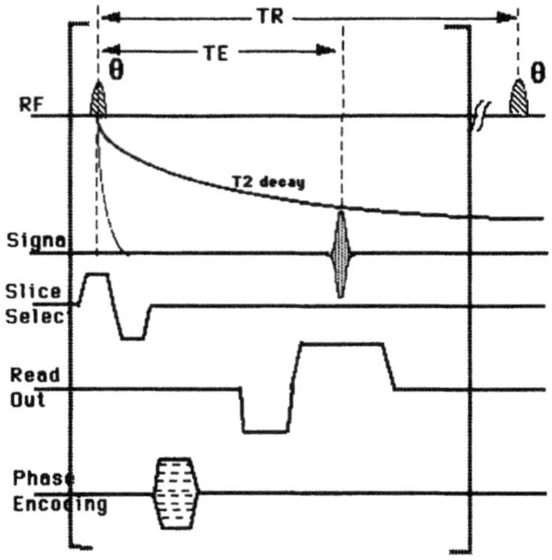

Figure 2. Pulse sequence diagram for a gradient echo. Each line has the same connotation as those in Fig.1. It should be noted that in contrast to the spin-echo, (i) the RF pulse is a fraction of 90°, i.e., a θ° pulse; (ii) there is no 180° pulse; and (iii) the echo is formed by an extra amount of gradient in the readout direction.

1.3. Fast-Spin-Echo Imaging

The method that has turned out to be the most effective in reducing the total scan time without compromising diagnostic quality of the image is fast-spin-echo (FSE) or turbo-spin-echo imaging. First suggested by Hennig et al (4), it needed significant improvement in the hardware performance characteristics of the scanner, particularly the gradient power supply and the gradient coils for routine clinical implementation. The conventional spinecho steps through one phase-encoding gradient level for each repetition time TR (Fig.1). The FSE method, on the other hand, traverses through several phase-encoding levels within each TR, collecting several echoes, as shown in Fig.3. In other words, in each TR, about 16 to 32 *k*-space lines (7 as shown in the figure, depending on the gradient performance characteristics of the scanner) are traversed. The total scan time is thereby reduced by a factor of the number of echoes collected in each TR, termed the *echo train length* (ETL). Thus, with respect to the example given above, if the conventional SE takes 8 minutes of scan time, an FSE with an ETL of 32 will take 8/32 minutes or a mere 15 seconds. Of course, the implementation of this scheme is associated with a great degree of sequence design complication (Fig.3), notably the necessity of applying a gradient of opposite polarity to the

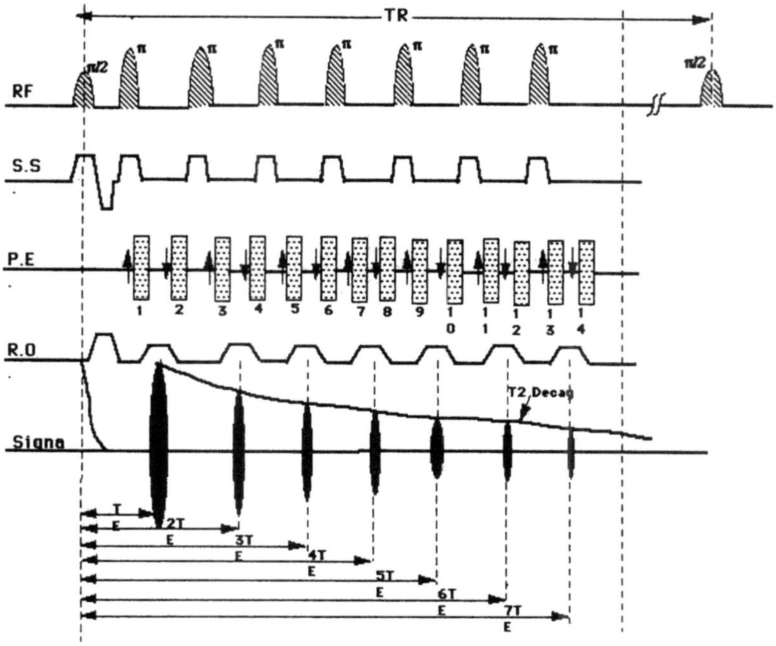

Figure 3. Pulse sequence diagram of a fast (or turbo-) spin echo. The first line shows a train of 180° pulses (7 shown here) following the 90° pulse, within one TR. This allows the collection of a corresponding number of echoes, shown in the last line, all within one FID curve. Each echo is individually phase (and readout) encoded, shown in the third and fourth lines, respectively. Each phase-encoding pulse has to be annulled before the next 180° pulse, shown by downward arrows as lobes 2, 4, 6, ... The number of echoes sampled in one TR can vary depending on the scanner and on the application, all the way up to 128 for the entire matrix. This is known as the single-shot FSE or HASTE.

phase-encoding gradient after the acquisition of the echo (pulses 2, 4, 6, ... in the third line of Fig.3), to annul the effects of that level before the next PE level of that echo train is applied. The requirement of rapidly switching the gradients in all three directions imposes considerable strain on the power supplies—one of the main reasons why its clinical implementation had to wait some time. The significant reduction of total scan time is generally used toward enhancing the image quality. Typically, longer TRs and TEs are used to get better T2-weighting, or a larger number of thinner slices are acquired. This is particularly important for imaging CNS lesions, since partial volume errors can be minimized. FSEs with multiple echoes are also available on most commercial scanners, as are FSEs with inversion recovery preparation. A special case of the latter, Fluid Attenuated Inversion Recovery (FLAIR), is described later. A few subtleties that accompany imaging with FSEs should be noted. The FSE technique is somewhat limited for short TEs which is necessary for both T1- and proton density weighted imaging. Since several echoes are collected from the same FID, the first echo is utilized to measure the lowspatial frequency of the K_y (phase-encoding) space, with later echoes sampling the higher frequencies. This is called centric reordering of the *k*-space, with the central lines being collected first. The result is a slight T2-weighting as a function of object size. Thus, in clinical situations, it should be noted that FSEs are ideally suited for T2-weighted imaging.

While it is true that T2-weighted images from conventional spin echoes and fast spin echoes are similar, there are several points of departure.

1. One factor that influences image intensity in FSE is incidental magnetization transfer contrast. This contrast, explained in detail later, arises from the large number of RF pulses used in this sequence (the first line in Fig. 3). Pulses that excite and refocus adjacent slices act as off-resonance pulses for saturating macromolecular protons, leading to magnetization transfer contrast (5). Compared to FSE, the much smaller number of RF pulses in spin echo results in negligible magnetization transfer contrast.
2. Fat generally appears bright onT2-weighted FSE as opposed to dark in conventional SE images, where the presence of *J*-coupling of fat-bounded hydrogen leads to a more rapid decay of the FID and a consequent dark appearance. The closely packed 180° pulses in FSE imaging break this *J*-coupling, leading to a confounding hyperintense appearance of fat (6).
3. An important point for imaging CNS lesions is that the signal void often observed in the conventional SE images proximal to hemorrhagic blood products is less conspicuous in FSE images because the closely packed RF refocusing pulses cause decreased sensitivity to susceptibility differences (7).

1.4. Echo Planar Imaging

Echo planar imaging (EPI) extrapolates the concept of FSE imaging to the case where, instead of traversing typically only 8 to 32 phase-encoding lines in *k*-space in one TR, the technique collects the entire matrix in one TR (Fig. 4). In other words, using a single excitation (shown as the magnetization preparation period in Fig.4) and multiple phase-encoding gradient echoes, all 256 or 128 lines are collected in one FID, allowing total acquisition times of less than 100 ms. A 90°–180° pulse pair is used to generate a spin echo, but here the similarity with spin echo imaging ends. The subsecond acquisition is enabled by a series of ultrafast oscillations of the frequency encoding gradient (fourth line in Fig.4) that generates a train of gradient echoes centered around the spin echo from the 90°–180° excitation. The phase encoding can be incorporated by either of the two schemes shown on the bottom two lines. As suggested originally by Mansfield (8), phase encoding can be applied by a constant gradient switched on in that direction; this results in an angular traversal of the *k*-space, but is not as demanding on the gradients as the next scheme. In the second method, a series of blips of phase encoding is applied for each echo, resulting in a more tractable, horizontal traversal of the *k*-space, but with a concomitant degree of extraordinary strain on the gradient system. The preparation for an EPI sequence can be either a spinecho (a 90°–180° RF pair) or a gradient echo ($\theta < 90°$) sequence.

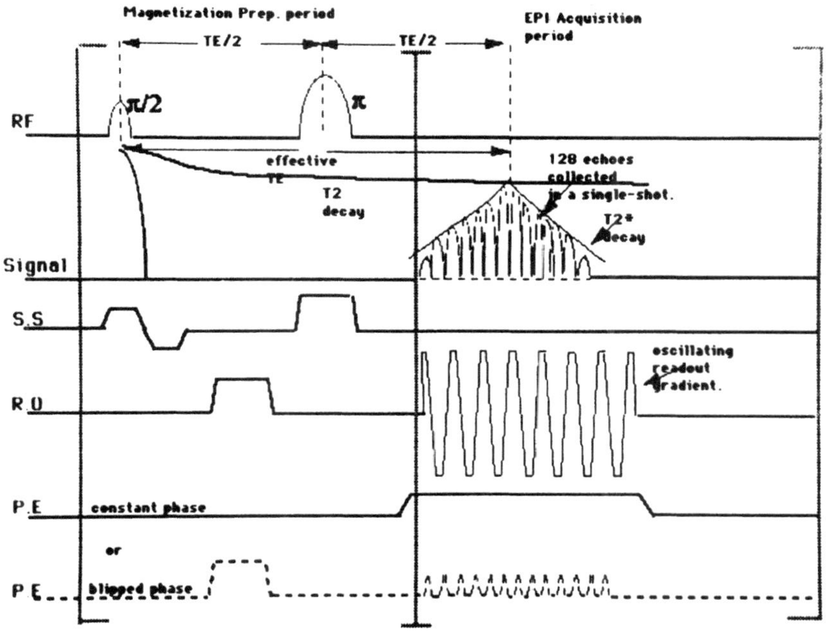

Figure 4. Echo planar imaging. All 128 echoes are collected within one FID or under the envelope of one spinecho (line 2). Different types of preparation can precede this detection of signal. Shown on the first line is a conventional spinecho, creating a spin-echo EPI. This preparation can be substituted by other types, such as gradient echo or diffusion weighted or inversion recovery. The echoes are collected on top of rapidly oscillating readout gradients (line 4). Phase encoding can be implemented by either of two methods shown on the last two lines: either a constant encoding (less strain on the gradients) or blipped (last line, less prone to artifacts). The entire acquisition can be less than 100 ms, with a TR of seconds

In Fig.5, the first three images show different contrasts. Panel a shows proton density, panel b, T1 contrast obtained using an inversion recovery preparation, and panel c, T2 contrast. All were obtained using spin echo preparation. The image quality of SE-EPI is superior to that of GE-EPI (panel i), but the latter is often used for T2* weighting as is required in BOLD imaging for functional MR imaging. The last six images of Fig.5 illustrate the use of EPI images for such fMRI experiments. Panel d shows the typical position of slices required for visual stimulation fMRI experiments, while the next two images (e, f) show good-qualityT2-weighted SE-EPI images obtained at two of those locations in less than a second each. Panels g–i show similar images obtained in auditory stimulation fMRI experiment. Panel g shows the location of the oblique slice necessary to visualize the third sulci. Such an oblique sagittal slice, however, encounters the air spaces of the inner ear and results in characteristic susceptibility artifacts shown in the SE-EPI image in panel h. The corresponding GE-EPI image had a much larger prohibitive extent of artifact. Panel i is a GE-EPI image. These are the best suited for visualizing T2* contrast but are relatively poorer in image quality. The penalty for the rapid image acquisition is in image quality and signal-to-noise. Multishot variations of the EPI technique have also been proposed to reduce susceptibility-induced artifacts. While EPI imaging may not be ideal for routine diagnostic scanning, it can be used for a variety of applications with the proper preparation. The preparation can be inversion recovery, which with centric reordering allows T1-weighting. The preparation can also be made diffusion-sensitive to yield diffusion-weighted images as explained in a later section. Some of the typical characteristics of EPI images are illustrated and discussed in the section on diffusion imaging. Rapid subsecond imaging has also made possible the use of EPI for perfusion imaging.

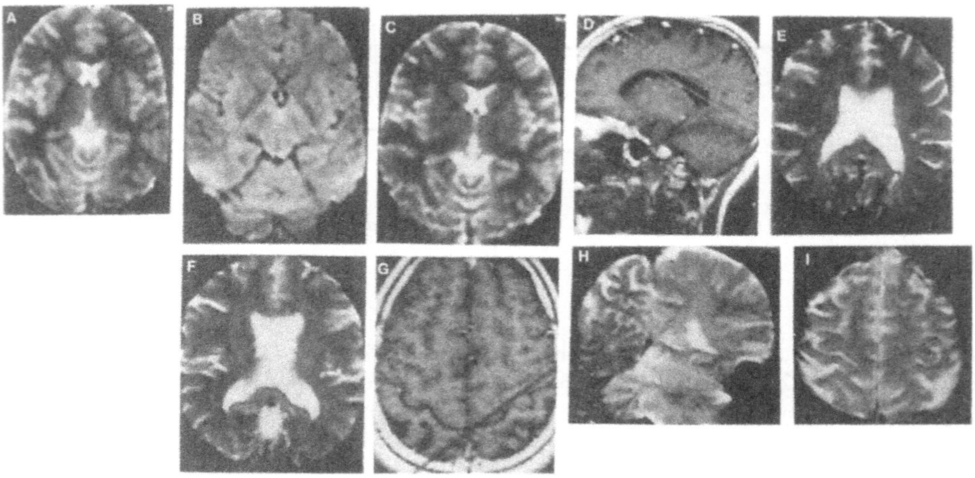

Figure 5. Images obtained using the echoplanar technique. (a) Proton density weighted, (b)T1-weighted, (c)T2-weighted. All were obtained using the SE-EPI mode, with panel b being obtained using inversion recovery preparation. Panel d shows the location of slices for imaging visual stimulation during an fMRI experiment. Panels e and f show two of those slices, obtained using SE-EPI technique; note the good quality and excellent T2-weighting of the images acquired in less than a second each. The bottom row shows similar images for an auditory stimuli activation fMRI experiment. Panel g shows the slice location while panel h shows the susceptibility artifact characteristic of EPI images even when collected in SE mode. The last image panel i shows a GE-EPI image; it has excellent T2* weighting suitable for visualizing BOLD effect, but has lower SNR and quality than the corresponding SE image.

2. CONTRAST MECHANISMS AND IMAGING STRATEGIES

2.1. Proton Density (ρ)

The overall strength of the signal is decided by the proton density (PD or ρ) that makes up the macroscopic magnetization. The proton density relevant here is the number of mobile protons per unit volume in contrast to the total number of protons (including stationary protons). Typically, for example, protons that belong to the organic matter in dense bone do not contribute to the signal that comprises the MR imaging experiment. They produce an MR signal, except that its characteristics are such that it is not relevant to the MR imaging process—more specifically, the signal from protons belonging to solids decays at too fast a rate (~ tens of microseconds) compared to mobile protons (hundreds of milliseconds) for any meaningful data to be collected. For this reason, solids, with an absence of mobile protons, appear dark in an MR image. The dearth of signal, however, does not render such information useless—a lot can be surmised from the appearance of the anatomy surrounding the solid structure. Typical structures that appear dark due to a lack of protons are cortical bone, sinus passages, lung parenchyma, and calcium deposits. Changes in proton density occur as a result of changes in the cellular composition in tissues that accompanies disease and will be reflected in the signal intensity. Nevertheless, proton density weighted images are inherently low contrast, with ρ in the brain varying from 0.71 in white matter, to 0.85 in gray matter, to 1.00 in CSF (9). This implies a maximum possible elevation in intensity from enhancement in ρ in white matter of about 41%, mitigated by partial volume effects and contamination from T1 and T2 to reduce the PD contrast further. In fact, if the contrast mechanism of MRI were to rely solely on ρ, MRI would not be much more superior an imaging modality than the more conventional modalities such as CT and ultrasound.

Proton density contrast can be obtained from conventional SE or FSE (explained later) by a combination of short TE and long TR. The physics behind this choice of combination of pulse sequence parameters can be briefly explained as follows. The long repetition time TR, before the spins are hit again by the next 90° pulse, allows all (or most) of the differences in signal intensity arising, from T1 differences in the tissues to be minimized. Only differences in proton-density between different tissues then decide the initial value of the magnetization to be flipped by the next 90°. Similarly, collection of the echo at a short value of TE allows only a minimal amount of T2 effects to be admixed into the echo that is collected. A simple calculation reveals that TR needs to be 3*T1 for contamination from T1 effects to be reduced to less than 5%, while TE needs to be 0.05 * T2 for T2 effects to be reduced by a similar amount. These values (3 s for TR and 4 ms for TE) are generally prohibitive in routine scanning. Instead, PD- weighted images are generally acquired as part of a double-echo sequence, in which the first echo, collected at short TE (~20 ms) and long TR (~2000 ms), yields PD-weighted images at the same time as the acquisition of theT2-weighted images at the same long TR but at a much longer TE (~80 to 120 ms). The image characteristics of these proton density images are illustrated a little later on, for better comparison with images with predominant contrast such as T1, T2, and others.

2.2. Spin–Spin Relaxation Time (T2)

As already explained, before the RF pulse is switched on, the individual magnetic spins (protons) are on the average aligned with the main magnetic field B_0, and in random distribution of phase with respect to each other. The RF pulse not only tips the magnetization away from the *Z*-axis but also puts them in phase coherence with each other. This coherent magnetization tipped into the transverse or *X–Y* plane, precesses about the B_0 (*Z*-axis), inducing the MR signal at the Larmor precessional frequency. The process starts with all of the spins bunched together, i.e., in phase with each other. Very soon, however, the phase coherence disappears. The latter occurs because the spins interact with each other and thereby are exposed to static as well as slowly varying local magnetic field fluctuations. These fluctuations cause a spread of the magnetic field around the central value of B_0, causing each to be exposed to a slightly different magnetic field, which in turn causes them to precess at slightly different angular frequencies, some faster than ω_0, others slower. The net effect is that the ordered phase-coherent state slowly disappears, with the spins fanning out in the *X–Y* plane. This causes the induced MR signal to decay at a rate governed by a factor called T2, i.e., the spin–spin relaxation time, since it is caused primarily by intrinsic fluctuations in magnetic field arising from interactions between adjacent spins, or alternatively, the transverse relaxation rate, since it takes place in the transverse plane. In this process of dephasing, it should be understood that there is no exchange of energy, only loss of coherence. The process is governed by the Second Law of Thermodynamics, namely, that any system starting in an ordered state tends naturally to go into a disordered state. The spin–spin relaxation time T2 arises primarily from static or slowly varying magnetic field fluctuations and hence does not depend on the magnetic field.

Molecules or ions in a solid tend to "sway" at a relatively slow rate, thereby generating a relatively large component of magnetic field fluctuations at low frequencies relevant to T2 relaxation. This results in solids having very short T2 values. This fact is dealt with in some detail in Section 2.8. Large molecules such as those of protein tend to behave the same way and consequently have short T2 values. In contrast, smaller molecules such as water, at least at body temperature, have very rapid Brownian motion that averages out any field fluctuations, or produces high-frequency components. The result is that water has very large T2 values, which implies that once a transverse spin–spin coherence is built up in an ensemble of spins, this coherence takes a long time to dephase—water and fluids such as CSF and edema have long T2s.

Another interesting consideration regards spin–spin interaction between protons in tissue that is mediated by the presence of minute amounts of paramagnetic substance. These are atomic particles having a relatively large amount of magnetic moment of their own, and occur among others in iron deposits in the liver, in deoxyhemoglobin, and in ferritin. In each of these occurrences, the iron ions facilitate the spin–spin interaction, cause the coherent proton spins to "talk" to each other and thereby lose their spin coherence much more rapidly, i.e., have very short T2s.

The differences between the T2 relaxation times between different tissues can be emphasized in what is known as aT2-weighted image by choosing a long repetition time, TR (~2000 to 4000 ms), and a long echo time, TE (~80 to 120 ms). The rationale is as follows: The long TR ensures that all the contrast between different tissues generated by differences between their T1 values have had time to minimize and the individual tissue species have relaxed along the longitudinal direction to their equilibrated magnetization values. Subsequently, after being flipped by the 90° pulse, detection at long echo times ensures that the intensity differences generated by differences in T2 relaxation times have maximized. Therefore, the longer the TR and TE, the greater is the T2 weighting of the image, with minimal admixture of T1 and proton density weighting. The downsides are that the longer the TR, the longer the total acquisition time, and the longer the TE, the smaller the signal-to-noise ratio (SNR) of the image. To circumvent the longer scan times at long TRs, FSE methods have become the mainstay of clinical T2 imaging. With improvements in hardware such as better receiving coils (quadrature and phased-array coils), the SNR of scanners are improving, and longer TEs (> 80 ms) are also becoming routine. EPI also yields excellent T2 weighting but artifacts related to image blurring (susceptibility artifacts) have made them less useful for routine diagnostic purposes. Similar problems are associated with T2 weighting using gradient echoes. Special sequences with steady state free precession (SSFP) have been designed specifically for T2 weighting but the images produced are susceptible to motion and susceptibility artifacts. For routine diagnostic scanning therefore, FSE imaging with TRs of about 2000 to 4000 ms and TEs of around 120 ms are the optimal solution.

2.3. Spin–Lattice Relaxation Time (T1)

The T1 relaxation time is a measure of the time taken by a proton, after being excited to the upper level, to come down to the lower ground state. This process involves an exchange of energy, with the excited spin exchanging energy with the surrounding molecular environment, termed *lattice* during the early stages of NMR. Since it is connected with the recovery of the equilibrium longitudinal magnetization, it is called *longitudinal relaxation* or spin–lattice relaxation. The relaxation of the excited spin requires an interaction with a time-dependent perturbation, typically a dipole–dipole interaction with a fluctuating electromagnetic field. Such perturbations or fluctuating field are produced by the tumbling and translation, inter- and intra molecular motion of the ions and molecules around the excited nuclei. The efficacy of the relaxation process depends closely on the correspondence between the energy associated with the Larmor precessional frequency of the excited spin and the energy of the tumbling motion of the surrounding molecules. Thus, water molecules which are relatively small in size tend to tumble at too rapid a rate to assist in effective relaxation and as a result have long relaxation times. At the other end are very large molecules such as those of proteins, which have too slow a motion to match the Larmor precessional frequency. Mid-sized molecules or water molecules whose motion is diminished by hydration bonds to larger molecules have the right correspondence and thereby short T1 relaxation times, appearing hyperintense on T1-weighted images.

In order to emphasize the differences in T1 values between different tissues being imaged, in other words to acquire a T1-weighted image, one needs to use a short TR (~300 to 500 ms) with a short TE (~20 ms) in the spin-echo sequence. The reason behind this choice of combination is as follows. Once the different species of spins belonging to different tissues, which are initially tipped to the *X–Y* transverse plane, start relaxing back to the longitudinal direction of the B_0 field, they do so at the rate of their individual T1s. If one waits too long, all of the different species would have relaxed to their equilibrium values, with no differences characteristic of their different T1 values registered. If they are instead "caught" at a time short compared to their T1 values, the extent to which they are relaxed back to the longitudinal direction would be different to the extent of the variation in their T1 values. Thus, if they are hit by the next 90° RF pulse at this stage, the initial heights of the subsequent FIDs or of the spin–echoes for the different species would be different and dependent on their T1s. Next, in the detection part of the pulse sequence, if one were to wait a relatively long time, such as 80 ms, the initial contrast created by the short TR and determined by T1 differences would disappear and be replaced by contrast created by T2 effects. Hence, in order to detect the initial contrast, one needs to use as short a TE as possible. Of course, the lowest minimum is decided by the hardware capabilities of the scanner, but 20 ms is a typical value.

The efficacy of relaxation can also be enhanced if paramagnetic ions bearing electronic charge or magnetic moments are artificially introduced into the tissue. This comprises the use of contrast agents. Gadolinium (Gd) ions are paramagnetic ions, with a large magnetic moment. They can be safely chemically bonded to inert chelates such as DTPA and introduced into the vascular system by intravenous injections. Gd has been used extensively throughout the world for about 15 years and has been found to be completely safe, excreted out in less than an hour with no residual presence. Other nonionic forms of Gd are also in trial. The contrast agent molecules of Gd-DTPA travel from within the vascular system to the extracellular space in regions that are compromised by disease and lower the T1 value of that region. In a T1-weighted postcontrast image, these regions are visualized by their conspicuity as hyperintense zones. In comparison to the other types of MR images such as the precontrast T1 or theT2-weighted images, the contrast-enhanced images are able to distinguish the lesions to the highest degree of conspicuity. This facility of selectively enhancing CNS and other lesions within the human body greatly enhances the strength of MR as an imaging technique. Images with different contrasts are discussed together in Section 2.6.

2.4. Inversion Recovery (IR)

The inversion recovery technique is an alternative method to the SET1-weighted sequence described in the previous section. The short-TE–short-TR sequence maps only a small fraction of the T1 differences in tissues to signal intensity differences. A more effective approach to mapping the T1 differences is the IR technique (Fig.6). In this pulse sequence, the typical SE sequence is preceded by a 180° pulse that inverts the protons in the slice of interest to the $-Z$ direction. This is followed by an evolution period, TI, called the inversion time, during which the inverted spins relax back along the Z-axis to the positive Z-axis at a rate governed by the T1s of the different tissues. After waiting for this IR period of time TI, during which the initial $-M_0$ magnetization would have relaxed back to $M_0*\{1 - \exp(-TI/TR)\}$, the magnitude of magnetization is sampled by a spin-echo pulse sequence exactly as described in Section 2.1. The only difference is that T1 and TI now govern the initial magnetization. Generally, the spin-echo sequence following the IR preparation has a short TE and a long TR in order to suppress T2 and obtain T1 weighting of the tissues.

The T1 weighting obtained by the IR method yields much higher contrast than the equivalentT1-weighted SE image. The only drawback of this technique is the long TR, resulting in relatively long scan times, and associated motion artifacts. To a large extent, this has been circumvented nowadays with the advent of the FSE, which can also be preceded by an IR preparation. In other words, one can get the excellent T1 weighting of IR sequences in much shorter times, by sampling several k-space phase-encoding levels in one TR.

The IR sequence has several other rather interesting and important clinical applications. The first is suppression of the fat signal. Because of the relatively short T1s of fatty tissue, they tend to register with high intensity in T1-weighted sequences. Combined with motion artifact in, for example, epidural fat, the propagation of the motion artifact can often obscure relevant pathology. It is possible to suppress the signal from fat by choosing a value of TI when the fat signal goes through zero. The magnetization from the other tissues would be nonzero since they have different values of T1. This method of fat suppression is called Short TI Inversion Recovery (STIR). The advantage of this method over the alternative of chemical shift saturation technique is that the latter is critically dependent on field inhomogeneities. The other application of the IR sequence is described in the next section.

2.5. Fluid-Attenuated Inversion Recovery (FLAIR)

Some particular types of CNS lesions such as those occurring at the cortical and subcortical regions and periventricular ones at the interface between brain and CSF are often difficult to visualize using typical SE T2 imaging because of the lack of contrast between the high intensity CSF and similar hyperintense cortex. FLuid-Attenuated Inversion Recovery (FLAIR) imaging, as its name implies, nulls or attenuates the bright signal from CSF while maintaining the T2 contrast of the rest of the surrounding anatomy by combining an IR preparation with a spin-echo T2 detection (10–13). This attenuation permits visualization of lesions adjacent to CSF. The basic design of the pulse sequence is

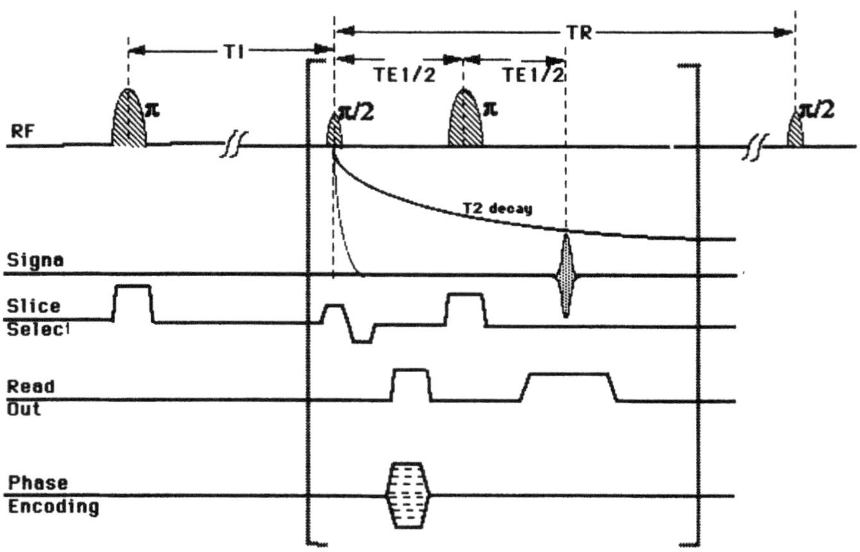

Figure 6. Inversion recovery sequence. A conventional SE sequence is preceded by a preparation period. The sequence is started with a slice selective 180° pulse that inverts the magnetization to the negative *Z*-direction. The inverted magnetization relaxes back to the positive *Z*-direction, to an extent determined by both the individual T1s of the different tissues as well as the evolution period, TI, called the *inversion time*. The detection is carried out by a short-TE/long-TR SE sequence. Presently, this detection is substituted by a fast spin echo, resulting in significant savings in scan times.

the same as that of the IR sequence (Fig. 6). In the preparation part of the sequence the fluid signal is nulled by choosing the correct value of TI (~1900 ms), such that the IR curve of the CSF protons is going through its null point. At that point of time, all of the protons are subjected to a conventional spin echo sequence with a choice of TR and TE (~6000 to 8000 ms for TR and ~120 ms for TE) that provides a high sensitivity to T2 differences between the tissues. It was obvious that even though the original implementation of this sequence with spin echo yielded remarkable results with almost complete fluid attenuation (10), the long values of TI and TR would render the acquisition times almost prohibitive. At present, therefore, FLAIR is almost exclusively used in combination with FSE mode of acquisition or FAST-FLAIR (13), in which for every repetition time TR, several phase-encoding levels are collected. Generally, echo train length (ETL) of 16 echoes (or larger, depending on the capabilities of the scanner) is utilized for routine clinical imaging. Significant improvements in lesion detectability, conspicuity, and lesion-to-CSF contrast have been shown (13).

2.6. Comparison of Images from Different Sequences

For a better appreciation of image quality and contrast obtained with different sequences and imaging schemes, a sample set of images is presented in this section. This set is by no means exhaustive but rather is chosen to highlight the strengths and weaknesses of different clinically relevant sequences.

The first of these is a series of images (Fig.7) obtained using gradient-echo imaging techniques in a normal subject. The first notable aspect of these acquisitions is that they are fast compared to standard spin echoes. Each scan took about 35s and yielded three slices, as opposed to SE acquisition times of 4 to 8 min. The next aspect of all of these images is that they look somewhat grainy (reflecting low SNR), with Fig.7b the lowest SNR and Fig.7a the best. The reason becomes obvious when one considers the pulse sequence parameters. All were acquired with a type of gradient echo called Fast Low Angle shot (FLASH). The TE, TR and flip angle for images a, b, and c of Fig.7 are 5 ms/200 ms/45°, 5 ms/200 ms/10°, and 15 ms/200 ms/15°. The SNR is therefore directly proportional to the flip angle—the larger the flip angle, the better the SNR. The contrast obtained depends on the three parameters. The only difference between images a and b was the flip angles, and yet they were able to produce completely different contrast, T1, and proton density, respectively. In general, in a low-TE sequence, a low flip angle will yield PD weighting. Then as the flip angle is increased, T1 weighting increases, as long as TEs are kept small. Increasing the TE and keeping the flip angle low increases T2 weighting. The signal intensity has a complicated dependence on TE, TR, and the flip angle. The use of short TRs allows transverse magnetization coherence to be present at the time of the next RF pulse, which can be either emphasized or spoilt. There are a number of variations of the original gradient-echo technique (GRASS) with acronyms that are vendor specific. Thus, long TRs are simulated by spoiling the transverse magnetization by randomizing the phase of successive

RF pulses—termed Spoilt GRASS (SPGR) or FLASH. Similarly, to simulate long TEs, the technique called Steady State Free Precession (SSFP) or FISP is used.

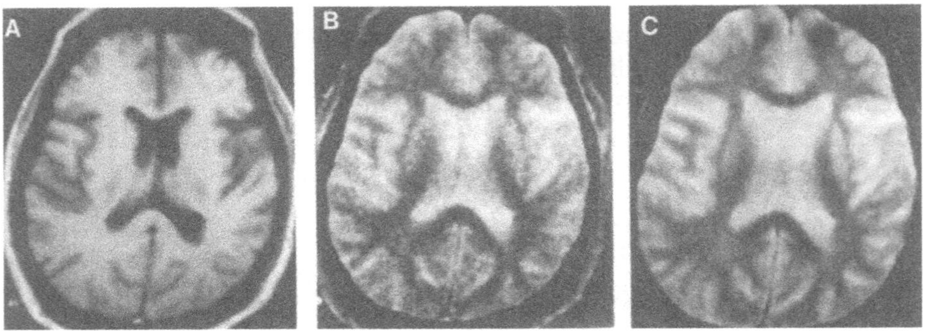

Figure 7. Images obtained using gradient echo technique of a normal subject.T1-weighted image (a) obtained using TE/TR/flip angle of 5 ms/200 ms/45°. Proton density image (b) obtained with TE/TR/flip angle of 5 ms/200 ms/10°. T2-weighted image (c) obtained using TE/TR/flip angle of 15 ms/200 ms/15° yielding T2-weighted image. Note that the signal-to-noise ratio of the images is proportional to the flip angle, and that the contrast depends quite critically on the combination of these three different sequence parameters.

Figures 8 and 9 show the contrast obtained from spin-echo images. The images are of a 12-year-old with HIV and associated CNS lesions. Figure 8a shows aT1-weighted scan obtained using a TE of 14 ms and a TR of 640 ms. As the TR is small, there is no need to resort to an FSE sequence since the total acquisition time is small in any case. Note that T1 contrast dictates that CSF is the darkest; white matter is white and gray matter gray. Quite good demarcation is observed between the white and gray matter. Images b and c in Fig.8 are obtained simultaneously as two images of a double echo sequence. Since an FSE sequence is utilized, the relatively short acquisition time permits an increase in the TR to 3800 ms. The TE for the proton density image was 22 ms and for the T2 image, 90 ms. Note the extremely good quality of the scan, with clear delineation of the different tissues of the brain. The contrast in theT2-weighted scan is flipped in sign compared to the T1 scan, with CSF bright, gray matter gray, and white matter the darkest. The contrast between tissues is lower in the proton density image than in the T2 image. Finally, in spite of the good quality of the T2 scan, the lesion is almost not visualized. However, when one uses FLAIR, the lesion becomes quite distinct. The CSF signal is suppressed but the rest of the contrast is similar to T2. Very large values of TR and TI are used (7000 ms and 2500 ms). These would not have been clinically practical but for the use of the FSE technique which brings down the scan time to reasonable values with excellent image quality.

One other type of contrast that needs to be considered in the imaging of CNS lesions is the enhancement subsequent to exogenous contrast agent infusion. Figure 9 shows the utility of such an approach. Images b and c are FLAIR and T2 scans with the same parameters as in the previous figure (they are on the same patient during the same study). The lesion conspicuity is quite low even in FLAIR. However, subsequent to contrast administration, the lesion can be detected immediately in the postcontrast scan.

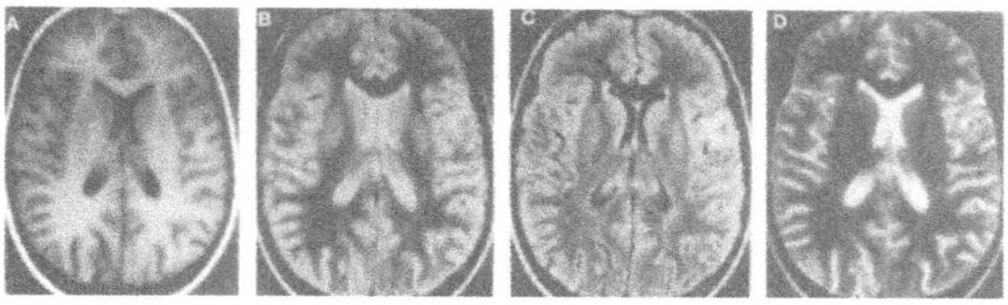

Figure 8. Images of a 12-year-old with HIV and associated CNS lesions. (a) T1-weighted image with TE/TR 14 ms/640 ms using a conventional spin echo. (b) Proton density weighting with TE/TR 22 ms/3800 ms using a turbo-spin echo. (c) A FLAIR image with a turbo-IR, with TE/TR/TI 105 ms/7000 ms/2500 ms, and (d) T2-weighted image obtained as the second image in a double echo sequence with the proton density image in (b), but with TE 90 ms. Note how the lesion becomes more conspicuous in the FLAIR compared to the T2, and how the CSF is rendered entirely dark.

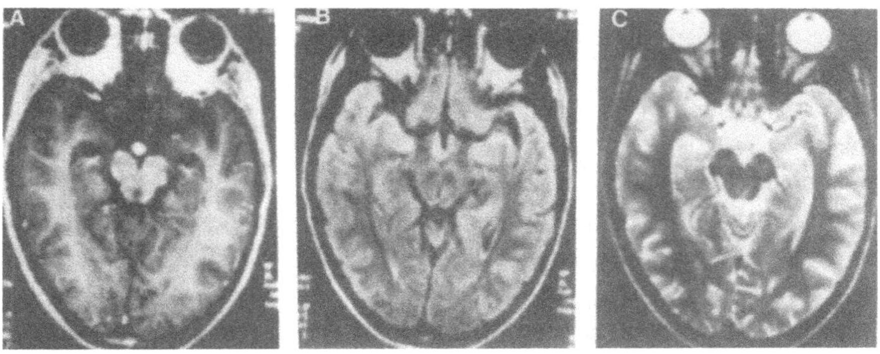

Figure 9. Images from the same subject as in Fig.8, but in different slices to demonstrate the efficacy of Gd-DTPA contrast enhancement. (a) Postcontrast image after intravenous infusion of Gd-DTPA. (b) FLAIR and (c) T2 weighting, with the same parameters as the corresponding scans in Fig. 8. Note the high conspicuity of the lesion in the contrast-enhanced scan in (a), better even than in the FLAIR. The visibility of the lesion is minimal in the T2 scan.

2.7. Magnetic Resonance Angiography (MRA)

Gradient echo imaging was designed to reduce the scan times in MRI. While the reduction in scan times was never quite enough to offset the degradation in image quality, gradient echoes have found widespread use in a variety of other imaging applications, perhaps one of the strongest being its use in MR angiography. Gradient echoes are used to image the vasculature within the human body, with and without contrast agents, either in the time-of-flight (TOF-MRA) or phase-contrast (PC-MRA) mode. The former is utilized to visualize the structure of the vasculature while the latter is used to quantify velocities within the vessels. Modern enhancements in MR imaging hardware and software have presently allowed angiographic images of extremely good quality to be produced in clinically relevant times.

The physical basis of TOF-MRA is as follows. A slice perpendicular to the vessel (hence generally in the axial plane) is irradiated repeatedly by RF pulses, causing saturation of all protons within that slice, except for protons flowing into it through the vessels. The flowing protons are subjected to basically only one RF pulse during the imaging. This results in the stationary tissue being rendered dark by saturation of the protons within it while the flowing protons within the vessels are hyperintense. A stack of such slices acquired along the direction of the vessel is added together to form the image of the vessel tree. Subsequent image processing includes reducing all pixels below a threshold intensity value to zero, further increasing the conspicuity of the vasculature. This processing is called *maximum intensity projection* (MIP). The set of slices, with hyperintense vessels and surrounding stationary tissue dark, after MIP, image can be viewed from the side from any of several angles (Fig.10). Figure 11 shows image views, generated from a 3D stack at 18° intervals, after processing by MIP. Such MRA images are of increasing value in the diagnosis of CNS disease, and are routinely acquired within the protocol. Other improvements that have been incorporated include magnetization transfer saturation of the tissue surrounding the vessel, thereby increasing the contrast of the vessel registration (this has been used in acquiring the images in Fig. 10). Slices with traveling saturation bands allow the visualization of only arterial or venous flow. Thinner slices or 3D imaging and better slice profiling allow reduction of step artifacts arising from the finite width of the slices. Ultrashort TEs and TRs allow reduction of artifacts from turbulent flow that caused overestimation of the degree and extent of stenosis. Other artifacts include reduced signal intensity in parts of the vessel that are in plane or have tortuous path, caused by saturation effects.

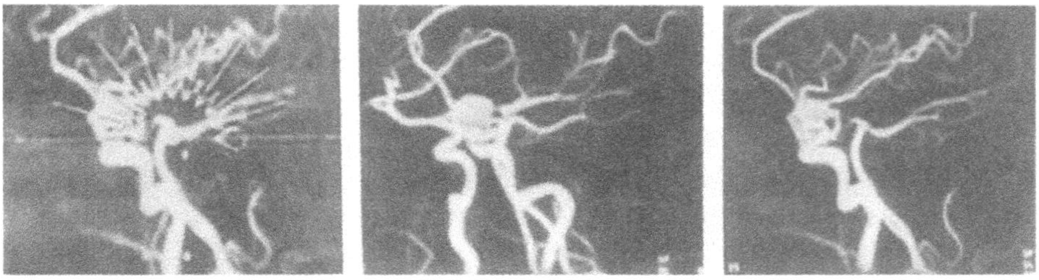

Figure 10. Magnetic resonance angiographic images acquired using the time-of-flight technique. (a) Sagittal projection image. The arrows demarcate the 18° interval at which views can be generated from the same stack of axial slices. (b) A sagittal view of intracranial vasculature. (c) A view from a rotated angle of 36° from sagittal to coronal. The ability to obtain such multiple views from different angles helps remove any "shadow" effect from overlapping vessels. Magnetization transfer effects were used to enhance the contrast between flowing spins and stationary tissue, as were saturation slabs.

Recent advances in hardware allow extremely rapid acquisition of these stacks of slices, in what is known as 3D volume acquisition covering the entire volume of interest in as short a time as 13 to 37 s, depending on the number of slices acquired. This allows sequential rapid imaging of an extent of vasculature repeatedly pre- and immediately postinfusion of Gd-DTPA contrast agent using a power injector. This allows imaging the bolus passage of the contrast agent as it passes through the vasculature, and using several image processing steps, yields angiographic images of excellent quality. The two main strengths of this technique are the rapidity with which the images can be obtained and the relatively noninvasive and nonionizing character of the procedure.

2.8. Magnetization Transfer Contrast (MTC)

Besides the more familiar T1, T2, and proton density (ρ) parameters, magnetization transfer contrast (MTC) is one more MR contrast mechanism by means of which tissues can be differentiated based on the differences between their intensity values on an MR image. This availability of a large number of intrinsic contrast parameters makes MR imaging inherently superior to other more radiologic imaging modalities. MTC is based on the fact that protons in tissues can be considered to exist in two distinct "pools." In the first pool, protons such as those in free water are relatively more mobile. A consequence of their high mobility is a relatively long T2, spin–spin or transverse relaxation time. In other words, once put into a coherent state, proton spins belonging to this species take a long time to lose their coherence or to dephase, resulting in a signal (i.e., the free induction decay) that lasts for a long time. In the Fourier transformed space (i.e., in the MR image), this manifests itself as a sharp spectral line, and represents the bulk of the signal detected by a typical MR scanner. In contrast, the other "pool" consists of protons that are "bound" to large macromolecules such as proteins and membrane and results in heavily restricted motion of these protons. The restricted motion in turn enhances the static or slowly varying (low frequency) components of the magnetic field fluctuations, which are the time-dependent perturbations responsible for T2-relaxation. In other words, the restricted motion enhances T2-relaxation of these protons significantly, leading to rapid decay of coherence or rapid dephasing, resulting in a very fast decay of the FID, too fast for the coil to detect any useful data, with T2 values as low as ~20 μs. In the frequency domain, this is manifested as a very broad line. The breadth of this resonance line renders it practically invisible in the MR imaging experiment which can only detect the sharp line resulting from the mobile protons in the first pool described above (though special experiments also can be devised in order to detect only these broad lines).

Even though the two resonances, one sharp (from the mobile protons) and the other broad (from the bound protons), may be centered at the same frequency, the broad spectrum of the latter allows one to saturate the bound protons selectively by means of a high-powered off-resonance saturation pulse, i.e., centered away from the resonant frequency, while leaving the mobile pool almost unaffected. Subsequently, after the bound protons are saturated, the magnetization of the bound protons can be transferred from this species to the species of mobile protons via chemical exchange or cross-relaxation (the mechanisms of dipole–dipole interactions between the spins). This transfer of magnetization causes the signal intensity of the free protons to decrease, thereby producing substantial changes in tissue contrast. This occurs via two pathways: (1) the equilibrium magnetization of the free protons is reduced from M_0 to M_s and is manifested as a reduced proton-density effect in a proton-density-weighted image, and (2) the relaxation time of free water is reduced from T1 to T1sat.

2.8.1. *Implementation of Magnetization Transfer Pulse Sequence*

The magnetization transfer ratio (MTR) is defined as the fractional reduction of M_0, i.e.,

$$\mathrm{MTR} = (M_0 - M_s)/M_0 \qquad (2)$$

Generally, two- or three-dimensional MR images can be obtained without and with MT saturation, regions of interest (ROIs) placed at the same anatomically relevant region, and MTR calculated by the above equation. One has to make provisions for misregistration from patient motion.

The implementation of MTC imaging sequence is relatively straightforward. Saturation pulses (MT pulses) at a well-defined frequency offset can be added on prior to the imaging pulses in most sequences. The main concern is the specific absorption rate (SAR) value, i.e., the maximum amount of RF power deposited increases due to the addition of the large flip angle MT pulses. A more detailed characterization of the MT effect goes beyond simple intensity changes to measurements of T1sat, the reduction in T1 value of the free protons from the MT effects. However, very few clinical studies have performed measured T1sat.

Figure 11 shows three images in a normal subject, the first image being acquired without MT saturation, the second with MT saturation, and the third, calculated image showing the MTR effect, using Eq. (2). MT effects have started to find increasing clinical usage. Interesting use is found in combining MT effect with Gd enhancement. MT effects do not suppress signal enhancement from T1 shortening caused by Gd infusion in extracellular spaces. Consequently, the two effects, parenchymal intensity suppression by MT and lesion enhancement by gadolinium, act synergistically to increase the lesion conspicuity (14). Gupta et al. (15) have found that an MT effect increases the detectability of CNS tuberculosis and helps in better assessment of the disease load. Alterations in MTR have been found in stroke (16) and multiple sclerosis (17).

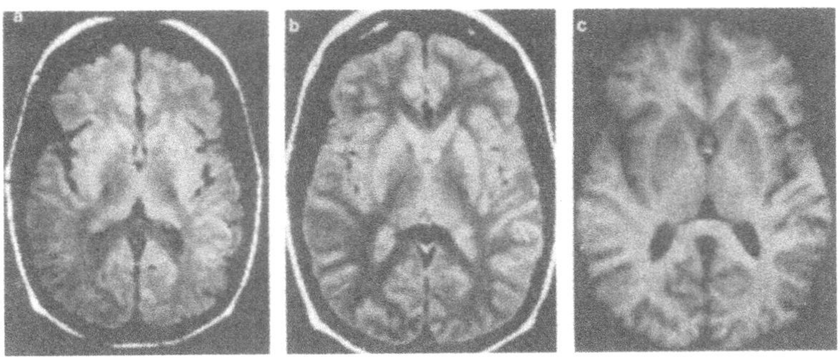

Figure 11. Magnetization transfer effects. The first image (a) was obtained without MT, the second (b) with MT, and the third (c) is the calculated magnetization transfer ratio, as per the equation MTR = $(M_0 - M_s)/M_0$. (Courtesy Drs. James Matiello and Nancy Sicote.)

2.9. Diffusion-Weighted MR Imaging (DWI)

2.9.1. Physical Basis of Diffusion

Molecular diffusion in matter arises from thermal energy and is manifested as random translational (Brownian) motion of the molecule along random paths, mediated by change in direction arising from collisions with other structures within the matter. The mean square distance, $\langle r^2 \rangle$, that a molecule travels in unit time-t is given in three dimensions by the famous Einstein equation, for unrestricted isotropic diffusion:

$$\langle r^2 \rangle = 6Dt \tag{3}$$

where D is the diffusion constant (mm^2/s). In a homogenous medium of infinite dimension, the direction of the diffusion path is totally random and the diffusion coefficient, a measure of the molecule's mobility, depends sensitively on the temperature of the medium and inversely on the viscosity. In tissues, this motion is severely restricted by the presence of cellular structures that pose as barriers to unrestricted motion. The presence of a greater amount of such barriers within the cell as opposed to the extracellular space results in smaller diffusion coefficient within the cell compared to outside. A direct consequence of this is that D is lower by a factor of about 10 in tissues compared to that in pure water, for which the diffusion coefficient is 3.4×10^{-3} mm^2/s at around body temperature. Further, in tissues that are highly organized, the barrier to free movement may be greater in one direction compared to the others, resulting in an anisotropic diffusion coefficient. Examples of tissues that have such directionally dependent diffusion coefficients are muscle fibers and myelinated white matter.

When the barriers to totally random motion are taken into account, such as occurs in normal tissue, the motion ceases to be a pure Gaussian function of distance, and the conventional definition of the diffusion coefficient breaks down. The time interval T_d, over which this diffusive motion is observed, becomes an important parameter to be considered. At very short values of T_d, the motion of the molecules is still likely to be random, because the molecules may not have had enough time to encounter the barriers, such as those within small capillaries. However, as T_d is increased, the probability of the molecule encountering some sort of a barrier that modifies its random motion increases. One then defines a modified or apparent diffusion coefficient (ADC), whose value depends critically on the time scale of observation. A few important points pertaining to the variation of ADC with T_d to be noted are: (1) the value of ADC for all tissues starts with the value of the diffusion coefficient of free water, when T_d is very small. (As explained above, at these short time scales, the motion of the molecules is still random). (2) Subsequently, as the motion of the molecule is observed at longer and longer time scales, the extent to which they diffuse in unit time (i.e., their diffusion coefficient) starts decreasing from their free water value, at a rate that is decided by the microscopic character of the particular tissue. Thus, for a fixed

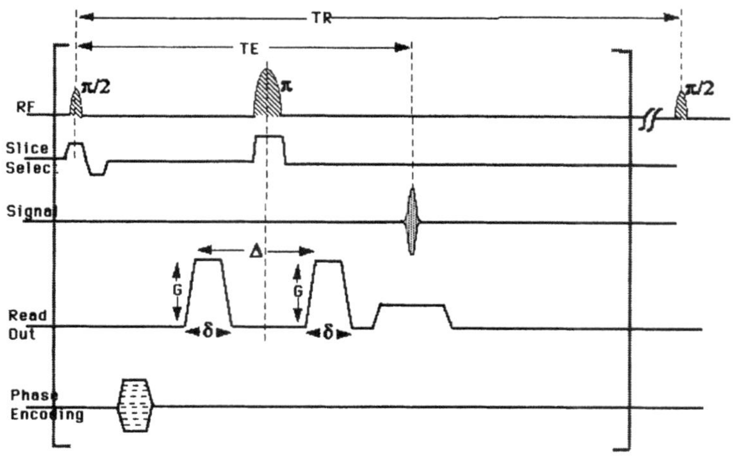

Figure 12. Pulse sequence diagram of a diffusion-sensitized spin-echo sequence. The characteristic part is the pair of diffusion gradient lobes, shown here in the read-out direction, of height *G*, individual width δ, and separation Δ, placed symmetrically about the 180° pulse. Symmetric placement about the 180° is not a necessity. The height *G* is stepped through various incremented values to the maximum gradient allowed by the scanner, in order to generate images at different *b* values. These gradient lobes can be placed on any or all of the three axes. Also, this diffusion preparation part, consisting of up to the end of the second diffusion gradient lobe, can be used as a unit to be attached to other types of sequences, for the detection part of the imaging sequence. The most successful one is an echo planar sequence, but one can also use single-shot fast spin echo and others.

value of T_d (which can be controlled by the MR pulse sequence design), the ADC value would be less than that of the free water value by an amount that would depend on the type of tissue, e.g., the ADC of normal brain parenchyma would be different from that which is compromised by a stroke. This thereby sensitizes the images to the diffusion parameter and provides a contrast mechanism that is unique to MR. (3) Once again, depending on the microscopic characteristics of the tissue observed, the ADC may or may not have a directional dependence that is a function of the symmetry of the barriers that the molecule encounters. Thus, the protons within the corpus callosum diffuse within a unit time to different distances in different directions, i.e., have anisotropic diffusion coefficients, compared with isotropic ADC for protons within for example the CSF.

Such random motion of the proton that constitutes the diffusion process within tissues results in irreversible dephasing or loss of phase coherence and, consequently, decrease of the MR signal intensity. The presence of magnetic field gradients such as those required for encoding in the slice select, readout, and phase-encoding directions in generating an MR image increases the extent of this dephasing. Dephasing that arises solely out of a combination of magnetic field gradients and regular flow such as those in vessels, can be reversed and any loss of signal gained back by special design of the gradient lobes. Loss of signal strength from the random diffusion process is, however, irreversible. Hence, a pair of magnetic field gradient lobes is used for diffusion imaging (Fig. 12). The first of these causes deliberate dephasing of the transverse spin magnetization. The second one refocuses "nonrandom" dephasing occurring in static spins, but is unable to regain the loss of signal arising from the random motion of diffusion. The loss of signal of the transverse spin magnetization detected in the manner above is a measure of the ADC in tissues. Inclusion of such diffusion gradient lobes in different directions allows one to measure the ADC in that direction. T_d, the time scale of observation described above, is decided basically by the time interval between these gradient lobes. Anisotropy of the ADC is revealed by image acquisition diffusion gradients along different directions.

The main factor that has restricted the use of diffusion imaging routinely until now has been the limitation in gradient strength attainable by the scanner, which is restricted by its hardware capabilities. Sensitization to diffusion requires fairly large gradients (~20 mT/m) that are switched on for relatively long durations (~20 to 30 ms). This also sensitizes the images to gross and physiological motion. Thus, conventional spin-echo sequences into which diffusion lobes are incorporated could not be used effectively in clinical imaging due to severe motion artifacts. Clinical realization of diffusion imaging has been feasible only with the implementation of ultrafast echo planar techniques. The recent rapid enhancement of MR scanner hardware has resulted in wider access to high gradient strengths (~23 mT/m) and rapid slew rates or the rate of change of the gradients (<250 μs), allowing echo-planar diffusion-weighted imaging.

2.9.2. Physiological Basis of Diffusion

Stejskal and Tanner had laid the foundation for diffusion imaging as practiced today as far back as 1965 (18,19). Two diffusion sensitization gradient lobes of magnitude G, width δ, and separation Δ (Fig.13) are generally placed on either side of the 180° RF pulse. The attenuation of the MR signal due to diffusion in the presence of such gradient lobes is given by

$$\mathrm{SB} = S_0 * e^{(-Db)} \tag{4}$$

here SB and S_0 are the signal strengths with and without the diffusion gradient lobes respectively, D is the diffusion coefficient, and b is a description of the lobes and is given by

$$b = (\gamma G\delta)^2 * [\Delta - \delta/3] \tag{5}$$

From Eq. (4),

$$\ln (S_0/\text{SB}) = D/b \tag{6}$$

or the diffusion coefficient,

$$D = b * \ln (S_0/\text{SB}) \tag{7}$$

In a typical diffusion imaging experiment, the gradient lobes are sequentially increased from a zero height to the maximum amplitude attainable by the scanner. A plot of D in Eq. (7) is that of a straight line, such that if a straight-line fit is made of the log of the S_0/SB, the slope of this line yields the diffusion coefficient. Generally a pixel-by-pixel fit of the entire MR image is processed in this manner to yield a calculated diffusion image of the anatomy, where a hypointense zone indicates a zone of low diffusion such as a zone compromised by a stroke.

As previously mentioned, the main impediment to diffusion imaging is the presence of physiological motion. These macroscopic components of motion tend to overwhelm this method of imaging which has been designed to be extremely sensitive to microscopic motion. The general solution to this problem lies in minimizing the overall acquisition time. Diffusion-weighted imaging using spin echoes in living systems is almost impossible because of problems associated with motion. Several strategies have been adopted toward minimizing the motion artifacts including adding diffusion sensitization to rapid sequences such as fast spin echoes, HASTE sequences, and turbo-gradient echoes. The most effective solution by far is diffusion-sensitized EPI, which therefore yields subsecond images at each b value. Unfortunately, the hardware upgrade necessary for EPI is relatively costly, in which case alternate techniques have to be implemented.

Figure 13 shows diffusion-weighted images using EPI. Two slices are shown. The susceptibility artifact of EPI images is prominent in the inferior row, in areas of the anatomy proximal to air–tissue interfaces. The difference in susceptibility of these two dissimilar materials causes inhomogeneities in the magnetic fields, resulting in rapid T2* decay. The resulting mismapping of pixels causes the distortion of images characteristic of susceptibility artifacts.

The first column shows the images of these slices at $b = 0$, i.e., the diffusion gradient lobe has zero area. Under these conditions, the prominent contrast in the images will be that of T2, arising from the long TE required to accommodate the diffusion lobes. The second

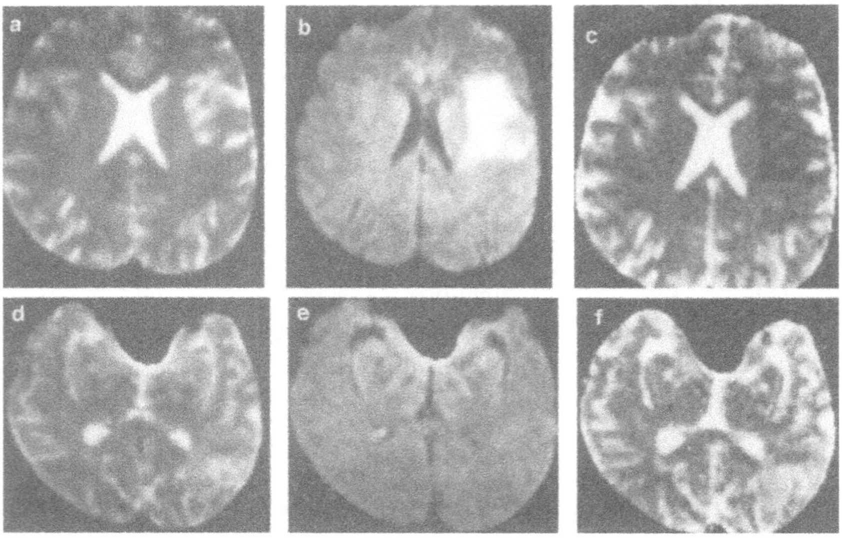

Figure 13. Diffusion-weighted echo-planar images of two slices of a patient with cerebral stroke. The first column (a,d) is acquired with $b = 0$, the second with $b = 1000$ (b,e), and the third is a calculated ADC map (c,f), obtained as explained in the text. Note the compromised zone stands out with high conspicuity in comparison to surrounding parenchyma. The second column shows the susceptibility artifacts that accompany most EPI images, arising in this case from air–tissue interface. (Courtesy Drs. James Matiello and Jeffery Alger.)

column shows the images of the two slices acquired at $b = 1000$ s/mm^2. Several points are to be noted. One can immediately observe the bright zone compromised by a cerebral stroke in the left image. A hyperintense zone indicates a region where the protons are diffusing less than in the zones with lower intensity. A smaller value of diffusion coefficient (as in the stroke region) causes lesser dephasing of the spins and consequently more MR signal than that arising from normal tissue zones where the protons are able to diffuse to a greater extent. This hyperintensity in a diffusion-sensitized image is the characteristic signature of strokes and other pathologic conditions that increase cellularity. Another point to be noted in these two images is that even though they are alsoT2-weighted images, CSF appears dark. The reason for this is that the protons within CSF have the larger diffusion coefficient, and therefore the consequent dephasing reduces its signal more than any other tissue in the region. The third column shows the calculated ADC images. A pixel-by-pixel calculation is made of the D value using Eq. (7), employing the log of the two intensities at the two b values. Each pixel intensity is a direct measure of the ADC value of the protons at that point. The compromised zone now appears dark due to a reduced value of ADC in comparison to the surrounding parenchyma. In particular, the CSF appears bright because of its near-water-like ADC values.

Given the present-day status of hardware in most MR scanners, diffusion weighting using EPI has become a clinical reality. Even if the relatively costly upgrade required of EPI is not available, other less costly alternatives are available, such as combining diffusion sensitization with single-shot FSE, or just FSE, or with certain turbo fast gradient echoes. Each of them has its associated caveats. These sequences are not single-shot acquisitions like EPI (with the exception of single-shot FSE), and thus have a greater degree of motion artifact registered on the image. Recent work at several research centers shows great promise for diffusion-weighted EPI technique. Applications such as detecting the presence and extent of stroke in the acute stage have become established as routine clinical practice. The potential of its application in other diseased conditions of the CNS is also being rapidly realized.

3. NUCLEAR MAGNETIC RESONANCE SPECTROSCOPY

Spectroscopy is the art of eliciting information without appreciably changing the source of information. The information is obtained by perturbing the phenomenon under study and then observing its behavior as it attempts to cope with the perturbation. Most spectroscopies deal with the interaction of electromagnetic radiation with matter. The different frequencies of electromagnetic radiation used as the probes affect matter differently and reflect different characteristics of matter, e.g., X-ray spectroscopy gives us information on the energy changes in the inner electrons of atoms or molecules, visible and ultraviolet spectroscopy informs us of the transitions in valence electrons, whereas microwave and infrared spectroscopy tells us of molecular and vibrational energy transitions of molecules. The Heisenberg uncertainty principle, however, ensures that the closer we approach the microscopic or quantum world, the less likely it is that our assumption of negligible effect will hold. Yet, as the state of the art in biomedical physics improves, increasingly we are realizing that to alter the story of any disorder or disease, understanding and intervention must take place at the molecular and genetic level. *In vivo* NMR spectroscopy is one such tool employed to understand the biochemical mechanisms of the human body in normal and diseased states. The technology provides a basis to model and interpret the macroscopic clues manifested as clinical and laboratory findings.

The RF region of electromagnetic radiation corresponds to the transition in energy levels of the magnetic states of the atomic nuclei. Because the same nucleus in different molecular structures may show different magnetic states, MRS can be applied to determine molecular structures.

3.1. Of Things Nuclear

Often material phenomena or objects are identified and studied through their attributes or properties. While these properties may qualify a uniqueness of the object and, hence, aid in categorization, they are not the whole object. For example, while charge (positive or negative) may aid in distinguishing an electron from a proton, it is not the whole electron or proton.

The two most familiar properties of nucleons are charge and mass. Possession of charges may or may not be associated with electric fields as in the case of neutrons where positive and negative charges cancel each other out. However, moving charges lead to a magnetic field and charges in orbital motion describe a third property of nucleons, i.e., spin. A neutron with no net charge can also possess spin, indicating that it must contain separated charges in motion at least part of the time even though its total charge is zero.

This spin property, also called *angular momentum* is associated with the induction of a magnetic dipole with a north and south pole. The nucleons within the nucleus combine to give a total spin property for an isotope. Thus, in the case of ^{1}H with its one unpaired proton, the nuclear spin is ½, whereas ^{2}H with its one unpaired proton and an unpaired neutron has a total nuclear spin of 1. Two or more particles with spins having opposite signs can cancel the individual spins with no net spin as in the case of ^{4}He. Only those isotopes that have nonzero magnetic spin are of importance in NMR.

The theory establishing the quantum nature of matter also dictates that the angular momentum (spin) can take on only certain definite values, i.e., the spin of a nucleus is an integral multiple of $h/2\pi$ where h is Planck's constant. Physicists also discovered another property of nuclei having nonzero angular momentum. When placed in an external magnetic field, these nuclei experience a certain torque or twisting effect, their magnetic moment, which could assume only certain discrete orientations with respect to the external magnetic field. Thus, a proton with a spin of ½ can exist in only two states, parallel or antiparallel with respect to the external magnetic field. Nuclei with higher spins can take on additional orientations, but still only a few distinct ones.

Table 1. Magnetic Properties of Some Common Nuclei

Isotope	Z	Spin	Natural abundance	Sensitivity at 1.5 T Relative[a]	absolute[b]	NMR frequency at 1.5 T (mHz)	Isotope
H-1	1	½	99.98	1.0000	1.000	1.0	63.89
H-2	1	1	0.016	0.0154	0.016	2.4×10^{-6}	9.80
C-13	6	½	1.108	0.0226	0.023	2.5×10^{-4}	16.06
F-19	9	½	100.00	0.85	0.850	8.5×10^{-1}	60.08
P-31	15	½	100.00	0.083	0.083	8.3×10^{-2}	25.85

[a] At constant field for equal number of nuclei.
[b] Product of relative sensitivity and natural abundance.

This twisting effect in the presence of the external magnetic field creates a cone-shaped path called *precession*, which is distinguished from the orbital motion in the absence of the field. Each isotope with a nonzero angular momentum has a unique precession rate in the presence of the same, constant external magnetic field strength. This unique property is determined by γ, the gyromagnetic ratio of an isotope. Thus, the precessional frequency, ω_0 (also called the Larmor frequency), for each isotope in the presence of a given external magnetic field, B_0, is given by Eq.1. Table 1lists the magnetic properties of some common nuclei of importance.

3.2. The Hierarchy of Energy States

As previously mentioned, the presence of an external magnetic field imposes a discrete orientation on the nuclear spins. In the case of H-1, nuclei that find themselves aligned anti-parallel versus the ones that are parallel to the external magnetic field also find themselves in different energy states. This hierarchy of energy states is brought about by the Larmor frequency. The energy gap between the two spin states (-1/2 and +1/2) is given by

$$\Delta E = \frac{\gamma\, h B_0}{2\pi} \tag{8}$$

For ^{1}H at 1.5 T with $\gamma = 267.512 \times 10^6$ rad/T-s, this energy gap is given by

$$\Delta E = \frac{(267.512 \times 10^6 \,\mathrm{rad/T\text{-}s})\,(6.63 \times 10^{-34}\,\mathrm{Js})\,(1.5\,\mathrm{T})}{2\,(3.14\mathrm{rad})}$$

$$\Delta E = 4.236 \times 10^{-26}\,\mathrm{J}$$

The population difference between the two states at 1.5 T and room temperature is given by the Boltzmann distribution

$$\frac{P_{m=-1/2}}{P_{m=+1/2}} = e^{-\Delta E / kT} \tag{9}$$

At 25°C (298 K), using the Boltzmann constant, $k = 1.381\times10^{-23}$ J/K, we have

$$\frac{P_{m=-1/2}}{P_{m=+1/2}} = e^{-(4.236\times10^{-26}\,\mathrm{J})/(1.381\times10^{-23}\,\mathrm{J\,K^{-1}})(298\mathrm{K})} = 0.999989$$

Since

$$P_{m=+1/2} = 1 - P_{m=-1/2}$$

$$\frac{P_{m=-1/2}}{1 - P_{m=-1/2}} = 0.999989$$

$$\therefore \quad P_{m=-1/2} = 0.4999974 \quad \text{and} \quad P_{m=+1/2} = 0.5000026$$

This small excess in the lower energy state (m=+1/2) is what determines the probability of a transition and, hence, the sensitivity of NMR spectroscopy. Thus, rather weak signals representing the hierarchy of energy states populated by the spin are detected in NMR spectroscopy.

3.3. Resonance: Portal of Energy Transfer

When the spinning ensembles of nuclear spins distribute themselves in two energy states (in the case of ^{1}H) in the presence of an external magnetic field, B_0, the macroscopic Boltzmann equilibrium magnetization can be represented by a vector **M** of magnitude M_0. This macroscopic vector stems from the individual magnetic moments of the nuclei that form the excess population in the lower energy state. The precessional motion of individual nuclei is not in phase and no component of this vector is in the transverse X–Y plane in the equilibrium state.

Resonance allows a portal to perturb this system and transfer energy to the nuclei in a controlled manner to deflect the spins from the *Z*-axis. This is achievable by alternating electromagnetic waves in the RF range even though the strength of this alternating RF magnetic field, B_1, may be relatively small compared to the static external magnetic field, B_0. However, to affect such a torque in the magnetization, **M**, and, hence, resonance, the alternating RF magnetic field must have the same Larmor frequency and must be perpendicular to the main static field. Let us see why this is so.

When an alternating RF magnetic field, B_1, is applied perpendicular to B_0, the magnetization **M** will at a particular point in its precessional path around the *Z*-axis experience a torque which would make it precess around B_1 and thus change the angle by an amount *d*2. Half a revolution further, the change would be -*d*2, so that the overall change is zero. When the frequency of B_1 is increased to compare with the Larmor frequency, T_0, each nucleus in the lower energy state will stay for some time under the influence of B_1 performing a precessional motion. When $T = T_0$, the angle 2 will be changed and the nucleus will absorb the energy due to resonance. The loss of energy from B_1 can be measured (20). The nuclear spins will also be in phase coherence.

The angle 2 is called the flip angle and describes the macroscopic net effect on the spins. The quantum spins can exist only at certain energy levels; therefore, there are no flip angles for individual spins. With a flip angle of 90°, the magnetization **M** is in the transverse plane with the result that the number of spins in the lower and higher energy states is the same. With twice the strength of the RF, the angle can be changed to 180° with the result that at the quantum spin level one has the opposite effect of the equilibrium state with more spins in the higher energy state than the lower energy (parallel) state (21).

3.4. Saturation and Relaxation

The transition of nuclear spins into the hierarchical equilibrium state in the presence of an external magnetic field is not instantaneous but rather is governed by a first-order rate process. Once a nucleus reaches a higher energy state whether it is due to the external magnetic field or an RF pulse as discussed in the previous section, there would be no further absorption of energy and soon the population difference between the two states would be zero. The system would thus reach saturation. The only way the nuclei can remain accessible to further energy absorption is if there is a mechanism of energy emission or loss. However, how can this "relaxation" take place? Nuclei do not collide easily and therefore, cannot lose energy as molecular, vibrational, or translational loss. A spontaneous transition to a lower energy state is a possibility but has very long decay times (10^{13} years). Nuclei lose their energy through resonance with the molecular motion of the lattice. Rotational and translational motions of molecules in liquids create fluctuating time-dependent magnetic fields. When the frequency of this fluctuating magnetic field equals the Larmor frequency, a magnetic torque is exerted to flip the orientation of the nuclei and thus their energy states.

The magnetic energy received by the lattice is transferred into thermal energy. Since this process of energy loss or relaxation is quite dependent on the molecular environment, different molecular environments encountered for ^1H will yield different rates of relaxation governed by their relaxation time, T1. Thus, gray and white matter would have different rates of recovery of protons in the longitudinal plane.

The time dependence of the magnetization in the transverse plane after the RF resonance is affected by another relaxation time, i.e., the spin–spin relaxation time, T2. As in the case of spin–lattice relaxation, the local fluctuating magnetic field due to the molecular motion also affects T2. The phase coherence of the individual nuclear spins when exposed to the fluctuations along with the inhomogeneities in the external magnetic field results in a spread of their Larmor frequencies seen as a "fanning out" or loss of phase after which the spins move back to the axis of the external field as magnetization in the transverse plane goes to zero. This type of relaxation is different from longitudinal relaxation since the total energy of the spin system does not change.

3.5. The Chemical Shift

Earlier it was stated that the resonance frequency for nuclei is dependent on the magnetic field experienced by the nuclei. One would hence expect the same nuclei in a compound to resonate at a single frequency. This, however, was found not to be the case soon after the discovery of NMR. Even a pure compound such as ethanol was found to have multiple frequency peaks. Since the external magnetic field was homogeneous within the liquid, the explanation lay in differences in microscopic magnetic environments within the molecules as a result of which nuclei experienced variations in the effective magnetic field. These microscopic field variations arise from the electrons orbiting the nucleus and producing their own weak magnetic field, B_i. The s-electrons are spherically symmetric and circulate producing current, which in turn produces a magnetic field at the nucleus, which opposes the external magnetic field. Thus, in order to obtain the resonance condition, it is necessary to increase the applied field over that for the isolated nucleus. If one nucleus is more shielded than another in the same external field, its signal will be shifted to a lower frequency. If B_0 is the applied field and B_{eff} the field at the nucleus, then

$$B_{eff} = B_0 - B_i \qquad (10)$$

$$B_i = \sigma B_0 \qquad (11)$$

$$B_{eff} = B_0(1-\sigma) \tag{12}$$

Because the strength of the induced field due to nuclear shielding is proportional to that of the applied field, we can define a constant σ, called the shielding or screening constant, which is sensitive to the chemical environment of the nuclei. Therefore, nuclei in different chemical environments experience different fields and hence produce signals at different frequencies. This upfield shift of the nucleus is called a *diamagnetic shift* and is present in every molecule since each has s-electrons. The electrons in p-orbital with no spherical symmetry produce magnetic fields that are large which when averaged over molecular motion give low-field shifts. This deshielding effect is called the *paramagnetic shift*. Of course, for protons (^{1}H) with no p-electrons there is no paramagnetic term (22).

Thus, the practical implication is that the differences in the resonance frequencies between different nuclei range in millions of hertz (Hz) and the spread of resonance frequencies for a single nucleus is in the range of a few thousand of hertz due to this chemical shift. The electronic screening is due not only to the electrons from the nucleus's own electronic cloud but also to differences in chemical bonds between various atoms and their different electron distributions. Thus, hydrogen bonded to carbon experiences a different "electron field" than hydrogen bonded to oxygen. In addition, the types of bonds (single, double) influence the field.

Spectroscopy is then this extraction of the shift in frequencies from the NMR signal, which clues one to shifts or differences in the molecular structure of tissues. While the chemical shift may introduce artifacts in NMR imaging, it is the basis of spectroscopy!

The chemical shift denoted by δ is measured with reference to a substance at a given magnetic field and is defined as

$$\delta = \frac{\nu_x - \nu_{ref}}{\nu_{ref}} \times 10^6 \text{ ppm} \tag{13}$$

where ν_{ref} is the resonance frequency of the reference nucleus at a given magnetic field and ν_x is the resonance frequency of a certain nucleus at that field. Thus, δ is dimensionless. One can approximate the ν_{ref} to ν_0, the spectrometer operating frequency, and can rewrite the above equation as

$$\delta = \frac{\nu_x - \nu_{ref}}{\nu_0} \times 10^6 \text{ ppm} \tag{14}$$

In this way, the chemical shift is normalized with respect to the operating frequency and is independent of the external magnetic field, making comparison of data easy and simple. In high-resolution spectroscopy, ν_{ref} is the resonance frequency of a stable compound, namely, tetramethylsilane (TMS). TMS is added to each compound in high-resolution MRS, but it is impractical in whole-body MRS.

3.6. Spin–Spin Coupling

If one observes a high-resolution spectrum of, say, dichloroacetaldehyde ($CHCl_2CHO$), one would expect to see two peaks due to the two different protons in the molecule. However, one finds four peaks, i.e., the two peaks are split into two peaks each a few hertz apart. This is because the magnetic field at each proton is made up of contributions due to the circulating electrons (nuclear shielding) plus the effect of the other nuclear magnet in the molecule. Thus, the magnetic field at nucleus A is

Magnetic field at A (H_A) = nuclear shielding + magnetic field due to nucleus B (H_B)

Nucleus B has two orientations in the external magnetic field, $m_B = +½$ which will produce a small field $+\Delta B$ and $m_B = -½$ which will produce an equal and opposite field $-\Delta B$ at H_A. Thus, H_A gives two equally intense peaks from the two orientations of H_B and similarly H_B gives two equally intense peaks with the same separation.

This explanation of the observation is called *indirect spin–spin coupling* and its intensity of coupling is defined by a constant, J. Thus, J_{AB} is the coupling between nuclei A and B, independent of the applied field and has units of hertz. Unlike the chemical shift, J-coupling is independent of the external magnetic field. The nuclear coupling constant can be positive or negative. If the coupling J_{AB} is positive, then the spin states with opposed spins (i.e., $m_A = +½$, $m_B = -½$ and vice versa) will have lower energy than the states with parallel spins. If J_{AB} is negative, the converse is true.

The internuclear couplings are transmitted via the bonding electrons, i.e., they are electron-coupled interactions. There are three possible mechanisms:

1. The nuclear moments interact with the electronic currents produced by the orbiting electrons.
2. There is dipolar interaction between the nuclear and electronic magnetic moments.
3. There is an interaction between the nuclear moments and electronic spins in s-orbitals because electron wave function has a finite value at the nucleus. This is called the *contact* term.

For all couplings involving hydrogen, the contact term is dominant and the other terms may be neglected (22). Note that this coupling can be either homonuclear (between the same nuclei) or heteronuclear (between different nuclei). When $J \leq \delta$, the chemical shift separation, the spectrum is called the *first-order NMR spectrum* and the number of lines in which a signal is split can be determined from

$$x = 2nI + 1 \tag{15}$$

here x is the number of lines of the multiplet, n is the number of equivalent coupled nuclei and I their nuclear spin. Similarly, the relative intensities of the lines forming that multiplicity are predictable because of Pascal's triangle.

Number of equivalent coupled nuclei

0
1
2
3
4

Multiplicity and relative intensity of the lines

1
1 1
1 2 1
1 3 3 1
1 4 6 4 1

Figure 14 shows an ethanol ^{1}H MR spectrum at low resolution and at high resolution. The methyl (CH_3) resonance is split in a triplet because of the coupling with two equivalent nuclei of the methylene group (CH_2), and the methylene resonance is split in a quartet of doublets due to the coupling with three equivalent protons of the $-CH_3$ group and one of the hydroxyl protons (23).

3.7. The Dialogue: Transmitting and Receiving the Resonant Signal

In our situation of NMR spectroscopy, so far we have learned that the presence of an external magnetic field creates the split of nuclei into energy levels, which in the case of ^{1}H numbers two. This population difference between the two energy levels is very minute but terribly important because our agent of perturbation relies on there being this difference. Our agent is the oscillating magnetic field, B_1, from the RF range.

3.7.1. *Rotating Frame and the B_1 Field*

In the presence of the static magnetic field, B_0, the nuclei precess with angular frequency $\omega_0 = \gamma B_0$. If one considers a rotating frame with angular frequency ω_r relative to the static laboratory frame, the nuclei in the rotating frame will appear to precess with angular frequency $\omega_0 - \omega_r$. If $\omega_r = \omega_0$, then the nuclei will appear not to precess at all and the static magnetic field will appear to be zero. In such a frame, if one were to transmit a magnetic field B_1 along the *X*-axis to perturb the nuclei, the nuclei will precess about B_1 with angular momentum γB_1 just like the application of B_0 caused them to precess about B_0 with angular momentum γB_0. This is because B_1 is the only field experienced in the rotating frame. If

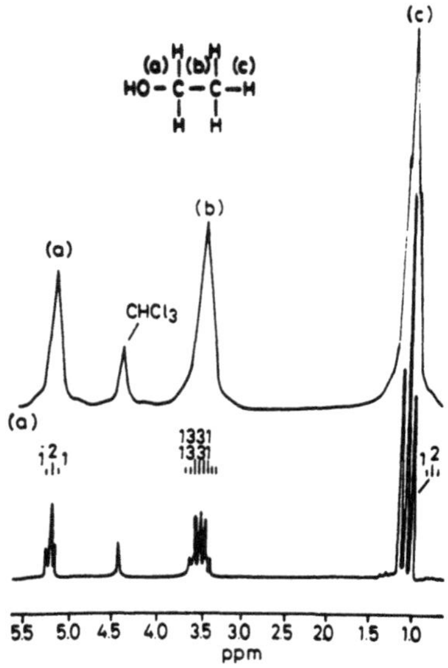

Figure 14. ^{1}H NMR spectrum of ethanol.

the field B_1 is applied for time t_p, the nuclei will rotate through an angle $\theta = \gamma B_1 t_p$ ($\omega_1 = \theta/t_p$). If t_p is such that $\theta = \pi/2$, this will tilt the nuclei in the transverse plane. In the laboratory frame, B_1 must correspond to a field that rotates about the *Z*-axis with angular frequency ω_0. The simplest method of generating this rotating B_1 field is to apply an oscillating field along a direction in the *X–Y* plane for such a field can be broken down into two components rotating in opposite directions. The component rotating in the opposite direction to the nuclear precession can be shown to have a negligible effect in NMR and so the oscillating B_1 field is effectively equivalent to the required rotating field (24).

3.7.2. The Perturbation or Transmittance

The transmitter irradiates with RF fields of appropriate frequency, power, and timing. Two parameters must be understood well when setting up the transmission signal, i.e., duration of each pulse, t_p (interval per pulse or pulse width), and the interval between pulses, T_R. The pulse width t_p is related to the pulse angle of magnetization, θ, by the equation given in the previous section (3.7.1).

If we assume that magnetization before the application of the RF field B_1 is M_0 along the Z-axis, then the pulse B_1 generates a component M_{xy} in the X–Y plane given by

$$M_{xy} = M_0 \sin\theta \tag{16}$$

$$M_{xy} = M_0 \sin(\gamma B_1 t_p) \tag{17}$$

A compromise has to be made between t_p and the interval between pulses. If the interval between pulses is too small, then the nuclei will not recover their magnetization (saturation effect) and consequently the signal received will diminish. If one waits too long between pulses, then in a given time few scans will be collected and since the SNR increases according to (number of scans)$^{-1}$, this will adversely affect the SNR. The optimum angle for a given TR (interval between pulses) is known as the Ernst angle, α, and is given by

$$\cos\alpha = \exp(-TR/T1) \tag{18}$$

here T1 is the spin–lattice relaxation time of the signal of interest.

In practice, the parameter that is adjusted is the pulse width, t_p. A sample scan can be used to measure the t_p for a 90° pulse or a 180° pulse. For different values of degrees t_p, the signal received, M_{xy}, will vary sinusoidally as predicted by the Eq. 17 and the amplitude will be maximum for the 90° pulse.

Another parameter that needs mentioning is the effective bandwidth of the RF pulses. The Heisenberg uncertainty principle states that the appropriate relationship between frequency and time is

$$\Delta\nu\,\Delta t \approx 1 \tag{19}$$

Thus, in our case, if the pulse lasts for a width t_p, the effective frequency bandwidth will be no better than $\Delta\nu \sim 1/\Delta t$. To optimize the sensitivity of the technique, one would like to simultaneously excite all of the nuclei, the resonance frequencies of which may range over, say, 5 kHz. It is clear that this can only be done by sufficiently powerful yet short pulse widths, i.e., much less than 200 μs (25).

3.7.3. *The Response or Reception*

After the RF pulse is turned off, the nuclei precess freely at their own characteristic Larmor frequencies about B_0. The coherence of their precession gradually disappears because nuclei in different environments and locations experience microscopic field variations differing by ΔB from the mean, which, in turn, introduces variations in their precessional rates of $\gamma\Delta B$ from the mean. The term *free induction decay* (FID) is often used to describe the signal resulting from an RF pulse; it describes the decay of the induced signal arising from free precession of the nuclei in B_0.

The FID is detected (usually nanovolts) by a receiver consisting of a preamplifier and an RF amplifier, one or two phase-sensitive detectors that convert the RF frequency to audiofrequency. The audiofrequency is filtered and amplified to required levels (~ 2 V peak to peak). This signal is sampled, digitized, and fed into a computer for further processing.

3.7.4. *Collection of Data*

The signals from the receiver are continuous. This received voltage is converted into a digital form by sampling the signal amplitude at equal time intervals, Δt_s (dwell time), using a stable gating frequency and an analog-to-digital converter (ADC). According to the Nyquist theorem, in order to adequately represent the actual frequency of the wave, each wave must be sampled at least twice per cycle. Thus, for a sampling frequency of ν_s, the frequencies that will be unambiguously represented lie in the interval $-\nu_s/2 < \nu < +\nu_s/2$. Frequencies outside this interval will be aliased into it by being shifted with an integer multiple of the sampling frequency.

If the spectral width is W, with quadrature detection, the reference frequency can be in the center of the spectrum, which ranges from $-W/2$ to $W/2$ and can be sampled at a frequency of W. But since there are two detectors, the computer will receive $2W$ samples per second. The sampling or acquisition time is given by

$$T_s = N_s \Delta t_s \tag{20}$$

here N_s is the number of sampled data points. The spectral resolution $\Delta\nu$ is related to the acquisition time by

$$\Delta\nu = 1/T_s \tag{21}$$

Therefore, if one increases T_s, the spectral resolution improves. However, in practice because of $T2^*$ decay, the NMR signal decays exponentially [exp $(-t/T2^*)$] and after 3 $T2^*$ seconds of sampling, in *in vivo* NMR mainly noise is acquired. The resolution (1/3 $T2^*$) is then approximately equal to the linewidth at half-maximum height, $\Delta\nu_{\frac{1}{2}} = 1/(\pi\ T2^*)$, which corresponds to the linewidth associated with the decay time $T2^*$ (26).

In order to improve the sensitivity of the NMR technique, successive scans are summed which reduces the level of noise relative to that from coherent signals. A coherent signal will increase linearly with the number of scans and since noise is random, its amplitude increases only with the square of the number of scans. The ratio of signal to noise after *n* scans to that of a single scan is (27)

$$S / N \propto \sqrt{n} \tag{23}$$

The ADC converts continuous time signals into a sequence of binary numbers. Each binary number is a sample of the signal. The accuracy of the ADC is linked to the number of digits representing the amplitude of the signal sample. For an ADC of *B* bits, the dynamic range is 2^B and the accuracy is $1/2^B$.

3.8. Water Suppression

The hydrogen nucleus in water serves as a useful marker in NMR imaging because of its large concentration (~100 moles) in soft tissues. However, the same feature makes it unattractive in 1H NMR spectroscopy because other metabolites containing 1H have concentrations on the order of micro to millimoles. Thus, in order for them to be NMR visible, the huge water peak has to be suppressed. A 12-bit ADC has a dynamic range of (211 - 1) or - 2000:1 (with one bit for the sign). If the water signal fills the digitizer, then a metabolite signal that is less than 2000 times that of water is undetectable. Sixteen-bit ADCs help in expanding this range to 33,700:1 but still require good water suppression techniques (28).

3.8.1. Presaturation

The most common approach to suppression of water is to presaturate the water resonance. This means that the water in the sample is irradiated with the water resonance frequency to achieve equally populated spin states before the sampling pulse is transmitted and acquisition started. This is illustrated in Fig.15. Figure 16 shows the effect of water suppression on a spectrum. In practice, this procedure yields suppression factors between 100 and 1000. The RF field strength B_2 needed to saturate water is usually double the linewidth of water and is achieved with long pulses of a few hundred milliseconds. One would want to keep B_2 as small as possible to keep the frequency range small.

3.8.2. Selective T1 Relaxation

An inversion recovery scheme can be used for water suppression since the water T1 may be significantly different from other metabolites. The pulse sequence can be

$$(\pi - \mathrm{TD} - \pi/2 - \text{acquisition} - \mathrm{T})_n$$

The magnetization is first inverted with a 180° pulse. After a time TD, the longitudinal magnetization Mz recovers. If TD is so chosen that TD = T1 (H_2O) ln 2, then the water magnetization is null at a time when other molecules are partially or completely relaxed and the sampling pulse can be generated. This sequence, however, has the drawback of requiring long TD since water T1 is long. In addition, it is also difficult to quantitate since the metabolites may be only partially relaxed. One could refine the technique by making the pulse frequency-selective water inversion as shown (28):

$$(\pi_{\mathrm{H2O/sel}} - \mathrm{TD} - \pi/2 - \text{acquisition} - \mathrm{T})_n$$

3.8.3. Selective Excitation

In this scheme, the water signal is left undisturbed while selective ^{1}H molecule frequencies are excited. Various combinations of pulses and shapes have been designed such that the Fourier transformation gives a null at the solvent frequency.

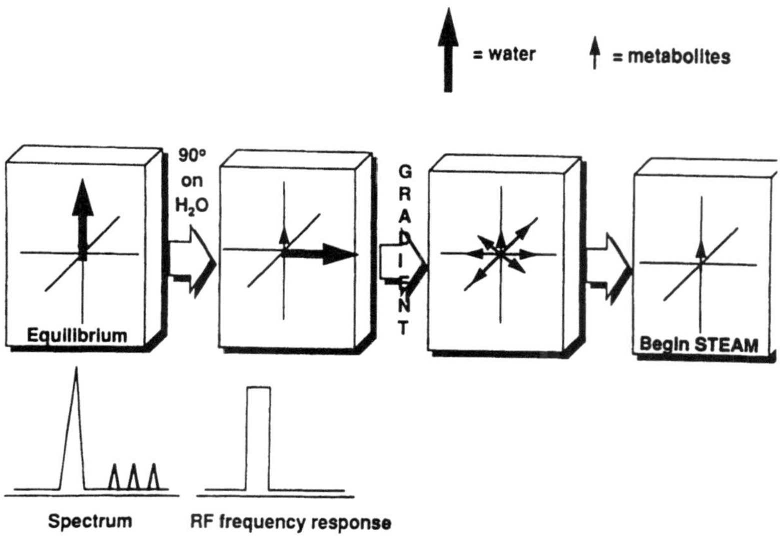

Figure 15. Presaturation technique for water suppression.

3.8.4. Other Methods

Several other methods of water suppression exploit certain features of the solvent or solute molecule. Some schemes may work very well in *in vitro* or biologic fluid samples or 2D NMR sequences but may not work in 1D NMR in *in vivo* spectroscopy. In addition, there are some postprocessing routines to filter out the water peak and its effect on metabolite signals.

3.9. Localization

Spatial localization methods in spectroscopy can be divided into two broad categories, single-volume or a voxel methods and multiple-volume methods. The single-volume method obtains spectrum from an area of interest while the multiple-volume method obtains a spectrum from every point in an image.

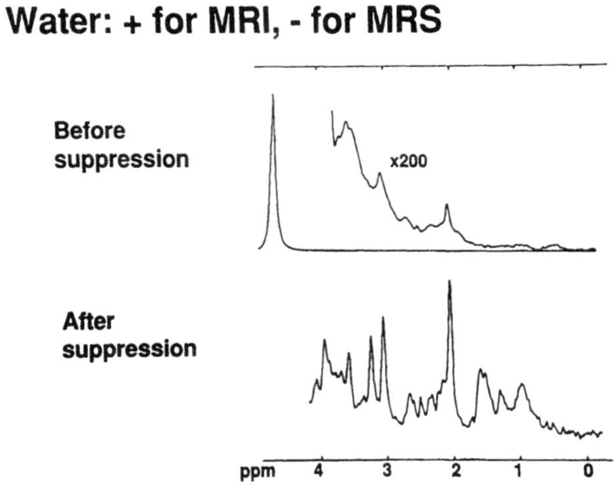

Figure 16. Effect of water suppression on a ^{1}H spectrum.

The single-volume methods can be further subdivided into (1) selective excitation methods which excite the area of interest only and attempt to leave all of the other volumes untouched; (2) selective deexcitation methods which excite all of the spins outside the volume and saturate or dephase them so that only the spins in the volume of interest are left; and (3) encoding schemes which use multiple acquisitions and addition and subtractions of spectra in order to obtain the volume of interest (29).

The NMR signal has two basic parameters into which information about the spins may be stored: the frequency (and, hence, phase) and amplitude. Spatial localization can be achieved by making either of these two paraeters dependent on the locations of the spins within the magnet. In practice, this process is achieved by sampling the NMR signal in the presence of an RF gradient (B_1 methods) or through the use of techniques similar to those employed in NMR imaging where B_0 gradients are used for spatial selection and encoding. We will discuss the two most popular methods employing the latter technique with direct excitation.

3.9.1. Point-RESolved Spectroscopy (PRESS)

PRESS uses a double echo as shown in Fig.17. A volume in space is localized by applying three selective pulses each along one of the orthogonal axes in the presence of the appropriate gradients. Although the second and third RF pulses are designed to give a tip angle π over the volume of interest, the actual tip angle may deviate outside and near its limits, producing undesired transverse magnetization. This can be dispersed or dephased by the application of a gradient pulse called the "TE-crushers." Because the spins of interest are in the transverse plane during the entire sequence, TE-crushers can be placed directly around the refocusing π pulses. Note also that the gradient direction of the TE-crusher pairs around the two refocusing pulses should be

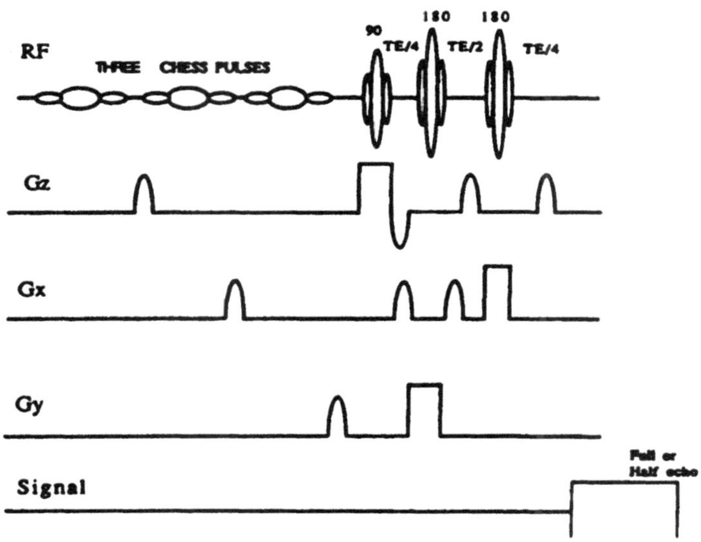

Figure 17. The pulse sequence of point-resolved spectroscopy (PRESS).

orthogonal or differing in magnitude by a factor of two in order to avoid rephasing of a possible unwanted stimulated echo. Minimum gradient dephasing power of the TE-crushers depends on B_1 and B_0 homogeneity over the sensitive region of the receiver coil. Water suppression is achieved by the application of a chemical shift-selective RF pulse followed by a dephasing gradient pulse (CHESS). These repeated CHESS sequences are constructed to avoid occurrence of unwanted echoes and interference with the localization scheme (30). One of the characteristics of this sequence is that the minimum echo delay achievable is quite long (more than 30 ms) on most clinical scanners. This feature may be a serious drawback in detecting resonances with short T2s. A conventional spin-echo sequence excites an echo after applying a pulse sequence of 90° – TE/2 – 180° – TE/2. A second 180° pulse after the conventional spin echo sequence will give rise to a spin-echo called *double echo*.

3.9.2. Stimulated Echo Acquisition Mode (STEAM)

Figure 18 explains the STEAM sequence. This sequence contains three 90° pulses, each of which is slice selective in one of three perpendicular directions. The stimulated echo occurs only in the volume, which was excited by each of the pulses.

Gradients are appropriately placed to dephase any spins, which are excited by the pulses but are not contained in the desired volume. The three CHESS pulses followed by dephasing gradients, which are applied before the three slice-selective 90° pulses, achieve solvent suppression. There have been a number of different reports dealing with the optimization of the CHESS pulses, the order of the dephasing gradient pulses and their durations. Using some of these approaches, excellent solvent suppression factors can be obtained for very short echo times (30).

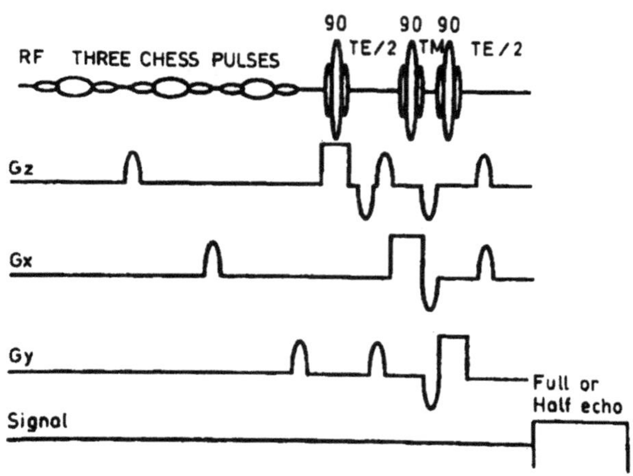

Figure 18 .The stimulated echo acquisition mode spectroscopy (STEAM) pulse sequence.

The TE-crushers in STEAM and PRESS cause a large phase dispersion to avoid unwanted transverse magnetization originating from regions outside the volume of interest. Because of the nearly complete dephasing by the first TE-crusher, the second $\pi/2$ RF pulse in STEAM rotates only half of the spins of interest to the *Z*-axis. This part of the magnetization will eventually lead to the stimulated echo. The other half will be further dispersed by the TM-crusher and will not lead to rephased magnetization at the time of acquisition. There is an additional signal loss of 50% due to the TE-crushers. The net loss of 75% of the potential signal is the distinct disadvantage of STEAM (30).

4. DATA PROCESSING

After the FID is collected in the time domain, it needs to be conditioned, corrected, and transformed into the frequency domain to see the chemical shift of the metabolites. This transformation is brought about by a mathematical operation called Fourier transformation.

4.1. Zero Filling

Some manipulations are done to the FID in the time domain before performing Fourier transformation. For example the NMR FID signal can be written as

$$S(t) = \sum_j A_j e^{i(\omega_j t + \phi_j) - t/T2_j^*} \qquad (t \geq 0) \tag{24}$$

here A_j, ω_j, ϕ_j and $T2_j^*$ are the amplitude, frequency, phase, and effective decay time of the spectral component j. The goal is to increase sensitivity (SNR) and improve spectral resolution. Fourier transformation contains the same number of data points as were transformed. For example, transformation of decays collected in quadrature dual channels, each containing $2K$ points, produces real and imaginary parts each containing 2K points. Since the frequency separation between adjacent points (spectral resolution) in the transformed spectrum is equal to 1/acquisition time, this implies that this separation would be similar in magnitude to the width of the signal if the acquisition time were equal to $4T2^*$. It is preferable to increase the number of complex data points by a technique called *zero filling*. It consists of extending the experimental time domain data set by a number of zeroes. This has a similar effect as increasing the sampling points after the signal has decayed. It results in a smooth interpolation between the spectral data points and improves the appearance of the spectrum after Fourier transformation (26).

4.2. Apodization

Apodization means "cutting the feet off." It indicates a mathematical operation by which all of the data points of the FID are multiplied by a function characterized by a starting value of 1 and that decays regularly to zero at the end of the acquisition time. There are many such mathematical functions used in the time domain. The selection of the proper one may depend on the application.

4.3. Phase Correction

This manipulation is done in the frequency domain. In general, phase correction is necessary for the following reasons:

1. The spectrometer phase-sensitive detector phase is not set correctly.
2. There is a delay between the excitation pulse and the start of the data acquisition; the FID is truncated.
3. There are off-resonance effects when the RF amplitude is not large compared to the spectral width.
4. The filter bandwidth is about equal to or smaller than the spectral width.

4.4. Baseline Correction

The spectral baseline is usually nonuniform and is corrected by fitting a line or polynomial to a selected part of the spectrum. A major contributor is the residual water due to incomplete suppression of water. This procedure is quite critical and must be applied with care because peak heights and areas may be affected and may lead to errors.

4.5. The Significance of Observed Metabolites

After the procedures described in the previous sections, one obtains a spectrum. Figure 19 displays ^{1}H NMR spectra recorded in the occipitoparietal white/gray matter region of a normal control subject (b) and a patient (a) with diagnosed with hepatic encephalopathy (31). Several peaks are identified in the spectrum. Their assignments at 1.5 T are 2.01 ppm = *N*-acetyl-L-aspartate, 3.0 ppm = creatine, 3.2 ppm = choline, and 3.54 ppm = myoinositol. There is a complex pattern of coupled resonances between 2.1 and 2.5 ppm together with a further group of resonances around 3.7 ppm that are assigned to glutamate (Glu) and glutamine (Gln). Since the multiplets due to Glu and Gln are unresolvable, these peaks are denoted by Glx.

4.5.1. N-Acetyl-L-aspartate (NAA)

In 1984, Koller et al. showed that NAA existed in higher concentration in gray matter than in white matter, and disappeared when the neurotoxin kainic acid was injected into neurons (32). Arnold et al. showed a decrease in brain NAA in patients with chronic multiple sclerosis (MS), a disease where there is a loss of axons. Conversely, in acute MS, where axons have not yet disappeared, there was a normal concentration of NAA (33). These studies suggest that NAA might be largely present within the neurons and can be used as a neuronal marker. A methylgroup of NAA has a singlet at 2.02 ppm, methylene multiplets at 2.5 ppm, and methine multiplets at 4.35 ppm.

The role of NAA in the brain is unknown. *N*-Acetyl-aspartyl-glutamate (NAAG) is colocalized with NAA in neurons and releases NAA and glutamate when it is cleaved by *N*-acetylated alpha-linked dipeptidase (34–36).

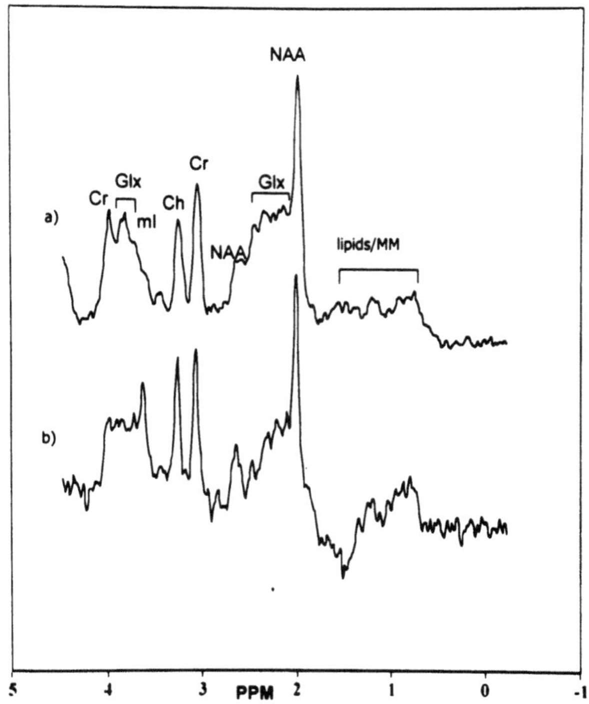

Figure 19. Proton spectra from 8 ml volume localized in the occipitoparietal region of (a) an HE patient and (b) a healthy control.

4.5.2. *Creatine*

Creatine is taken in the diet and is synthesized within the liver, kidneys, and pancreas. The precursor molecules for creatine are arginine, glycine, and *S*-adenosylmethionine. Approximately 120 g of creatine and creatine phosphate is present in human muscle and neurons and about 2 g per day is replaced by diet and *de novo* synthesis (37). In ^{1}H MRS, both creatine and creatine phosphate are measured as a single peak. Creatine phosphate serves as a reserve for high-energy phosphates in the cytosol of muscle, neurons, and buffers cellular ATP/ADP reservoirs. Creatine kinase converts creatine to creatine phosphate utilizing ATP. Tissues like muscle and brain where the largest changes in energy metabolism occur have the largest concentrations of creatine kinase (38). Creatine has two groups of peaks, one at 3.0 ppm and the other at 3.9 ppm.

4.5.3. Choline

Choline (Cho) is present in most foods and absorbed in the diet. It is the precursor for two important molecules, acetylcholine (ACho) and phosphatidylcholine (PtdCho). While all cells use Cho to synthesize PtdCho, which is a constituent of the cell membrane, ACho is found only within cholinergic neurons (39). ACho is a neurotransmitter that is critical for many aspects of memory, cognition, and mood. Changes in ACho also occur in Alzheimer's disease. There is some uncertainty as to which of the Cho-containing compounds contribute to the peak. Not all of the major water-soluble Cho-containing compounds add up to the concentration seen in the spectrum. However, the relatively immobile lipid-soluble PtdCho has been shown to contribute to the *in vivo* spectrum as well (40). The dominant methyl group appears at 3.15 ppm.

4.5.4. Myoinositol

Myoinositol was discovered more than 100 years ago and is a cycle sugar alcohol. Its function is still uncertain. Inositol diphosphate (IP_2) and diacylglycerol, both liberated from phosphatidylinositol by phospholipase, are active in releasing Ca^{2+} from endoplasmic reticulum and from mitochondria. Thus, IP_2 performs a vital role in hormone–receptor binding with its intracellular activity, i.e., these are second or third messengers for type IIb hormone actions, including vasopressin, thyrotropin, alpha-adrenergic activity, angiotensin II, and gonadotropin-releasing hormone (41). However, intracellular messengers IP_2 and IP_3 are present only in micromolar concentrations and cannot contribute to the more than 5 mM observed peak. It may be that myoinositol is the "storage" form of the IP_n messenger system (42). The observable peak at 3.5 ppm is used as representative of myoinositol.

REFERENCES

1. Lufkin RB. The MRI manual. St. Louis: Mosby, 1997.
2. Haacke M. Magnetic resonance imaging: Basic principle and sequence Design. New York: Wiley, 1999.
3. Stark DD, Bradley WG. Magnetic resonance imaging. St. Louis: Mosby, 1996.
4. Hennig J, Nauerth A, Friedburg H. RARE-imaging: A fast imaging contrast in fast spin-echo MR imaging. J Magn Reson Med 1986; 3: 823–833.
5. Melki PS, Mulkern RV. Magnetization transfer effects in multislice RARE sequences. Magn Reson Med 1992; 24:189–195.
6. Henkelman RM, Hardy PA, Bishop JE, Poon CS, Plewes DB. Why fat is bright in RARE and fast spin-echo imaging. J Magn Reson Imging. 1992; 2:533–540.
7. Jones KM, Mulkern RV, Mantello MR, et al. Brain hemorrhage: Evaluation with fast spin-echo and conventional dual spin-echo images. Radiology 1992; 182:53–58.

8. Mansfield P. Multi-planar image formation using NMR spin-echoes. J Phys C 1997; 10:L55–L58.
9. Tourtellote WW, Parker JA. Some spaces and barriers in postmortem multiple sclerosis. Prog Brain Res 1968; 29:493–525.
10. Hajnal JV, DeCoene B, Lewis PD, et al. High signal regions in normal white matter shown by heavily T2-weighted CSF nulled IR sequence. J Comput Assist Tomogr 1992; 16:506–513.
11. DeCoene B, Hajnal JV, Gatehouse P, et al. MR of the brain using fluid-attenuated inversion recovery (FLAIR) pulse sequences. Am J Neuroradiol 1992; 13:1555–1564.
12. DeCoene B, Hajnal JV, Pennock JM, Bydder GM. MRI of the brain stem using fluid attenuated inversion recovery pulse sequences. Neuroradiology 1993; 35:327–331.
13. Rydberg JN, Hammond CA, Grimm RC, et al. Initial clinical experience in MR imaging of the brain with a fast fluid attenuated inversion recovery pulse sequence. Radiolgy 1994; 193:173–180.
14. Tanttu JI, Sepponen RE, Lipton MJ, Kuusela T. Synergistic enhancement of MRI with Gd-DTPA and magnetization transfer. J Comput Assist Tomogr 1992; 16:19–24.
15. Gupta RK, Kathuria MK, Pradhan S. Magnetization transfer MR imaging in CNS tuberculosis. Am J Neuroradiol 1999; 20: 867–875.
16. Prager JM, Rosenblum JD, Huddle DC, Diamond CK, Metz CE. The magnetization transfer effect in cerebral infarction. Am J Neuroradiol 1994; 15:1497–1500.
17. Dousset V, Grossman R, Ramer KN, et al. Experimental allergic encephalomyelitis and multiple sclerosis: Lesion characterization with magnetization transfer imaging. Radiology 1992; 182:483–491.
18. Stejskal EO, Tanner JE. Spin diffusion measurements: Spin echoes in the presence of a time dependent field gradient. J Chem Phys 1965; 42:288–292.
19. Stejskal EO, Tanner JE. Use of spin echo in pulsed magnetic field gradient to study anisotropic restricted diffusion and flow. J Chem Phys 1965; 43: 3579–3603.
20. Kasler F. *Quantitative analysis by NMR spectroscopy.* New York: Academic Press, 1973.
21. Chakeres DW, Schmalbrock P. *Fundamentals of magnetic resonance imaging.* Baltimore: Williams & Wilkins, 1992.
22. Abraham RJ, Fisher J, Loftus P. *Introduction to NMR spectroscopy.* New York: Wiley, 1988.
23. Spisni A. 1D spectrum analysis. In *Magnetic resonance spectroscopy in biology and medicine,* JD deCertaines, WMMJ Bovée, F Podo, eds. New York: Pergamon Press, 1992.
24. Hoult DI. The NMR receiver: A description and analysis of design. Prog NMR Spectrosc 1978; 12: 41–77.
25. Bovée WIM, Creyghton J. Acquisition and processing of the NMR data. In *Magnetic resonance spectroscopy in biology and medicine,* JD deCertaines, WMMJ Bovée, F Podo, eds. New York: Pergamon Press, 1992.
26. Lindon JC, Ferrige AG. Digitization and data processing in Fourier transform NMR. Prog in NMR Spectrosc 1980; 14: 27–66.
27. Leibfritz D. Water suppression. In *Magnetic resonance spectroscopy in biology and medicine,* JD deCertaines, WMMJ Bovée, F Podo, eds. New York: Pergamon Press, 1992.
28. Bolinger L, Lenkinski RE. Localization in clinical NMR spectroscopy In *Biological magnetic resonance; vol. II. In vivo spectroscopy,* LJ Berliner, J Reuben, eds. New York: Plenum Press, 1992.
29. Moonen CTW, von Kienlin M, van Zijl PCM, et al. Comparison of single-shot localization methods (STEAM and PRESS) for in vivo proton NMR spectroscopy. NMR Biomed *1989*; 2: 201–208.
30. Thomas MA, Yue K, Binesh N et al. Localized two-dimensional shift correlated MR spectroscopy of human

brain. Magn Reson Med (in press).

31. Thomas MA, Huda A, Guze B, et al. Cerebral 1H MR spectroscopy and neuropsychological status of patients with hepatic encephalopathy. Am J Roentgenol 1998; 171:1123–1130.
32. Koller KJ, Zaczek R, Coyle J. N-Acetyl aspartyl-glutamate: Regional levels in rat brain and the effects of brain lesions as determined by a new HPLC method. J Neurochem 1984; 43: 1136–1142.
33. Arnold DL, Mathews PM, Francis G, Antel J. Proton magnetic resonance spectroscopy of human brain in vivo in the evaluation of multiple sclerosis: Assessment of the load of disease. Magn Reson Med 1990; 14: 154–159.
34. Passani L, Tsai G, Slusher BS, Simmons M, Coyle JT. Immunocytochemical co-localization of N-acetyl-aspartate and N-acetylaspartylglutamate in monkey. Soc Neurosci Abstr 1990; 16: 1190.
35. Simmons ML, Frondoza C, Coyle JT. Monoclonal antibodies to N-acetyl-aspartate. Soc Neurosci Abstr 1990; 16: 1190.
36. Rothstein JD, Tsai G, Kuncl RW, et al. Abnormal excitatory aminoacid metabolism in amyotrophic lateral sclerosis. Ann Neurol 1990; 28: 18–25.
37. Rodwell VW. Conversion of aminoacids to specialized products. In *Harper's review of biochemistry*. Los Altos, CA: Lange Medical Publications, 1981.
38. Miller BL. A review of chemical issues in 1H-NMR spectroscopy: N-acetyl-L-aspartate, creatine, and choline. NMR Biomed 1991; 4: 47–52.
39. Freeman JJ, Jenden DJ. The source of choline for acetylcholine synthesis in brain. Life Sci 1976; 19: 949–962.
40. Riedy G, Miller BL. Phosphatidylcholine contributes to the choline peak in 1H-NMR spectroscopy. Bull Clin Neurosci. (in press).
41. Berridge MJ. Inositol triphosphate and diacylglycerol as second messengers. Biochem J 1984; 220: 345–360.
42. Ross BD. Biochemical considerations in 1H spectroscopy. Glutamate and glutamine; myo-inositol and related metabolites. NMR Biomed 1991; 4: 59–63.

Chapter 2

PYOGENIC INFECTIONS

Sudhakar K. Venkatesh and Rakesh K. Gupta

Pyogenic bacterial infections of the central nervous system are dynamic processes influenced by multiple interactions between microorganisms and host defenses. Brain and spinal cord are relatively well protected from infective agents by the skull bones and vertebral column, the meninges and blood–brain barrier. Thus, bacterial species must overcome considerable barriers to produce disease. However, once the bacterial pathogens enter the CNS, host defense mechanisms are frequently suboptimal regarding control of the infection. In addition, preexisting immunosuppressive conditions in the host are frequent in modern clinical practice. This may account for the continuing high mortality and morbidity rates from bacterial infections of the CNS despite the advances in diagnosis and treatment that have been made in recent years.

Various clinico-pathologico-radiological features may be found depending on the site of the infection and the type of bacterial agent. In this chapter, we will focus on meningitis, brain abscess, spinal infections, infective spondylitis, and epidural and subdural abscesses.

Sudhakar K. Venkatesh and Rakesh K. Gupta • Department of Radiodiagnosis, Sanjay Gandhi Post-Graduate Institute of Medical Sciences, Lucknow 226014, India.

MR Imaging and Spectroscopy of Central Nervous System Infection, edited by Gupta and Lufkin. Kluwer Academic / Plenum Publishers, New York, 2001.

1. PYOGENIC BRAIN INFECTIONS

1.1. Pyogenic Meningitis

Adam et al. described the pathology of bacterial meningitis in 1948 based on autopsy material from patients with *H. influenzae* meningitis (1). Since then, experimental models of bacterial meningitis, knowledge of host defense mechanisms, and pathophysiology of associated complications have allowed a more complete understanding of the pathologic processes operating in this entity (2).

Bacteria reach the meninges through one of the following pathways: (1) hematogenous dissemination from a distant site (e.g., lung, skin, genitourinary tract). (2) spread from an adjacent suppurative focus of infection (e.g., otitis media, sinusitis, mastoiditis), (3) vegetations from a congenital or an acquired structural cardiac defect. Once bacteria gain access to the subarachnoid space, an inflammatory process ensues. Neutrophils migrate into the subarachnoid space producing purulent exudate. On gross examination, the exudate has a grayish yellow or yellowish green appearance. The purulent exudate extends to the convexities of the hemispheres and to the various cisterns in the base of the brain. It may also extend to the posterior surface of the spinal cord, arachnoid sheaths of the cranial nerves, and into the perivascular spaces of the cortex. A small exudate may also be found in the ventricular fluid and attached to the ventricular wall and choroid plexus (3,4).

Microscopic examination of the exudate in the early stages of infection demonstrates a large number of neutrophils and bacteria. The presence of a large number of bacteria in the exudate suggests that phagocytosis by neutrophils is incomplete as a result of deficient opsonic activity in the CSF. Within the first 48–72 hours of infection, there is evidence of inflammation in the walls of the small and medium-sized subarachnoid arteries. The endothelial cells swell and multiply, narrowing the lumen. There is an infiltration of the subintimal layer and adventitia by neutrophils and lymphocytes. The meningeal veins become distended and develop mural inflammation during bacterial meningitis. There may be focal necrosis of the vessel wall along with mural thrombus formation in the lumen of the vein or in the dural sinus. Hemorrhagic cortical infarction is the result of cortical venous and dural sinus thrombosis (2–4).

Toward the end of the first week of meningeal infection, neutrophils begin to degenerate and are removed by macrophages. Lymphocytes and fibroblasts proliferate in the exudate.

Microscopic changes in the brain may present in the form of astrocyte swelling, loss of myelinated fibers in ischemic/hypoxic injury, suggesting that hypoxia may contribute to the pathologic changes from bacterial meningitis at this stage. As the infection progresses, the subarachnoid exudate continues to accumulate and the flow of CSF may be obstructed. It may result in communicating or obstructed hydrocephalus. The development of diffuse cerebral edema and increased intracranial pressure further complicates the pathologic changes. The cerebral edema results in increased intracranial pressure and adversely affects cerebral perfusion pressure. Loss of autoregulation of the cerebral blood flow in bacterial meningitis may be another potential contributing factor to the development of cerebral ischemia. The cerebral edema may lead to cerebral herniation and result in apnea, hemodynamic instability, coma, and death (2–6).

The clinical symptoms in bacterial meningitis depend on the age of onset of the illness. The initial symptoms of bacterial meningitis in children and adults may be fever, neck stiffness, headache, lethargy, irritability, nausea, vomiting and photophobia. Although the signs are nonspecific, the combination of one or more of these symptoms with other signs of meningeal irritation suggests the diagnosis of meningitis. In the elderly, the symptoms and signs are usually fever, confusion or coma, headache, nuchal rigidity, altered consciousness and seizures (4–6).

Diagnosis of meningitis is usually made by examination of CSF. The CSF of a meningitis patient typically shows low glucose (<40 mg/dl), high protein (>50 mg/dl), pleocytosis with predominant neutrophilia, and positive Gram's stain (70–90% of untreated cases). Culture of CSF is positive in approximately 80% of cases of bacterial meningitis; 48 hours is usually required for accurate identification. Patients who are treated with antibiotics before admission have a significant decrease in CSF neutrophils, protein, and positive Gram stain. In general, bacteria should not be seen on Gram's stain or grow in culture from CSF examined 24 hours after treatment has begun with appropriate antibiotic therapy. Polymerase chain reaction has been utilized to amplify bacterial DNA from patients with meningitis caused by various meningeal pathogens, including *H. influenzae*, *N. meningitidis*, streptococci, *S. pneumoniae*, *S. aureus*, gram-negative bacilli, anaerobes, and *L. monocytogenes*. The sensitivity and specificity are 91%, although there are problems with false positive results (5–6).

Conventional MR imaging without contrast injection in patients with uncomplicated acute bacterial meningitis may be unremarkable (7). Postgadolinium MR studies may demonstrate leptomeningeal enhancement similar to what is observed on contrast enhanced CT (Fig.1). However, this enhancement is often better demonstrated by MR, with its multiplanar capability. In experimental studies performed on dogs with *S. aureus* meningitis, T1-weighted MR images with gadolinium demonstrated abnormal meningeal enhancement that more closely approximated the extent of inflammatory cell infiltration than CT (8). Postcontrast magnetization transfer (MT) T1-weighted images are shown to be more sensitive than conventional T1-weighted images in delineating the abnormal meningeal enhancement and are found to correlate with degree of meningeal inflammation on histopathology (8). Distention of the subarachnoid spaces on MR imaging is an early sign of severe meningitis. The complications of meningitis described above are better demonstrated with conventional pre- and postcontrast MR imaging. Patients with focal neurological abnormalities show a variety of abnormalities on MR that have helped in prognosis and better understanding of the pathophysiology of the disease. Pyogenic meningitis may be associated with subdural effusion, hydrocephalus, cerebral edema, cerebritis, and infarction (6,7,9,10).

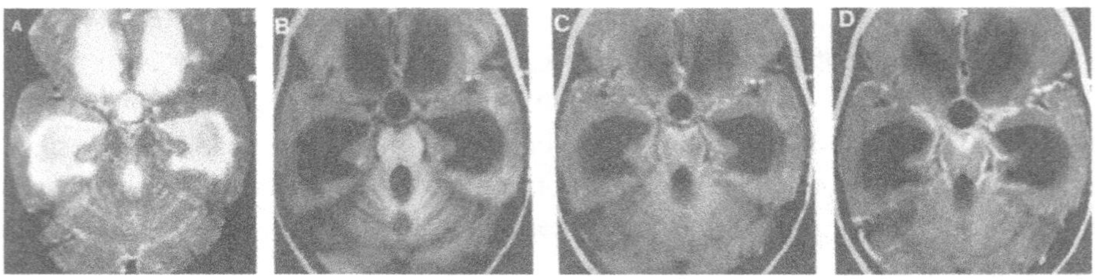

Figure 1. Pyogenic meningitis. T2-weighted axial image (a) at the level of midbrain shows evidence of dilated lateral and third ventricles.T1-weighted axial image (b) shows no other abnormality. MT T1-weighted axial image (c) shows hyperintense thick meninges in the basal cisterns and along the Sylvian fissures and anterior interhemispheric fissure which show enhancement in postcontrast MT T1-weighted image (d).

1.1.1. Subdural Collections

On MR, subdural effusions can often be distinguished from empyema because effusions are isointense to CSF signal, whereas empyema, with higher protein content, is often of greater signal intensity than CSF on T1- and proton density -weighted images. Subdural effusions, such as those that occur in association with *H. influenzae* meningitis in children, may be bilateral and are usually crescent collections adjacent to the frontal and parietal lobes. Calcification of the margins of an effusion is an uncommon late sequel (11). Postgadolinium images should show an enhancing membrane associated with empyema but not with a simple effusion (11). The presence of signal alterations in the cortex subjacent to an extra-axial collection more likely indicates empyema, with associated cortical inflammatory changes, than a simple effusion (9–11).

1.1.2. Hydrocephalus

Hydrocephalus may be either communicating or obstructive, and is a complication seen more frequently in children than in adults (6). On MR, T1- weighted images in the sagittal and coronal planes permit direct evaluation of midline structures and better delineate the site of obstruction. Periventricular CSF accumulation secondary to ventricular obstruction is seen on proton density- and T2-weighted images as areas of hyperintense signal surrounding the ventricular system (11). Similar periventricular hyperintensity may be seen in cases of ependymitis (Fig. 2) without hydrocephalus; however, in those cases contrast MR would demonstrate the enhancement (11). One sequel of ependymitis is the trapped fourth ventricle, which has the appearance of an expanding posterior fossa cystic mass resulting in obstructive hydrocephalus (11).

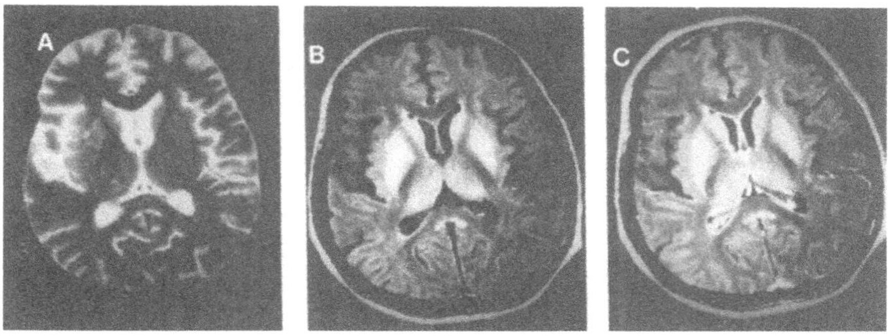

Figure 2 Pyogenic meningitis with ependymitis. T2-weighted axial image (a) at the level of thalami shows no obvious abnormality. T1-weighted MT image (b) shows hyperintense ventricular walls of bilateral frontal and right occipital horn which show enhancement in postcontrastT1-weighted image (c) consistent with ependymitis.

1.1.3. Cerebritis and Edema

Cerebritis appears as areas of increased signal intensity on T2-weighted images. Differentiation of cerebritis from infarction should become apparent with time because cerebritis typically evolves to a well-formed abscess, whereas the signal changes present with arterial infarction will conform to a distinct vascular distribution (6,11). However, these entities may be difficult to distinguish. The incidence of cerebral edema is variable; recent studies have shown it to occur in 2–5% of cases (6, 12). Cerebral edema determined primarily by changes in brain water contents is a consistent finding in most animal studies of meningitis and has been considered important in the pathogenesis of the disease, primarily by increasing the intracranial pressure and reducing the cerebral blood flow.

Edema is most prominent in the frontal and parietal lobes, with follow-up imaging frequently demonstrating some degree of atrophy.

1.1.4. Cerebral Infarction

Infarction is relatively common (2–19%) in bacterial meningitis and is a major determinant of outcome (6). Cortical and subcortical ischemic infarctions are evident as areas of increased signal intensity onT2-weighted images. Occlusion of small perforating

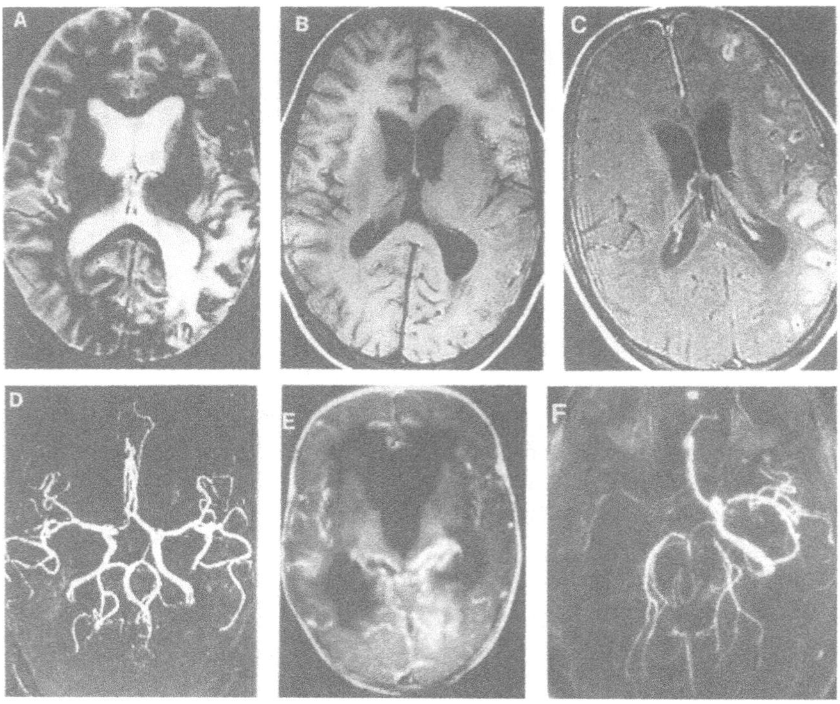

Figure 3. Pyogenic meningitis sequelae.T2-weighted axial image (a) at the level of third ventricle shows a hyperintense area in the left middle cerebral artery territory with dilatation of ipsilateral ventricular system.T1-weighted axial image (b) is unremarkable except for dilatation of ventricle. PostcontrastT1-weighted image (c) shows gyral enhancement in the left parieto-occipital region. MR angiogram (d) shows no obvious vascular abnormality. In another patient with pyogenic meningitis, post-contrastT1-weighted axial image (e) shows general enhancing meninges with hydrocephalus. MR angiogram (f) shows no flow signal in right internal carotid artery suggestive of occlusion.

arteries results in focal infarcts of the basal ganglia, whereas spasm of anterior or middle cerebral arteries may lead to massive infarction. Cortical venous occlusion with or without dural venous thrombosis, results in hyperintense areas in the subpial cortex and underlying white matter, which do not confirm to the expected arterial distribution. A favored site for this complication is near the vertex, secondary to superior sagittal sinus thrombosis (13). Hemorrhagic infarction has a characteristic appearance on T2-weighted images. Depending on the age of the infarct, areas of hemorrhage appear decreased in signal intensity acutely and of increased intensity when subacute, on T2-weighted images. OnT1-weighted images, the hemorrhage increases in signal intensity with evolution from the acute to the subacute phase. For subacute infarcts, the pattern of enhancement on MR mimics the gyriform pattern (11). MR angiography may show a focal decrease in signal in the affected arteries in the early part of the disease, which is shown to recanalize on treatment. If an MR angiogram is done during the recovery phase while the patient is in treatment, it may be normal even when the territorial infarction is visible on imaging (Fig.3). This is due to the recanalization of the vessels once the disease activity diminishes (11,14). Similarly, MR venograms may demonstrate the occlusion if there is venous infarct (11). Since complications of cerebral infarction influence the prognosis of meningitis, repetitive MR angiograms and venograms may be beneficial in patients with bacterial meningitis in order to evaluate the vascular lesion.

MR spectroscopy. Bacterial meningitis is associated with a decrease in CSF pH and glucose concentrations and an elevation in lactate (15). These abnormalities are thought to be due to increased activity of leukocytes or bacteria and anaerobic glycolysis in brain cells.

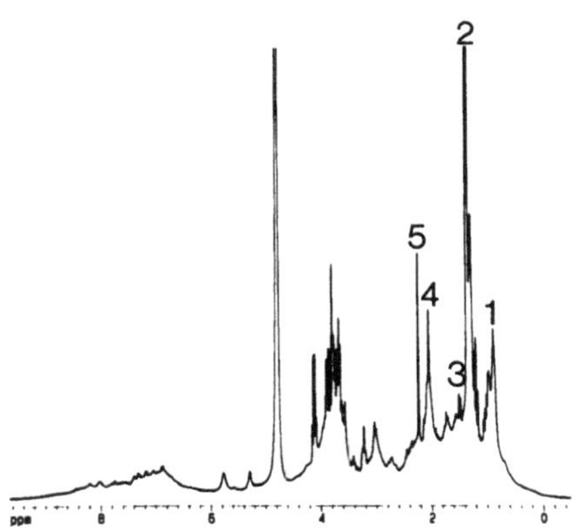

Figure 4. Ex vivo high-resolution proton MR spectroscopy of CSF in a proven case of pyogenic meningitis shows large amount of amino acids (1), lactate (2), Alanine (3), acetate (4), and acetoacetate (5).

It has been suggested that CSF lactic acidosis could also be caused by impaired cerebral perfusion (6). Other studies have found decreased brain pH without differences in cellular high-energy metabolites in the face of CSF lactic acidosis and hypoglycemia in animal studies and in humans using phosphorus magnetic resonance spectroscopy (16). On *in vivo* proton MR spectroscopy, only mild elevation of lactate with normal NAA, Cr, Cho, and inositols has been reported (6). CSF lactate accumulation therefore is unlikely to be caused by brain energy failure and suggests that the cause of encephalopathy associated with meningitis remains unknown (6).

Ex vivo high-resolution proton MR spectroscopy of CSF in patients with pyogenic bacterial meningitis shows large amounts of amino acids, lactate, acetoacetate and reduced glucose (Fig.4) (15).

1.2. Brain Abscess

Brain abscess is a focal suppurative process within the brain parenchyma that continues to be a diagnostic and therapeutic challenge to the clinician. Paradoxically, the mortality and morbidity from a brain abscess remained high until relatively recently despite potent, specific antimicrobial therapy and advances in neurosurgical technique. The incidence of brain abscess has remained relatively stable in the antibiotic era: It varies from 0.18 to 1.3% in the developed world, but remains a significant problem in the developing world. Predisposing causes usually include otitis media and mastoiditis, frontoethmoidal and sphenoid sinusitis, dental sepsis, congenital heart disease, bacterial endocarditis, immunosuppressive conditions, lung abscess, empyema and bronchiectasis. No predisposing factors are recognized in 15–20% of reported cases (17). Aerobic or microaerophilic streptococci are frequently found in mixed infections and the total number of organisms isolated is greater than the number of abscesses in a given series. The majority of the aerobes in different series responsible for the brain abscess are *Staphylococcus aureus*, gram-negative bacilli, and streptococci. The anaerobic organisms include *Bacteroides* and *Prevotella* species such as *B. fragilis*, anaerobic streptococci, *Clostridium* species, *Fusobacterium* species, *Actinomyces* species and others (17).

Brain abscess occurs most commonly in association with one of three distinct clinical settings: (1) a contiguous focus of infection (Fig.5), (2) hematogenous spread from a distant focus, and (3) cranial trauma. The majority of the patients with brain abscess demonstrate a contiguous focus of infection, usually sinusitis or otitis. The majority of the otogenic brain abscess is located in the temporal lobe. The cerebellum is the next most commonly affected area, and it has been observed that 85–99% of the cerebellar abscesses are secondary to otogenic infections. Most otogenic abscesses are solitary lesions.

Hematogenous brain abscesses often share the following characteristics: (1) a distant focus of infection, most often within the chest; (2) location in the distribution of the middle cerebral artery; (3) initial location at the gray matterwhite matter junction; (4) poor encapsulation; and (5) high mortality. These abscesses are more commonly multiple and multiloculated as compared with those that have an origin in foci of contiguous infection (17). Brain abscess is the least common intracranial complication of either traumatic injury to the brain or neurosurgery (17).

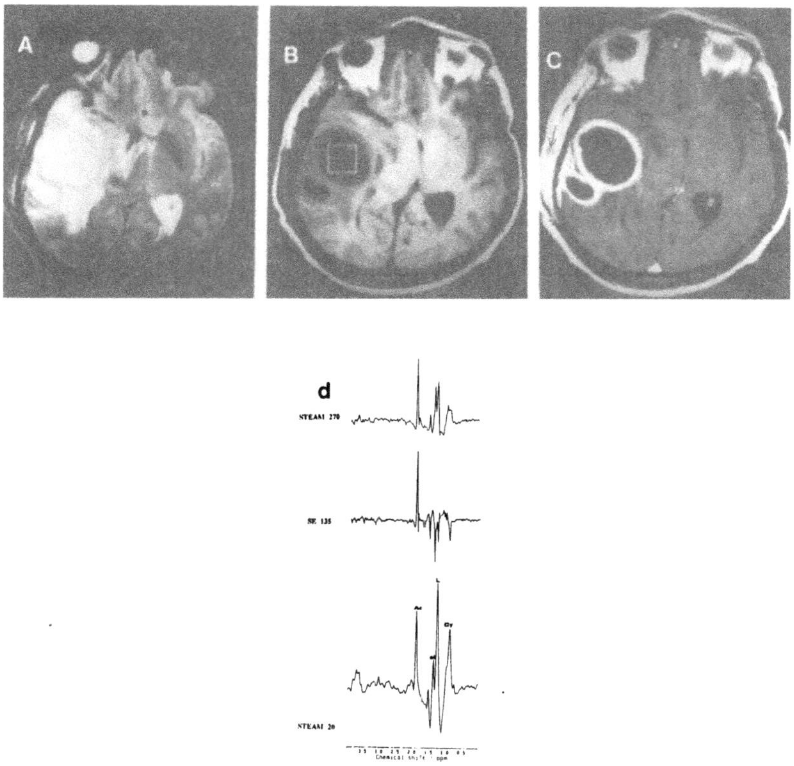

Figure5. Pyogenic abscess secondary to a skin abscess: T2- weighted axial image (a) at the level of thalami shows presence of two hyperintense areas with thin hypointense rim and perifocal edema and mass effect. Note the hyperintensity in the scalp of the right temporal region. T1-weighted image (b) shows isointense rim of the lesion. Note a small daughter lesion lying inferolateral to the lesion. Postcontrast T1-weighted image (c) shows rim enhancement of the lesion along with adjacent dura. Note the enhancement of the adjacent scalp with a small hypointense area (arrow) suggestive of an abscess. *In vivo* proton MR spectroscopy (d) from the cystic cavity with STEAM 20 ms sequence shows presence of cytosolic amino acids (Cy), lactate (L), alanine (al), acetate (Ac) characteristic of an abscess. Note the phase inversion of Cy, L, al at SE 135 ms sequence with no change in acetate and normal phase of all metabolites at STEAM 270 ms.

The final common pathway for brain abscess development in the preceding conditions appears to require a compromised area of the brain. Experimental data suggest that infection is extremely difficult to establish in normal brain tissue. Brain abscess may develop from a contiguous infection via two major mechanisms: (1) direct extension through areas of associated osteitis or osteomyelitis and (2) retrograde thrombophlebitic spread via diploic or emissary veins into the intracranial compartment. Additional possibilities in the case of otogenic infection include spread through preexisting channels (the internal auditory canal, cochlear and vestibular aqueducts) or between temporal suture lines (17). The development of brain abscess in the setting of neonatal meningitis deserves special comments. The pathogenesis appears to follow the initiation of a necrotizing vasculitis leading to subsequent hemorrhagic necrosis and liquefaction of the subcortical white matter, an area highly susceptible to changes in cerebral perfusion. An alternative hypothesis is initial ventriculitis followed by ependymal disruption and subsequent direct extension of the infection into the brain parenchyma (17) (Fig.6).

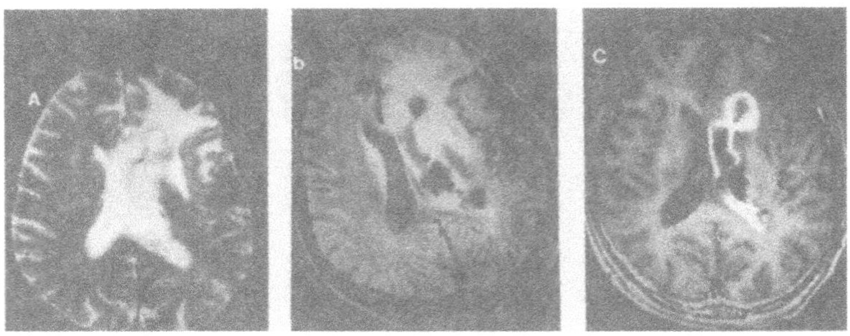

Figure 6. Pyogenic abscess with extension into ventricle. T2-weighted axial image (a) at the level of ventricles shows evidence of a hyperintense area in the left frontal region and a hypointense rim with perifocal edema. Note the heterogeneous intensity of left lateral ventricle. MTT1-weighted image (b) shows hyperintense rim of the lesion blending with the perifocal edema. Note the hyperintense left ventricular wall with irregularity especially posteriorly. Postcontrast T1 image (c) shows rim enhancement of the frontal lesion with enhancing ventricular wall suggestive of ventriculitis.

In approximately decreasing order of frequency, a solitary abscess may involve the following brain regions: frontal, temporal, frontoparietal, parietal, cerebellar, and occipital lobes. Intrasellar, brainstem, basal ganglia, and thalamic abscesses are rare. Once the infection is established in the brain, acute inflammatory cells are recruited and local vascular permeability is altered. The evaluation of an abscess includes four histopathologic stages. This staging process described in animal models of brain abscess, correlates with human brain abscess evolution (18,19). The first stage is an early cerebritis (days 1–3 following intracerebral inoculation in animals), which progresses to a perivascular inflammatory response surrounding the developing necrotic center by the third day. Profound edema in the surrounding white matter develops concurrently. In the second stage of late cerebritis, development of a wellformed necrotic center reaches its maximum size.

Additionally, fibroblasts appear, setting the stage for capsule formation and a marked increase in neovascularity at the periphery of the necrotic zone. These newly formed capillaries lack tight junctions and leak proteinaceous fluid. Surrounding this is the beginning of a reactive astrocyte response, along with persistent white matter edema. The third stage, early capsule formation (days 10–13), is characterized by a slight decrease in the size of the necrotic center. At this point, there is a well-developed layer of fibroblasts, with significantly more reticulin deposition on the cortical side of the lesion than on the ventricular side. Outside this developing capsule is a region of persistent cerebritis and neovascularity, with a further increase in reactive astrocytes. These processes continue in the fourth and final stage, late capsule formation (days 14 and later). The capsule continues to thicken, with an abundance of collagen present by the third week. It was more recently shown that basic fibroblast growth factor beta is produced early in brain abscess formation by a continued influx of leukocytes and nonneuronal proliferating cells. A loss of growth factor production and fibrosis was only seen in chronic abscess (18,19).

The clinical manifestations of a brain abscess can vary greatly among patients and depend on several interdependent factors. These factors include the relative virulence of the infecting organism, the host immune status, abscess location, number of lesions, and the presence or absence of associated meningitis or ventricular rupture. Additionally, symptoms of the primary site of infection may predominate. The duration of symptoms may vary from 3 to 120 days (median 13.3 days). The usual clinical symptoms and signs are headache, fever, focal neurological deficit, nausea/vomiting, seizures, papilledema, and nuchal rigidity. Because the symptoms are nonspecific and frequently lack fever, brain abscess may be confused with several other clinical problems (17).

T2-weighted spin-echo MR images demonstrate early cerebritis as an area of increased signal intensity indistinguishable from or slightly hypointense to the surrounding edema (11,20). The inherent sensitivity of proton density- and T2-weighted images to alteration in tissue water enables earlier detection of cerebritis compared to CT. OnT1-weighted images, cerebritis appears isointense to slightly hypointense to adjacent normal brain parenchyma, with associated mass effect manifested by sulcal effacement or ventricular compression. Postcontrast study shows either no enhancement or minimal and inhomogeneous enhancement (11,20). Pyogenic abscess has characteristic MR features, sufficient to make an accurate diagnosis. In the usual case of a mature abscess with central liquefactive necrosis, the center of the cavity is slightly hyperintense to CSF whereas the surrounding edematous brain is slightly hypointense to normal brain parenchyma on T1- weighted images. On T2-weighted images, the signal intensities are quite variable depending on the TE chosen and the protein composition and fluidity of the material in the central cavity. With a long TE, the differences in intensity between CSF, cavity fluid, and edema usually diminish. This is because the contribution of T2 relaxation to the signal intensity on long TR images is maximized. The central proteinaceous fluid and brain edema have higher signal intensity than CSF due to the T1 shortening effect of the macromolecules present in these fluids. Without protein and other macromolecular components that shorten T1, the cavity fluid would be nearly equal to CSF in signal intensity. On unenhanced MR images, the mature abscess often has a rim with distinctive features (21). The rim is isointense to slightly hyperintense to white matter on T1 and is hypointense onT2-weighted images. The rim reportedly does not show the mesial thinning seen with the collagen capsule, nor is the rim identical to the rim enhancement seen on contrast study (11). The signal properties of the rim seen on MR have been attributed to collagen, paramagnetic free radicals released from the phagocytosing macrophages and probably microhemorrhages (11,21). The support for the hypothesis of the free radical release is based on the fact that a similar hypointense rim onT2-weighted images has been observed in granulomas which are not notably hemorrhagic and in some metastases, which have no pathologic evidence of iron deposition or hemorrhage, but do have abundant macrophages (22,23). We have observed calcification and hemorrhage in the wall of the chronic untreated abscesses, subsequently confirmed on histopathologic examination (Fig.7). Both hemorrhage and calcification appeared hypointense on T2-weighted images and showed bloom effect on T2*-weighted images. The hypointense rim resolves with successful surgical/medical treatment of the abscess with a reduction in phagocytic activity. We have observed that the abscess is replaced by residual hypointensity onT2-weighted images as patients are being followed up on conservative treatment and the rim may not really resolve as suggested by the earlier reports. Multiple hypointense concentric rims have been described in a case of multiple abscesses secondary to *Nocardia asteroides* (24). It is believed that *N. asteroides* is a slow-acting bacterium and could modify the defense mechanisms of the host and allow more time to develop multiple rims in an abscess (24). We have not observed such rims in a case of nocardial abscess suggesting that such features may not always be observed in brain abscess due to nocardia (Fig.8).

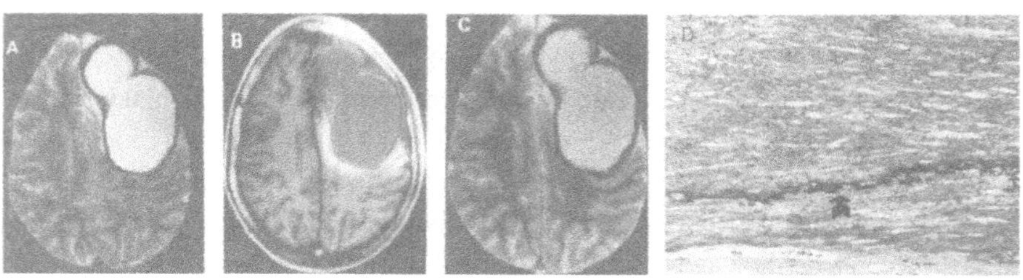

Figure 7. Pyogenic abscess with calcified wall.T2-weighted axial image (a) at the supraventricular level shows a lobulated hyperintense area in the left frontal region with hypointense rim.T1-weighted image (b) shows isointense rim with areas of hypointensity. Gradient echo low flip angle image (c) shows susceptibility effect in the peripheral rim. Patient underwent excision of the abscess. Histopathology showed calcification (arrow) in the abscess wall (d).

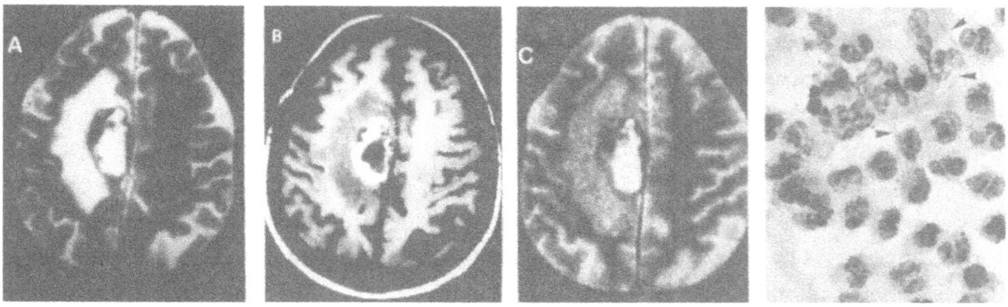

Figure 8. Nocardia abscess with hemorrhage in the wall. T2-, T1-, and T2*-weighted images from the supraventricular cerebral hemisphere show central hyperintense mass with peripheral incomplete hypointensity and perifocal edema onT2-weighted image (a) in left parasagittal region. It appears as hypointense center with peripheral irregular hyperintensity onT1-weighted image (b) with perifocal hypointense edema. The peripheral rim signal characteristics are consistent with subacute hemorrhage. Note the susceptibility effect on the periphery of the lesion on T2*-weighted image (c) suggestive of hemorrhage. Histopathology slide (d) showing evidence of long chain of bacteria (arrowheads) with a background of leukocytes and macrophages.

The ring enhancement of an abscess capsule on postcontrast study is similar to what is seen on a postcontrast CT. The rim is usually thin walled, but the thickness of the wall may vary depending on the chronicity of the abscess. Nodular or solid enhancement, incomplete thin rings or thick irregular rings are less often observed. Nodular enhancement along with rim enhancement may be occasionally observed in a chronic abscess and may simulate a neoplasm (11). Daughter abscesses appear as adjacent smaller enhancing rings, often along the medial margin of the parent abscess. Edema surrounding an abscess may be greater in volume than the abscess itself, and cause much of the mass effect. If an abscess ruptures into the ventricular system and ependymitis develops, there is enhancement of the ventricular margin in addition to the rim enhancement of the abscess, which heralds a poor prognosis. Purulent material within the CSF may show increased signal intensity on both T1- andT2-weighted images. Grossman and colleagues have shown in the experimental brain abscess that gadolinium-enhanced MR is better than contrast-enhanced CT in detecting small abscesses with necrotic center and in showing ring-enhancing lesions long after lesions there have evolved to nodular enhancement by CT (25). The differential diagnosis for a ring-enhancing lesion includes high-grade glioma, metastasis, infarction (bland or septic), radiation necrosis, AIDS-related lymphoma, and other inflammatory conditions. Helpful clues to the diagnosis of an abscess are pattern of enhancement, presence of daughter abscess, location, ependymal and/or meningeal enhancement, and extra-axial collection (11).

The standard treatment of mature brain abscess is surgical drainage and/or excision, whereas cerebritis and some cases of abscess may be managed with antibiotics. In cases with daughter abscess formation, intraoperative ultrasound guidance is recommended to ensure that all abscesses are identified and drained. After successful combined medical and surgical treatment, serial MR imaging reveals a decrease in the edema, mass effect, and degree of enhancement associated with the abscess. A small focus of enhancement may persist even after a full course of antibiotics, yet the lesions usually resolve within the succeeding 3 to 4 months (11). With healing, an area of gliosis and occasionally focal calcification is all that remains (11). Postoperatively, the MR appearance of resolving abscess may be complicated by the presence of subacute and chronic hemorrhage. Recent reports indicate that medical treatment alone can be effective in brain abscess especially when it is located in regions like the brainstem, thalamus, and basal ganglia. Even large abscesses have been treated conservatively with excellent results (26–30).

Proton MR spectroscopy. As stated above, it is possible to make a definitive diagnosis in most patients with brain abscess based on clinical and imaging features. However, in a small number of cases, it may not be possible to differentiate it from glioblastoma multiforme and a metastasis with necrosis. *In vivo* proton MR spectroscopy has been found to useful in the differentiation of brain abscess from other cystic mass lesions (31–37). In an untreated abscess, the usual metabolites observed are succinate (2.4 ppm), acetate (1.92 ppm), lactate/lipids (1.3 ppm), leucine/isoleucine/valine (0.85 ppm), glycine (3.5 ppm), alanine (1.5 ppm), and other amino acids (31–33) (Fig.5). Some of the metabolites (acetate, succinate) are considered to be of bacterial origin while others are secondary to the degeneration and death of the inflammatory cells and exudate (35, 38). The dead and degenerating inflammatory cells produce a large amount of proteolytic enzymes that degrade the proteins to free amino acids, visible on proton MR spectroscopy. Demonstration of these metabolites is considered specific of an abscess and has not been observed in a cystic neoplasm so far (31–38). Cystic glioma and metastasis show lipid/lactate, choline with a small resonance of creatine (33). As conservative treatment is being instituted in the brain abscess, it is essential to be sure of the diagnosis before starting the broad-spectrum antibiotics covering the aerobic–anaerobic bacteria. Imaging is the only objective criterion along with improvement in the clinical signs as an indicator of response to treatment. It may take 2-3 weeks before the response to conservative treatment can be assessed objectively. Recently, it has been observed that response to treatment can be assessed by serial spectroscopic examination (35). If a patient is treated with combined surgical and medical therapy, there is a disappearance of the bacterial metabolites along with a decrease in inflammatory metabolites. These changes in the metabolites have also been confirmed in the pus removed on surgery repeatedly using *ex vivo* highresolution MR spectroscopy (35). The response to treatment consisting of medical therapy alone has also been studied by proton MR spectroscopy (35,39). Dev et al. showed a decrease in the lactate (1.33 ppm)/amino acid ratio (0.85 ppm) after 1 week in two patients studied on serial medical therapy alone (35). It appears that *in vivo* MR spectroscopy is a useful noninvasive technique that may help in confirming the diagnosis and in objectively assessing the response to treatment (Fig.9) in patients with brain abscess (35,39).

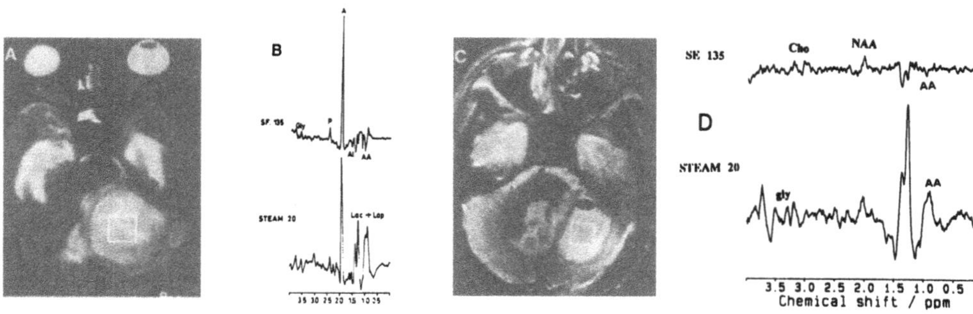

Figure 9. Left cerebellar abscess with combined medical and surgical management. A T2-weighted axial image (a) shows a large abscess cavity in the left cerebellar hemisphere, extending to the midline, with mass effect on the fourth ventricle and perifocal edema (inset, 8 ml volume from which the proton spectra were obtained). The proton spectrum obtained by using STEAM (20 ms) sequences shows resonances at 0.9, 1.3, 1.5, 1.92, 2.1 to 2.3, 2.4, 3.01, and 3.56 ppm. In the SE 135 ms spectrum, resonances at 0.9 and 1.3 ppm show phase reversal as well as marked reductions in signal intensity. The resonance at 1.5 ppm shows phase reversal only. T2-weighted axial image (c) obtained after 1 week of aspiration and antibiotic treatment shows a reduction in size with the air fluid level. *In vivo* spectroscopy reveals the presence of only lipids and lactate, small amounts of cytosolic amino acids and glycine. Note the marked decrease in the acetate signal, compared with panel b. Contamination from *N*-acetylaspartate (NAA) and choline (Cho) is the result of a partial volume effect.

Recently diffusion-weighted imaging has been used in differentiation of cerebral abscess from intracerebral necrotic tumors (40,41). The abscesses appear bright on diffusion-weighted images and show a very low apparent diffusion coefficient (ADC) (Fig.10), while the necrotic tumors appear hypointense on diffusion-weighted images and have a very high ADC. Ebisu et al. (42) performed *ex vivo* diffusion imaging of the aspirated pus and measured the ADC and showed that the pus imaged had a very high signal intensity and very low ADC consistent with the results of *in vivo* study. They concluded that the pus structure itself is responsible for low ADC values and the heavily impeded water mobility of pus may be related to its high cellularity and viscosity.

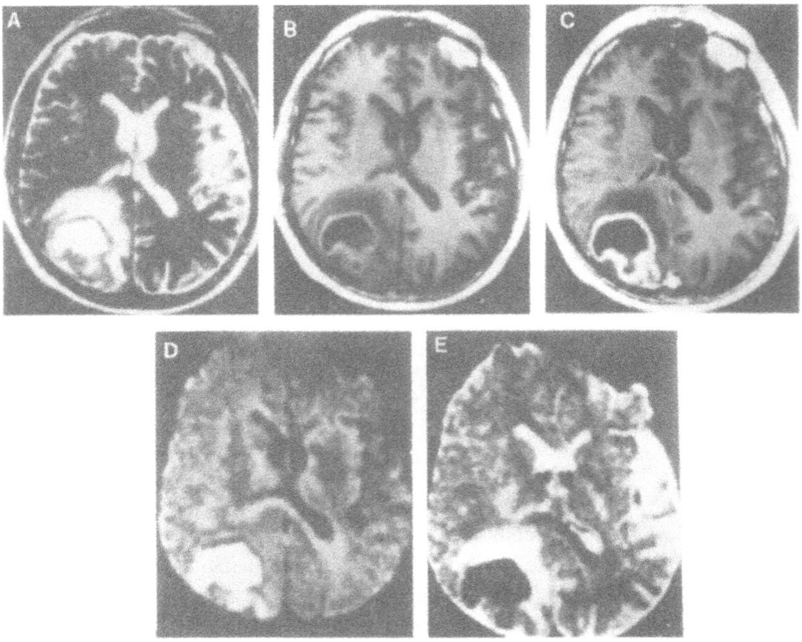

Figure 10. Diffusion imaging in a case of cerebral abscess.T2-weighted axial image (a);T1-weighted images without (b) and with (c) contrast show a cystic mass in right occipital region with perifocal edema with rim enhancement. Diffusion-weighted multishot EPI image (d) and the corresponding ADC map (e) show very low ADC values from the cavity (0.21–0.34 X 10^3 mm^2 /s) suggesting diagnosis of an abscess. (Reprinted with permission from Desprechins B, et al. Use of diffusion-weighted MR imaging in differential diagnosis between intracerebral necrotic tumors and cerebral abscesses. Am J Neuroradiol 1999; 20: 1252–1257.)

1.2.1. Multiple Brain Abscesses Secondary to Septic Emboli

Patients with a history of cyanotic heart disease, rheumatic heart disease (Fig.11), bacterial endocarditis, and intravenous drug abuse are at risk for the development of septic embolism leading to cerebrovascular occlusions. Depending on the size of the emboli, the MR findings may vary from major arterial branch infarction to multiple small abscesses at the gray–white matter junction, secondary to occlusion of small arteries and arterioles (17). Multiple small abscesses may mimic toxoplasma encephalitis and should be considered in the differential diagnosis when enhancing intracranial lesions are seen in patients with AIDS (11). The small abscesses are accompanied by surrounding edema and mass effect, well demonstrated on MR. Mycotic aneurysms may occur as a result of septic embolism, involve intermediate to small cerebral arteries, and thus are usually located in the more peripheral arterial branches than congenital aneurysms (11,17). Although mycotic aneurysms are usually small and thus difficult to detect a high index of suspicion should help pick up the lesions on MR, in at-risk patients. Presence of subarachnoid and parenchymal bleed in a patient with infection should suggest a mycotic aneurysm. MR angiography may be useful in detecting the mycotic aneurysms (11).

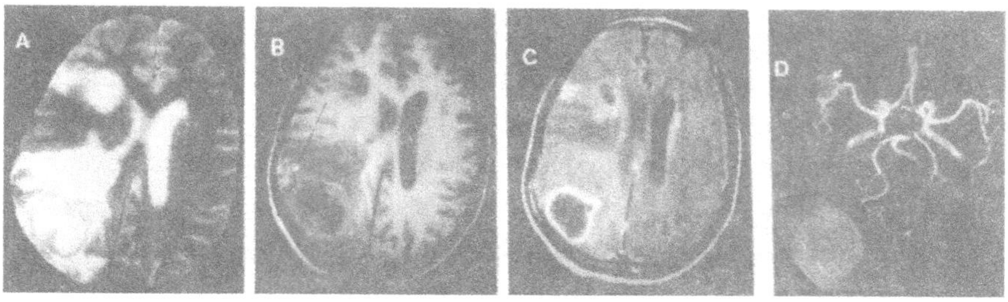

Figure 11. Brain abscess secondary to septic emboli in a young boy with rheumatic heart disease and infective endocarditis. T2-weighted axial image (a) at the level of ventricle shows evidence of a hyperintense area in the right parietal region with a hypointense rim with extensive perifocal edema and mass effect. Note another hyperintense area in the frontal area extending to the surface. T1-weighted axial image (b) shows isointense rim of the parietal lesion. There is hypointensity of the gyri in the right frontal region with a cavity medially. Note hyperintensity anterior to the parietal lesion (arrow) suggestive of hemorrhage. Postcontrast T1 image (c) shows rim enhancement of the lesion. Note also enhancement of gyri in the frontal region suggestive of infarct and rim enhancement of the cavity medial to the infarction consistent with early abscess formation. MR angiogram (d) shows irregularity with cutoff (arrow) of anterior division of right middle cerebral artery.

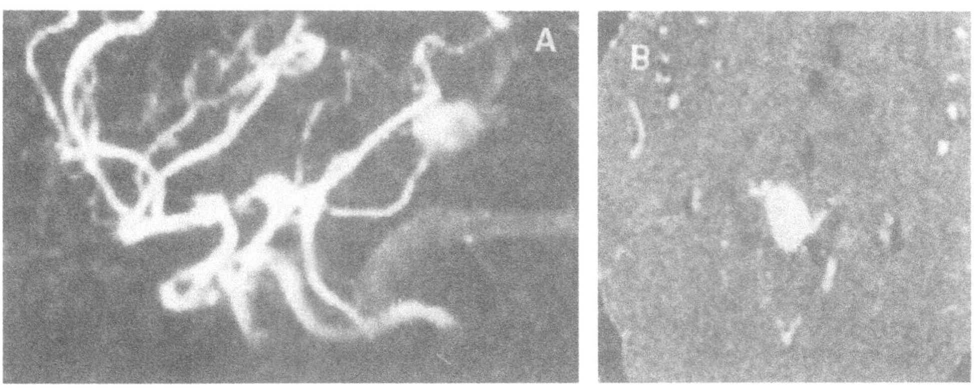

Figure 12. Mycotic aneurysm in a patient suffering from rheumatic heart disease with subacute bacterial endocarditis. MR angiogram (a) shows evidence of an aneurysm arising from right posterior inferior cerebellar artery. The image from the raw angiogram data (b) shows the aneurysm as a bright signal intensity area.

1.3. Cranial Subdural and Extradural Empyema

Purulent collections in the subdural and epidural space are uncommon and account for approximately 20–33% of all intracranial infections (43,44). The majority occur in the setting of sinusitis, have a fulminant clinical course, and require prompt diagnosis and urgent neurosurgical intervention (45–48). Less often, these develop secondary to infection of a posttraumatic extra axial hematoma or postcraniotomy cavity and have a prolonged indolent course, but nonetheless require surgical drainage for proper management (49,50). Bacteremia and meningitis are an infrequent cause of these collections and occur more so in children with meningitis. Subdural empyema should be considered a neurosurgical emergency and neuroimaging is essential to early diagnosis. When empyema results from sinusitis or mastoiditis, it is often associated with seizures, focal deficits, and rapid neurological deterioration.

Subdural empyema is most commonly located over the cerebral convexity and is frequently bilateral. A paratentorial location is the least common. MR is the modality of choice for the diagnosis of extra-axial empyema. The subdural empyema may be lentiform or crescentic in shape; epidural empyema is typically lentiform. It is possible to differentiate subdural from epidural empyema when the latter is continuous across the midline and/or when there is a hypointense rim on both T1- and T2-weighted images, representing medially displaced dura, seen at the interface between the collection and the brain. On T1-weighted images, convexity and interhemispheric subdural empyema appear hypointense relative to brain and hyperintense to CSF (51). On proton density- and T2-weighted images, the collections are hyperintense to brain and isointense to hyperintense relative to CSF. These signal characteristics are typical of proteinaceous fluid with T1 and T2 values intermediate between gray matter/white matter and CSF (51, 52). The signal intensity of the collection may not be markedly different from underlying brain if there is a significant edema secondary to ischemia and/or cerebritis (Fig.12). T2-weighted images, which are very sensitive to abnormal tissue water distribution, may show cortical hyperintensities that are reversible and are thought to represent edema related to transient ischemia produced by inflammatory vasospasm and venous stasis. MR venogram can demonstrate cortical vein and/or dural sinus thrombosis. Spin-echo and gradient-echo MR imaging can demonstrate the characteristic signal properties of acute and subacute hemorrhagic infarctions (51–53).

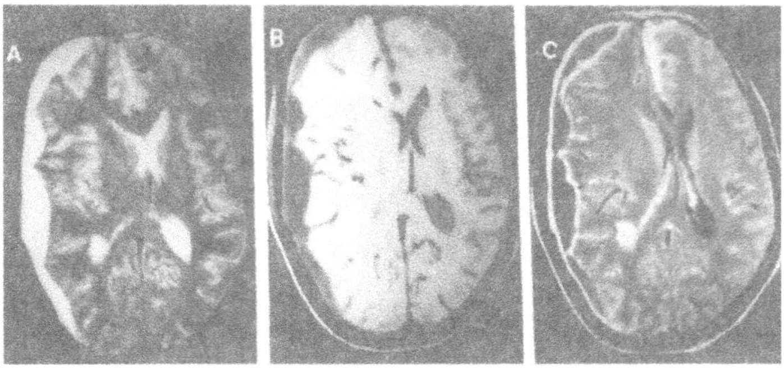

Figure 13. Subdural empyema with focal late cerebritis secondary to septic embolism. T2-weighted axial (a) image shows evidence of a hyperintense extra-axial collection in the right frontoparietal region with mass effect along with collection in the anterior interhemispheric fissure (arrow). Note the hyperintense area near the right occipital horn.T1-weighted image (b) shows hypointensity of the extra-axial lesion with areas of adhesions and isointensity of the occipital lesion. Postcontrast T1-weighted image (c) shows rim enhancement of the extra-axial lesion, interhemispheric collection with enhancing dura lying laterally suggestive of subdural empyema. The occipital lesion shows enhancement consistent with late cerebritis.

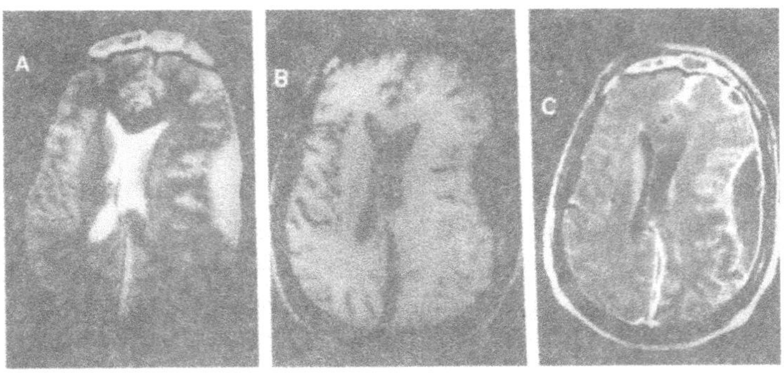

Figure 14. Subdural empyema secondary to frontal sinusitis. T2-weighted axial image (a) at the level of ventricles show evidence of extra-axial hyperintense area in the left parietal and frontal region. Note also hyperintensity in the posterior interhemispheric fissure and in frontal sinuses. The collection in interhemispheric fissure and loculated collection in the frontal region are well seen inT1-weighted image (b). PostcontrastT1-weighted image (c) shows rim enhancement of the pockets of collection in the interhemispheric fissures, frontal and parietal region with continuous dural enhancement on the left side. Note the enhancement of the frontal mucosa in close proximity with left frontal loculation.

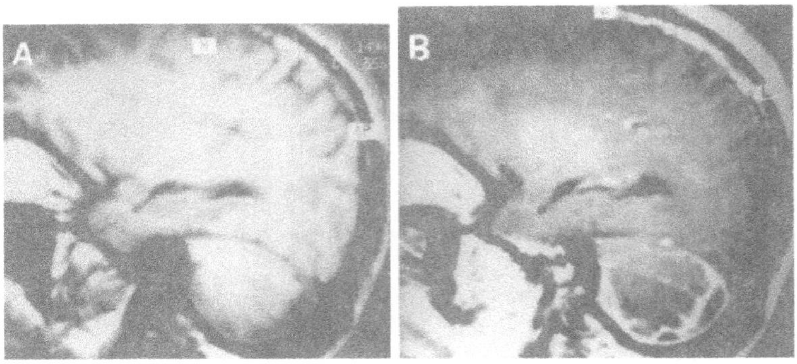

Figure 15. Retrocerebellar subdural empyema secondary to middle ear infection. T1-weighted sagittal image (a) shows evidence of a hypointense area posterior and inferior to left cerebellar hemisphere. Postcontrast image (b) shows peripheral rim enhancement with extension of enhancement in between the folia suggesting focal meningitis. Pus on culture yielded *Staphylococcus aureus*.

On MR, subdural and extradural empyemas show enhancement of the margins in postgadolinium studies (Figs.13–15). As the source of infection is usually otorhinologic, MR is quite sensitive in detecting inflammatory changes in the paranasal sinuses and mastoid air cells, usually as areas of increased signal intensity on T2-weighted images. Enhancement is usually prominent and peripheral. Mucopyocele may have decreased signal intensity on T2- weighted images related to chronic inspissated secretions (51–53).

1.4. Sellar Abscess

Primary pituitary abscess is a rare entity (54–56). Less than 100 cases have been reported in the literature (56). The pathogenesis of this disease is not entirely clear. Traditionally, pituitary abscess was thought to arise by direct extension from an infected sphenoid sinus or as a complication of a sellar CSF leak. Meningitis is associated with pituitary abscess in about 60% of the cases and can be recurrent. Other cases have been reported following transsphenoidal surgery, and occasionally within a tumor that has not been operated (57–60).

The diagnosis of pituitary abscess is elusive and is most often made at operation for a presumed pituitary adenoma (58). Often pituitary abscess manifests itself in a manner identical to other intrasellar mass lesions, with signs and symptoms of hypopituitarism. Preoperative diagnosis in such cases is particularly difficult. Presence of diabetes insipidus

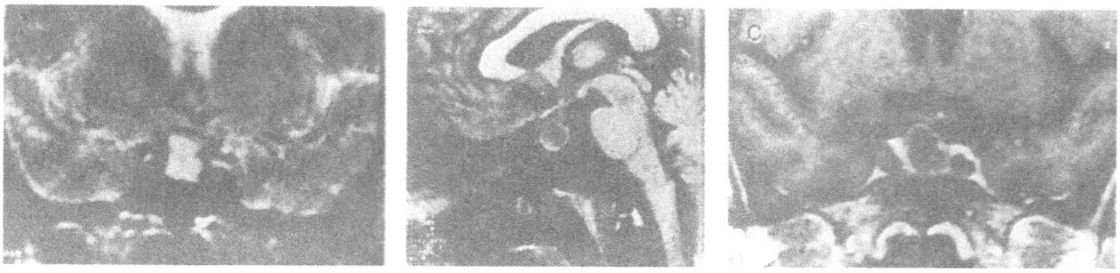

Figure 16. Sellar abscess. Midline sagittal T1-weighted image (a) shows low-intensity intrasellar mass with thin peripheral wall filling the sella. On coronalT2-weighted image (b) the intrasellar and the suprasellar part both show high signal intensity. PostcontrastT1-weighted image (c) shows enhancing margins of the abscess. Patient was operated and imaging at 9 months showed total regression of the lesion. (Reprinted with permission from Guigui J, Boukobza M, Taner I, et al. MRI of primary pituitary abscess. Clin Radiol 1998; 53: 777–779,W. B. Saunders.)

is of value in differentiating adenoma from an abscess, as diabetes insipidus is more commonly associated with an abscess (50%) than an adenoma (10%) (54–60). Signs of CNS infection, when present, consist fever and meningeal signs. Intrasellar abscesses on MR imaging appear as low signal intensity on T1- and hyperintense on T2-weighted images (Fig.16). This is probably due to the T1 and T2 shortening effect of the macromolecules in thick proteinaceous fluid (55,56). The posterior hypophysis is demonstrated as a hyperintense spot on T1-weighted images. Absence of the posterior hypophysis bright spot on MR imaging may be more likely to be seen in intrasellar abscess than pituitary adenoma (58,59). Postcontrast MR imaging may disclose a complete or an incomplete ring-enhancing lesion in the sella (54–60). The enhancement of the sella turcica may be associated with enhancement of the sphenoid and leptomeningeal enhancement.

1.5. Cavernous Sinus Thrombosis

The cavernous sinuses are dura-covered structures lying on either side of the sella and pituitary gland. They contain large venous channels, the internal carotid arteries, and the III, IV, and first and second divisions of the V and VI cranial nerves. The bone of the body of the sphenoid containing the sphenoid sinus is positioned both medially and inferiorly. The cavernous sinuses transmit cranial nerves to the orbits and receive blood draining from the orbit, paranasal sinuses, nasal cavity, nasopharynx, and face. The spread of bacterial infection may occur either directly from the sphenoid sinus through osteomyelitis or along a dehiscence in the bone, or retrograde extension of infection as a thrombophlebitis from the orbit or facial structures. Frequently, the intracavernous internal carotid artery is compressed or occluded. MR angiography and MR venography can demonstrate patency of the cavernous sinus and carotid artery. MR shows the thrombosed cavernous sinus as enlargement of the cavernous sinus by a clot in acute and subacute stages (61). The intensity of the clot is dependent on the chemical composition of the clotted blood. Hydrocephalus, an infrequent complication of cavernous sinus thrombosis, is thought to be due to extension of thrombosis to the major cerebral veins and sinuses (62). Mycotic aneurysm in the intracavernous portion of the internal carotid artery secondary to cavernous sinus thrombosis is rare (Fig.17) (63). Occlusion of the internal carotid artery and aneurysm formation secondary to cavernous sinus thrombosis have been reported separately as well as in combination. These abnormalities have been well demonstrated with MR angiogram and MR venogram (62).

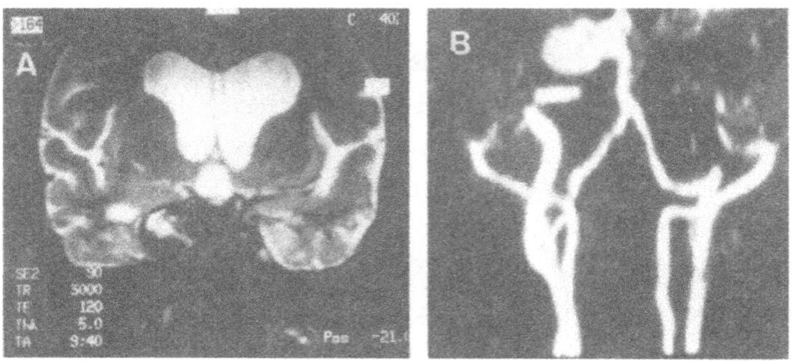

Figure 17. Cavernous sinus thrombosis with internal carotid aneurysm. AT2-weighted coronal image (a) through the cavernous sinuses shows enlarged right cavernous sinus with an irregular hyperintense area on the lateral aspect. MR angiogram (b) shows evidence of a pseudo-aneurysm arising from the cavernous portion of the right internal carotid artery. Note also the absence of flow signal of left carotid artery suggestive of thrombosis.

1.6. Infective Cranial Neuritis

Infective cranial neuritis secondary to bacterial infections is uncommon. This could be either secondary to bacterial meningitis or mastoiditis. Eighth cranial nerve is most frequently involved. On MR imaging the involvement of these structures is seen as enhancing areas (64) with or without evidence of enhancing meninges. We have observed a case of mastoiditis in which there was isolated involvement of right VIII cranial nerve that resolved on medical treatment (Fig.18). Isolated involvement of VIII cranial nerve by bacterial neuritis needs to be differentiated from small acoustic neuromas. Presence of focal infection in the vicinity of the involved cranial nerve helps in its differentiation (Fig.18).

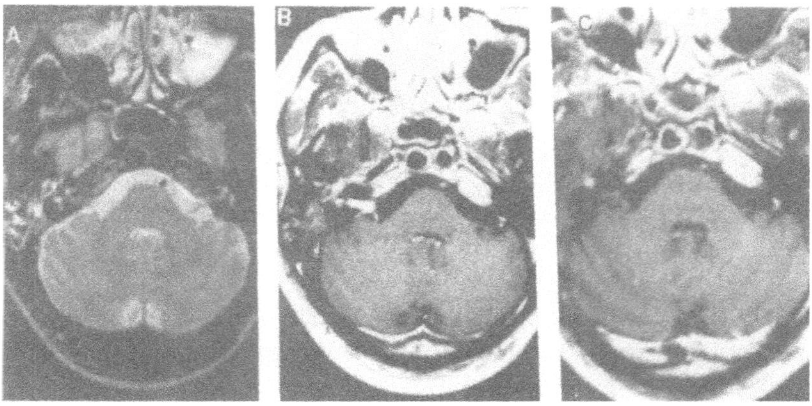

Figure 18. Infective eighth cranial nerve neuritis. T2-weighted axial image (a) at the level of pons shows evidence of hyperintensity in the right mastoid region suggestive of mastoiditis. Postcontrast T1-weighted image (b) shows enhancement of the intracanalicular portion of the eighth nerve and cochlea. Followup postcontrast T1-weighted image (c) shows no enhancing areas suggesting resolution of neuritis.

2. PYOGENIC INFECTIONS OF THE SPINE

2.1. Spondylitis

Pyogenic infectious spondylitis is rapidly suspected when it develops secondary to spinal surgery (65,66). However, bacteria may also enter the body at other sites and reach the intervertebral junction preferentially via the bloodstream (67,68). Such primary pyogenic infective spondylitis is rare with one estimate of two cases per year in an active tertiary care hospital. The low expectation rate and relatively uncharacteristic clinical symptoms frequently delays the diagnosis of pyogenic spondylitis. This may result in longstanding pain for the patient and the risk of spreading infection with complications that can lead to irreversible neurological deficits or even death (68–72).

The usual clinical presentation is focal backache followed by radicular pain. The mean interval from the first symptoms to the diagnosis is 2 months with range of 4 days to 8 months. The order of frequency of involvement is thoracolumbar, lumbar, and lumbosacral spine. The risk factors associated with pyogenic spondylitis are diabetes mellitus, old age, immunocompromise, drug abuse, previous nonspinal surgery, and other sites of infections and inflammation (68–72). *S. aureus* is the most common pathogen isolated in cases of pyogenic spondylitis; *E. coli*, *Salmonella*, *Pseudomonas aeruginosa*, and *Klebsiella pneumoniae* have also been implicated. Two major criteria are essential for establishing the diagnosis of pyogenic spondylitis: (1) the presence of characteristic imaging features of spine infection and (2) the isolation of the offending organism from blood or the affected site. In a majority of the cases, medical therapy is adequate and surgical decompression is rarely required (66–72).

The MR imaging findings of hematogenous pyogenic spondylitis are characteristic and correctly reflect the pathophysiologic changes. The increase in extracellular fluids within the vertebral marrow—a manifestation of inflammatory reaction—is responsible for the abnormal signal, which is relatively low on T1- and high on T2-weighted images (66–72). This abnormal signal is usually distributed throughout the whole vertebra. Vertebral collapse and spinal deformity are extremely rare and the affected vertebra does not seem to lose its cortical margins. Classically, two consecutive vertebrae and the intervening disk are involved. The disk is invariably reduced in size and shows high signal intensity on T2-weighted images. Pyogenic spondylitis is usually confined to the vertebral marrow with no significant extension into the paraspinal region (71). Moderate granulation tissue can be produced which is distributed around the vertebral body. Occasionally, the vertebral body may be hypointense or may show the hypointense bands intervening hyperintensity onT2-weighted images. This may be due to the sclerotic trabeculae occupying the marrow space (67). Epidural extension is infrequent and seen as an iso- to hypointense lesion obliterating the region between the vertebral bodies and the cord (Fig.19). The epidural lesions appear as hyperintense or of mixed intensity onT2-weighted images. Postcontrast study is essential for better delineation of epidural extension and meningeal involvement (71,72). The homogeneous enhancement of the epidural space is consistent with phlegmon and ring enhancement suggests abscess formation. The body of the infected vertebra and intervertebral disk show enhancement in a majority of the cases with pyogenic spondylitis on postcontrastT1-weighted sagittal and axial images (Fig.20)(71,72). Differential diagnosis usually includes other infective spondylitis like tuberculosis, mycosis, and brucellosis (70). Vertebral collapse and skip lesions are not a feature of pyogenic spondylitis and are classically seen in tuberculous spondylitis (70).

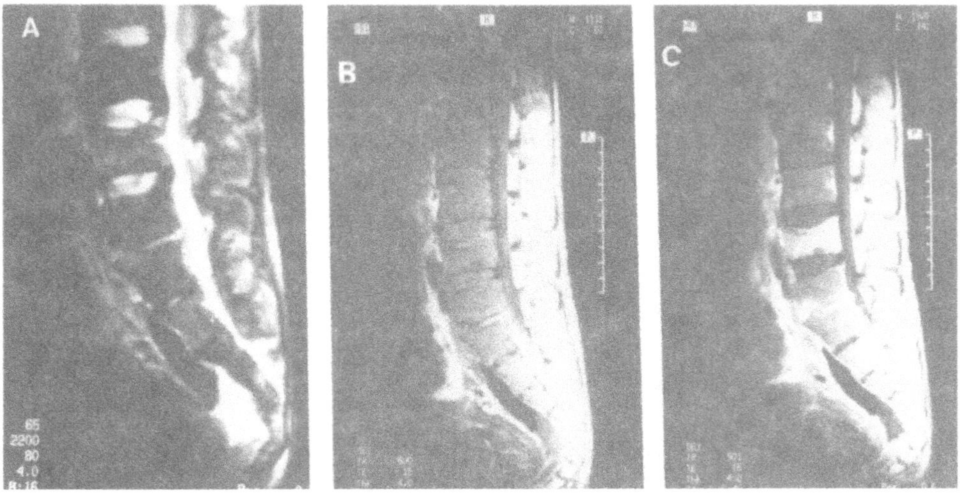

Figure 19. Pyogenic spondylitis with presacral collection. T2-weighted sagittal image (a) of lumbosacral spine shows evidence of subtle hyperintensity of third, fifth lumbar and first sacral vertebrae with reduced L5–S1 disk space along with an air fluid level in the presacral region (arrow).T1-weighted image (b) shows hyperintense L5–S1 disk. Postcontrast image (c) shows enhancement of the L3, L5, S1 vertebrae, intervening L5–S1 disk and rim enhancement of the presacral collection. Image guided pus aspiration was done and culture revealed anaerobic streptococci.

Follow-up in cases of pyogenic spondylitis can be effectively assessed with MRI. T2-weighted sagittal images may demonstrate a decrease in the abnormal high signal noted on the first examination. The presence of increased signal intensity onT1-weighted images from previously infected vertebrae reflects replacement of the inflammatory material by fat and indicates healing (70).

2.2. Spinal Epidural Abscess

An epidural abscess is a medical as well as a surgical emergency, requiring prompt and accurate diagnosis and treatment to prevent irreversible spinal cord dysfunction, paralysis, and death. It is a suppurative infection of the epidural space. Despite improvement in the diagnostic modalities and treatment, mortality and morbidity remain high due to the delay in diagnosis.

The epidural space is composed of two compartments: (1) a true space posterior and lateral to the spinal cord containing a cushioning layer of fat embedded with penetrating arteries and an extensive venous plexus and (2) a potential anterior space where the dura adheres to the posterior surface of the vertebral body. The epidural space is circumferential around the spinal cord distal to the second sacral segment. These anatomic considerations

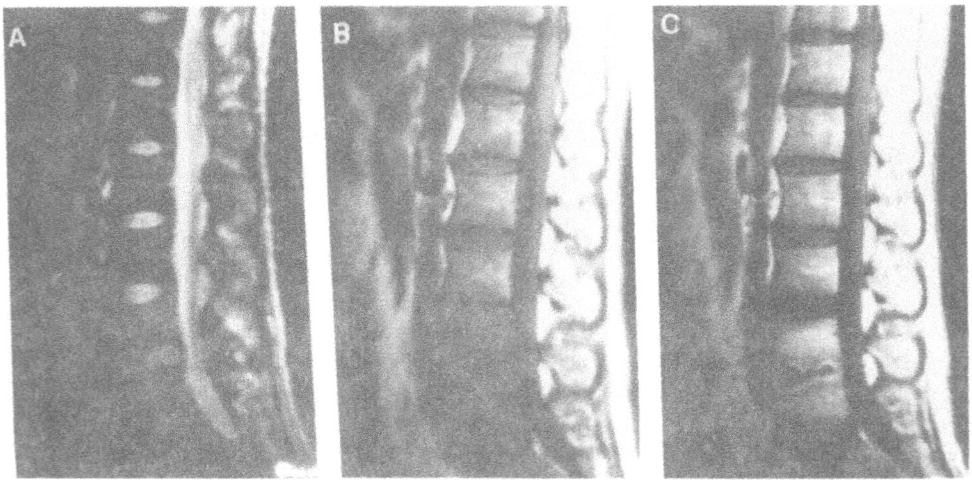

Figure 20. Pyogenic spondylitis. T2-weighted sagittal image (a) of lumbar spine shows evidence of mildly hyperintense L4 and L5 vertebrae with mildly swollen L4–L5 disk which shows heterogeneous signal intensity.T1-weighted image (b) shows hypointense L4 and L5 vertebrae. Postcontrast T1-weighted image (c) shows enhancing L4–L5 vertebrae with rim enhancement of the disk suggesting spondylitis with intervertebral disk abscess. Culture from the pus aspirated from the disk space yielded *Staphylococcus aureus*.

explain why the epidural abscesses are more commonly observed posteriorly in over 70% of the cases (73). The epidural abscess is most commonly seen in the thoracic spine, followed by lumbar and cervical spinal region. The epidural abscesses usually affect multiple adjacent segments; however, an infection of the entire length of the spinal epidural space has been reported (74,75).

Infection may be introduced into the epidural space by direct extension from a contiguous infection or by hematogenous or lymphatic seeding from a remote site. Contiguous infections include vertebral osteomyelitis; retropharyngeal, perinephric, and psoas abscesses; decubitus ulcers; and persistent dermal sinus tracts (76–78). Local invasion from superficial infections can also occur following penetrating injuries, prior surgery, or spinal procedures including epidural catheterization (79,80) and CT guided needle biopsy and lumbar puncture (73–75). In contrast to spinal epidural abscesses in adults, adjacent vertebral osteomyelitis is distinctly uncommon in children with this infection; hence, the majority of spinal epidural abscesses in children have been attributed to hematogenous seeding from a distant focus of infection (73–75). Underlying medical conditions associated with spinal epidural abscesses include degenerative joint disease, diabetes, intravenous drug abuse, alcoholism, cirrhosis, malignancy and renal failure (73,78,81).

The damage produced by the spinal epidural abscess to the spinal cord is out of proportion to the size of inflammation (76–82). This may be due to many factors including thrombosis and thrombophlebitis of the veins draining the spinal cord with resultant edema and venous infarction; compression of the arterial supply to adjacent cord segments with local ischemia and infarction; focal areas of vasculitis induced by adjacent inflammatory mass; or bacterial exotoxin production by *S. aureus*. *S. aureus* is the most common etiologic agent and is responsible for 52–95% of the bacterial abscesses seen in adults and children. Other bacilli seen are gram-negative aerobic bacilli, aerobic streptococci, and anaerobes. Multiple organisms are found in 5–10% of the cases. The initial signs and symptoms of a spinal epidural abscess characteristically include fever, backache, and malaise. The duration of symptoms is variable; they may be present for just a few days or up to several months before the patient presents for evaluation (73–75).

On MRI, an epidural abscess has low or intermediate intensity on T1- and high or intermediate intensity onT2-weighted images (76–87). The fluid portion of the abscess tends to have a markedly increased signal onT2-weighted images and the surrounding inflammatory edema and granulation tissue have inhomogeneous areas of mildly increased signal (Fig.21). T2-weighted images can be insensitive to the presence of an epidural abscess as both CSF and abscess may be hyperintense. Rarely, hemorrhage may accompany infection producing a heterogeneous high signal on T1- andT2-weighted images that have the appearance of an epidural hematoma (83). Axial and sagittal images obtained with MR can yield the extent and location of the abscess in the epidural space. MR can assess the degree of thecal compression and involvement of the spinal cord. The hyperintensity of the spinal cord may be due to either compression, inflammation of the cord, infarction secondary to vasculitis or due to venous edema secondary to venous thrombophlebitis (65). If the extent of involvement is more than the degree of compression, possibility of primary cord involvement is more likely. The enhancement pattern is variable depending on the age and consistency (pus or granulation tissue) of the abscess. Following intravenous injection of gadolinium, three patterns of enhancement have been described: (1) dense homogeneous, (2) inhomogeneous with scattered areas of sparse or no uptake, (3) thin peripheral enhancement. The third pattern represents a collection of liquid pus with a surrounding thin or thick enhancing rim. Gadolinium aids in the definition of the activity of the infective process and may be used to localize a potential area of biopsy and may determine the response to antibiotic treatment (83–86).

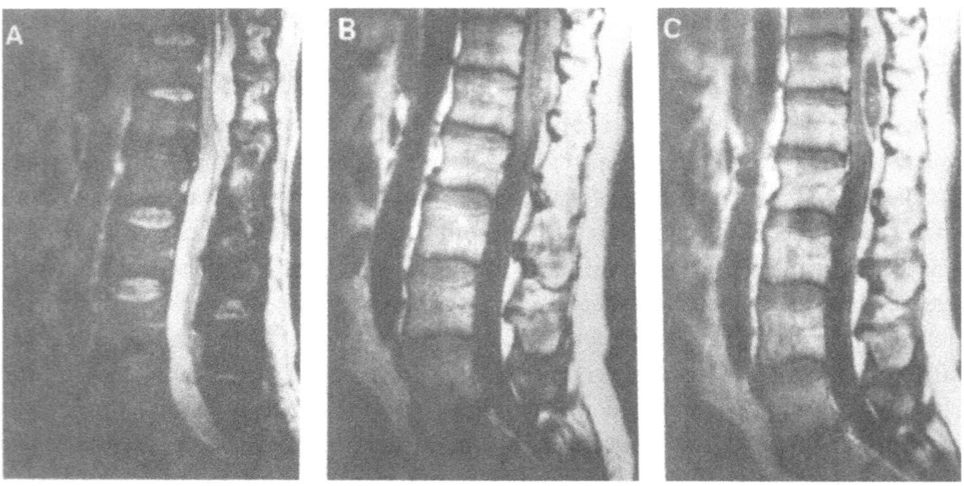

Figure 21. Pyogenic epidural abscess. T2-weighted sagittal image (a) of the thoracolumbar spine shows evidence of an extradural hyperintense area at the level of D11–L1 vertebrae with compression on the lower part of the cord. OnT1-weighted image (b) the collection appears iso- to hyperintense to the cord, and postcontrast image (c) shows peripheral thick enhancement and nonenhancing central area consistent with epidural abscess. Vertebrae show no obvious abnormality. Culture from the pus aspirated yielded *Staphylococcus aureus*.

The main differential diagnosis of an epidural abscess is metastatic disease. The finding of high signal intensity in the disk space onT2-weighted images adjacent to the eroded vertebral bodies and an epidural mass establishes spinal infection as the diagnostic choice over metastatic disease (65). The treatment is usually surgical, although there are reports of complete regression of the epidural abscess on medical therapy alone (79). The indications for the medical therapy are (1) identification of the offending organism, (2) stable neurological condition, and (3) appropriate neurosurgical backup. These patients should be followed up by repeated MR examination (86). The resolution of the previously noted anatomic distortion indicates a favorable medical response (80). The imaging appearance, however, does lag behind clinical response. The enhancement patterns may be variable while the patient is on a combination of therapy and abnormal enhancement does not necessarily indicate ongoing infection (83). Healed areas may be enhanced due to scar tissue (65).

2.3. Spinal Subdural Abscess

Spinal subdural abscess is a rare disease entity. The exact incidence is not known with less than 50 cases reported so far in the literature (88). The typical patient with spinal subdural abscess presents with fever, back pain, and symptoms of cauda equina or spinal cord compression, without the presence of spinal percussion tenderness (88–92). The spread of infection to the subdural space is probably due to hematogenous spread from a distant skin infection (88–91). Direct extension of a contiguous infection has also been noted occasionally (92). However, iatrogenic causes such as lumbar puncture, injection of a local anesthetic agent, and diskography have frequently been involved in the development of spinal subdural abscess (92). The lower incidence of spinal subdural abscesses in comparison with cranial subdural abscesses has been postulated to be due to several factors: the absence of sinuses in the spine; the possible role of the epidural space as a filter; and the centripetally directed blood flow in the spine in comparison with centrifugally directed cerebral blood flow in the brain (88–92).

S. aureus is the most common organism isolated in the cases reported in the literature (88). This is because cutaneous infections that have been noted frequently to be the peripheral infectious source of spinal subdural abscess are most often caused by *S. aureus*. MRI allows superior localization of the intraspinal process (88). MRI allows superior localization of the intraspinal process. Sometimes it becomes difficult to differentiate epidural from subdural abscess based on conventional T1- and T2-weighted images (88,92). Postcontrast study showed diffuse enhancement in a case described in the literature (Fig. 22). Based on the enhancement, it was difficult to differentiate abscess from lymphoma and it was confirmed on histopathology of an open biopsy (88). Surgical decompression is the best mode of treatment in a majority of the cases. There are cases that have responded completely on medical therapy following biopsy (88). A follow up MRI reveals dramatic resolution of infection in such patients.

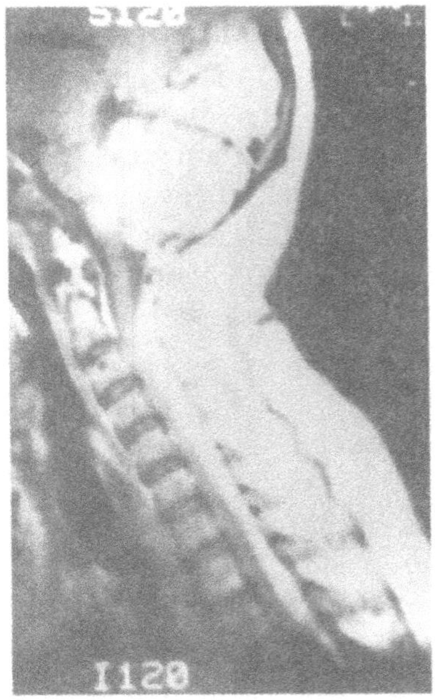

Figure 22. Spinal subdural abscess. Sagittal T1-weighted postcontrast image shows a large posterior spinal canal collection extending from C1 to the D2 interspace compressing and deforming the spinal cord. (Reprinted from Sathi S, Schwartz M, Cortez S, Rossitch E. Spinal subdural abscess: Successful treatment with limited drainage and antibiotics in a patient with AIDS. Surg Neurol 1994; 42:424–427 with permission from Elsevier Science.)

2.4. Myelitis and Intramedullary Abscesses

The occurrence of intramedullary infection is rare; only a few cases have been described in the literature (93). Bacteria such as staphylococcus and streptococcus are the most common organisms to invade the spinal cord; however, *Listeria* and *E. coli* (93–95) may also cause the infection. Hematogenous spread is the most common source of infection; however, lymphatic and direct extension may also occur. The mechanism of infection is considered to be the same as described for brain abscess. On MRI, bacterial myelitis appears as areas of intramedullary high signal on proton density- and T2-weighted images and hypointense on T1-weighted images (93,95). The cord diameter may show a marginal increase or may be normal. In a postcontrast study, the enhancement may vary from diffuse intense enhancement to minimal or no enhancement. Posttreatment MRI shows a decrease in T2 and T1 signal with better definition of the lesion after contrast administration (93,95). The cord may become completely normal on follow up with no cord atrophy (Fig.23). This

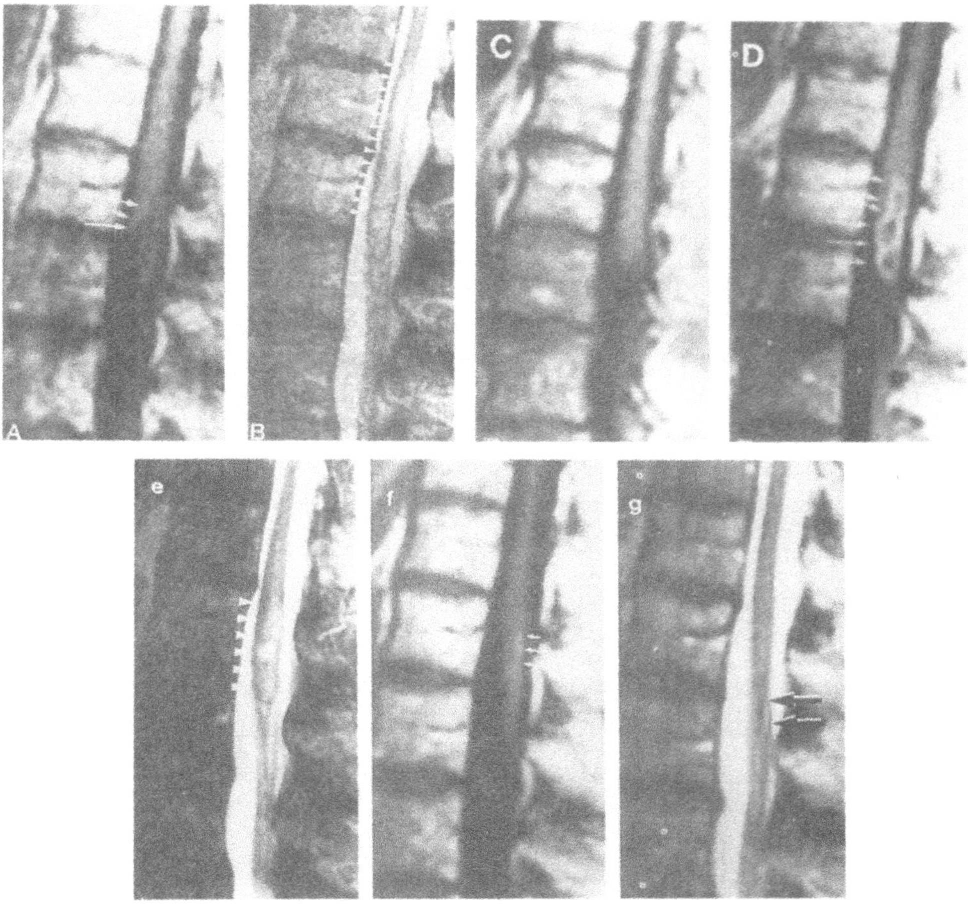

Figure 23. Intramedullary abscess. Sagittal contrast enhanced T1-weighted image (a) shows evidence of cord widening and low signal intensity in the conus. There is minimal enhancement (arrows) around the region of low signal intensity. T2-weighted image (b) shows high signal intensity in the central portion of the spinal cord (arrowheads). Follow-up MR study after 1 week, T1-weighted sagittal (c) image shows swollen cord. Postcontrast T1-weighted sagittal (d) image shows ring enhancement extending to the surface of the cord (arrows). T2-weighted image (e) shows intramedullary high signal at and above the conus (arrowheads). Follow-up after 3 months with sagittal contrast-enhanced T1-weighted image (f) shows subtle abnormal intramedullary enhancement (arrows) in the conus. TheT2-weighted image (g) shows resolution of edema with a central region of low signal intensity, possibly representing hemosiderin deposition (arrows). (Reprinted with permission from Murphy KJ, Brunberg JA, Quint DJ, Kazanjian PH. Spinal cord infection: Myelitis and abscess formation. Am J Neuroradiol 1998; 19: 341–348.)

has been described as the differentiating feature from spinal cord infarction secondary to anterior spinal artery occlusion (95). Differential diagnosis also includes infections from other agents such as tuberculosis, fungal, viral, and parasitic infections. It may not be possible to differentiate the other types of infective myelitis by imaging alone; microbiological and serologic evaluation of the CSF, blood, and tissue will help in correctly diagnosing the offending organism (95).

Intramedullary spinal cord abscesses are uncommon (96). When no demonstrable cause is present, they are referred to as primary abscesses; when there is detectable cause, like trauma, dermal sinus, or any other focus of infection in the body; they are referred to as secondary abscesses. Etiologic agents are the same as mentioned for myelitis. Abscess secondary to dermal sinus tends to extend a great distance up and down the cord rather than remaining localized to the site of dermal sinus (96,97). The abscess cavity appears as hyperintense on T2- and hypointense onT1-weighted images (Fig.24) (96). Signal intensity depends on the viscosity and protein contents of the fluid. Postcontrast study shows rim enhancement. While demonstrating the intramedullary abscess, it is important to look for a cause such as dermal sinus that needs to be excised along with drainage of an abscess cavity; otherwise, the abscess will recur (96,97).

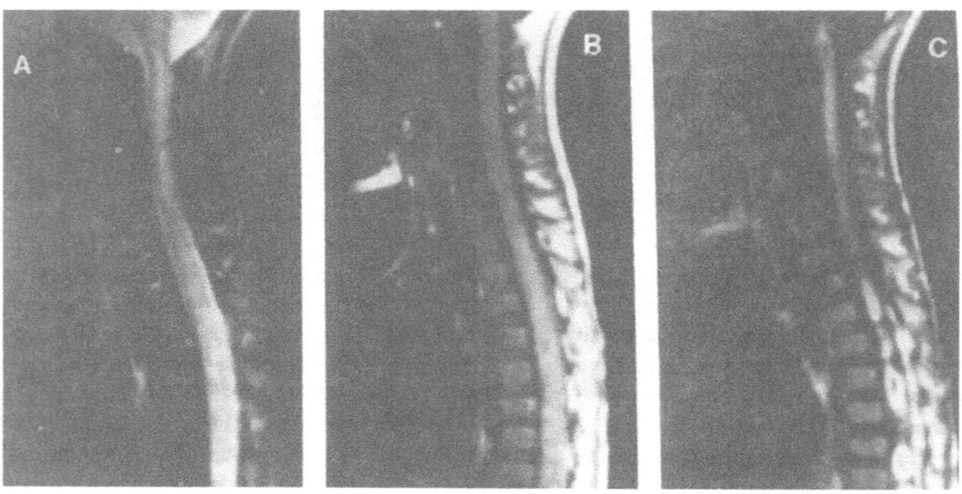

Figure 24. Intramedullary abscess. T2-weighted sagittal image (a) of the cervicothoracic spine shows swollen cord with hyperintensity especially in the thoracic part. T1-weighted image (b) shows heterogeneous, predominantly hyperintensity of the thoracic part of the cord. Postcontrast image (c) shows multiple rim enhancing areas in the thoracic and cervical cord suggestive of multiple intramedullary abscesses. Pus aspirated was positive for *Staphylococcus aureus*.

REFERENCES

1. Adam RD, Kubik CS, Bonner FJ. Clinical and pathological aspects of B influenzal meningitis, Arch Pediatr 1948; 65: 408–411.
2. Tunkel AR, Scheld WM. Pathogenesis and pathophysiology of bacterial meningitis. Clin Microbiol Rev 1993; 6: 118–136.
3. Tunkel AR, Scheld WM. Pathogenesis and pathophysiology of bacterial meningitis. In *Infection of the central nervous system*, WM Scheld, RJ Whitley, DT Durack, eds. New York: Lippincott–Raven, 1997; 297–312.
4. Roos KL, Tunkel AR, Scheld WM. Acute bacterial meningitis in children and adults. In *Infection of the central nervous system*, WM Scheld, RJ Whitley, DT Durack, eds. New York: Lippincott–Raven, 1997; 335–401.
5. Richardson M. Bacterial meningitis. Hosp Med 1996; 35: 685–688.
6. Shawl S. Neurologic evaluation of patient with acute bacterial meningitis. Neurol Clin 1995; 13: 549–577.
7. Castillo M. Magnetic resonance imaging of meningitis and its complications. Top Magn Reson Imaging 1994; 6(1): 53–58.
8. Runge VM, Welle JW, Williams NM, Lee C, Timoney JF, Young AB. Detectability of early brain meningitis with magnetic resonance imaging. Invest Radiol 1995; 30: 484–495.
9. Schroth G, Kretzschmar K, Gawehn J, Voigt K. Advantage of magnetic resonance imaging in the diagnosis of cerebral infections. Neuroradiology 1987; 29: 120–126.
10. Barnes PD, Poussaint TY, Burrows PE. Imaging of pediatric central nervous system infections. Pediatr Neuroradiol 1994; 4: 367–391.
11. Whiteman MLH, Bowen BC, Post MJD, Bell MD. Intracranial infection. In *Magnetic resonance imaging of brain and spine*, SW Atlas, ed. New York: Lippincott–Raven, 1996; 707–766.
12. Lebel MR, Hoyt J, Waagner DC, et al. Magnetic resonance imaging and dexamethasone therapy for bacterial meningitis. Am J Dis Child 1989; 143: 301.
13. Wolpert SM, Kaye EM. Intracranial inflammatory processes. In *MRI in pediatric neuroradiology*, SM Wolpert, PB Barnes, eds. St Louis: Mosby Yearbook, 1991; 151–176.
14. Tanabe T, Morimoto T, Kawasaki Y, et al. Usefulness of MRA in an infant with cerebral infarction due to streptococcal meningitis. No To Hattatsu 1996; 28: 520–524.
15. Roy R, Gupta RK, Kishore J, Taparia S, Poptani H, Bhakuni V. High resolution proton MR spectroscopy of the cerebrospinal fluid from children with tuberculous meningitis. Proceeding of annual meeting of ISMRM 1996; p 1157.
16. Matthews PM, Shoubridge E, Arnold DL. Brain phosphorus magnetic resonance spectroscopy in acute bacterial meningitis. Arch Neurol 1989; 46: 994.
17. Wispelmey B, Dacey RG, Scheld WM. Brain abscess. In *Infection of the central nervous system*, WM Scheld, RJ Whitley, DT Durack, eds. New York: Lippincott–Raven, 1997; 463–493.
18. Brit RH, Enzmann DR, Yeager AS. Neuropathological and CT findings in experimental brain abscess. J Neurosurg 1981; 55: 590–603.
19. Brit RH, Enzmann DR. Clinical stages of human brain abscesses on serial CT scans after contrast infusion: CT, neuropathological and clinical correlation. J Neurosurg 1983; 59: 972–989.

20. Hatta S, Mochizuki H, Kuru Y, Miwa H, Kondo T, Mori H, Mizunao Y. Serial neuroradiological studies in focal cerebritis. Neuroradiology 1994; 36: 285–289.
21. Haimes AB, Zimmerman RD, Morgello S, et al. MR imaging of brain abscesses. Am J Neuroradiol 1989; 10: 279–291.
22. Fleming CA, Zimmerman RD, Haimes A, Morgello S, Deck MDF. The diagnostic significance of rim intensity and edema patterns in the differentiation of intracranial mass lesions on MRI. Am J Neuroradiol (abstr) 1987; 8: 454.
23. Gupta RK, Jena A, Sharma A, Guha DK, Khush S, Gupta AK. MR imaging of intracranial tuberculomas. J Comput Assist Tomogr 1988; 12: 280–285.
24. Pyhtinen J, Paakko E, Jartti P. Cerebral abscess with multiple rims on MRI. Neuroradiology 1997; 39: 857–859.
25. Grossman RI, Joseph PM, Wolf G, et al. Experimental intracranial septic infarction: Magnetic resonance enhancement. Radiology 1985; 155: 649–653.
26. Spinnato S, Mazza C, Bricolo A. Nonoperative treatment of cerebellar abscesses. Child's Nerv Syst 1998; 14: 606–609.
27. Lutz TW, Landolt ET Wasner M, Gratzl 0. Diagnosis and management of abscesses in the basal ganglia and thalamus: A survey. Acta Neurochir (Wien) 1994; 127: 91–98.
28. Kondziolka D, Duma CM, Lunsford LD. Factors that enhance the likelihood of successful stereotactic treatment of brain abscesses. Acta Neurochir (Wien) 1994; 127: 85–90.
29. Fulgham JR, Wijdicks EFN, Wright AJ. Cure of a solitary brainstem abscess with antibiotic therapy: Case report. Neurology 1996; 46: 1451–1454.
30. Carpenter JL. Brainstem abscesses: Cure with medical therapy, case report and review. Clin Infect Dis 1994; 18: 219–226.
31. Remy C, Grand S, Lai ES, et al. ^{1}H MRS of human brain abscesses in vivo and in vitro. Magn Reson Med 1995; 34: 508–514.
32. Poptani H, Gupta RK, Roy R, et al. Characterization of Intracranial Mass lesions with in vivo proton MR spectroscopy. Am J Neuroradiol 1995; 16: 1593–1603.
33. Poptani H, Gupta RK, Jain VK, Roy R, Pandey R, Gujral RB. Cystic intracranial mass lesions: Possible role of in vivo MR spectroscopy in its differential diagnosis. Magn Reson Imaging 1995; 13: 1019–1029.
34. Martinez-Perez I, Moreno A, Alonso J, Aguas J, Conesa G, Capdevila A, Arus C. Diagnosis of brain abscess by magnetic resonance spectroscopy. J Neurosurg 1997; 86: 708–713.
35. Dev R, Gupta RK, Poptani H, Roy R, Sharma S, Husain M. Role of in vivo proton magnetic resonance spectroscopy in the diagnosis and management of brain abscess. Neurosurgery 1998; 42: 37–43.
36. Kim SH, Chang KH, Song IC, et al. Brain abscess and brain tumour: Discrimination with in vivo H-1 MR spectroscopy. Radiology 1997; 104: 239–245.
37. Harada M, Tanouchi M, Miyoshi H, Nishitani H, Kannuki S. Brain abscess observed by localised proton magnetic resonance spectroscopy. Magn Reson Imaging 1994; 12:1269–1274.
38. Damaerel P, Hecke PV, Oostende SV, et al. Bacterial metabolism shown by magnetic resonance spectroscopy. Lancet 1994; 344: 1197–1202.
39. Burtscher IM, Holtas S. In vivo proton MR spectroscopy of untreated and treated brain abscesses. Am J Neuroradiol 1999; 20: 1049–1053.

40. Kim YJ, Chang KH, Song IC, Seong SO, Kim YH, Han MH. Brain abscess and necrotic or cystic brain tumor: Discrimination with signal intensity on diffusion weighted MR imaging. Am J Roentgenol 1998; 171: 1487–1490.
41. Desprechins B, Stadnik T, Koerts G, Shabana W, Breucq C, Osteaux M. Use of diffusion weighted MR imaging in differential diagnosis between intracerebral necrotic tumors and cerebral abscesses. Am J Neuroradiol 1999; 20: 1252–1257.
42. Ebisu T, Tanaka C, Umeda M, et al. Discrimination of brain abscess from necrotic or cystic tumors by diffusion weighted echo planar imaging. Magn Reson Imaging 1996; 14: 1113–1116.
43. Danziger A, Price H, Schechter MM. An analysis of 113 intracranial infections. Neuroradiology 1980; 19: 31–34.
44. Blaquiere RM. The computed tomography appearances of intra- and extracerebral abscesses. Br J Radiol 1983; 56: 171–181.
45. Kaufman DM, Miller MH, Steigbigel NH. Subdural empyema: Analysis of 17 recent cases and review of the literature. Medicine (Baltimore) 1975; 54: 485–498.
46. Joubert MJ, Stephanov S. Computerized tomography and surgical treatment of intracranial suppuration. J Neurosurg 1997; 47: 73–78.
47. Sharif HS, Ibrahim A. Intracranial epidural abscess. Br J Radiol 1982; 55: 81–84.
48. Kaufman DM, Litman N, Miller MH. Sinusitis: Induced subdural empyema. Neurology 1983; 33: 123–132.
49. Luken MG, Whealan MA. Recent diagnostic experience with subdural empyema. J Neurosurg 1980; 52: 764–771.
50. Post EM, Modesti LM. "Subacute" postoperative subdural empyema. J Neurosurg 1981; 55: 761–765.
51. Weingarten K, Zimmerman RD, Becker RD, Heier LA, Haimes AB, Deck MDF. Subdural and epidural empyemas: MR imaging. Am J Roentgenol 1989; 152: 615–621.
52. Baum PA, Dillon WP. Utility of magnetic resonance imaging in the detection of subdural empyema. Ann Otol Rhinol Laryngol 1992; 101: 867–869.
53. Morikawa M, Rothman MI, Numaguchi Y. Infratentorial subdural empyema: MR findings. Am J Roentgenol 1994; 163: 1270.
54. Bossard D, Himed A, Badet C, et al. MRI and CT in a case of pituitary abscess. J Neuroradiol 1992; 19:139–144.
55. Wolansky LJ, Gallagher JD, Heary RF, et al. MRI of pituitary abscess: Two cases and review of the literature. Neuroradiology 1997; 39: 499-503.
56. Guigui J, Boukobza M, Taner 1, et al. MRI of primary pituitary abscess. Clin Radiol 1998; 53: 777–779.
57. Abs R, Parizel PM, Verlooy J, Neetens I, Arnouts P. Magnetic resonance characterisation of a long standing pituitary abscess. J Endocrinol Invest 1993; 16: 635–637.
58. Blackett PR, Bailgy JD, Hoffman HJ. A pituitary abscess simulating an intrasellar tumor. Surg Neurol 1980; 14: 129–131.
59. Post KD, McCormick PC, Bello JA. Differential diagnosis of pituitary tumors. Endocrinol Metab Clin North Am 1987; 16: 609–645.
60. Henegar MM, Koby MB, Silbergeld DL, Rich KM, Moran CJ. Intrasellar abscess following transsphenoidal surgery. Surg Neurol 1996; 45:183–188.

61. Zimmerman RA, Girard NJ. Imaging of Intracranial Infections. In *Infection of the central nervous system,* WM Scheld, RJ Whitley, DT Durack, eds. New York: Lippincott–Raven, 1997; 923–944.
62. Kumar S, Gupta RK, Kathuria M, Dev R, Husain M, Gujral RB. MR demonstration of a giant cavernous carotid aneurysm with occlusion of the contralateral intracranial carotid artery: An unusual complication of cavernous sinus thrombosis. Neuroradiology 1998; 40: 587–5 89.
63. Micheli F, Schteinschnaider A, Plaghos LL, Melero M, Mattar D, Parera C. Bacterial cavernous sinus aneurysm treated by detachable balloon technique. Stroke 1989; 20: 1751–1754.
64. Dichgans M, Jager L, Mayer T, Schorn K, Pfister HW. Bacterial meningitis in adults. Demonstration of inner ear involvement using high resolution MRI. Neurology 1999; 52: 1003–1009.
65. Post MJD, Quencer RM, Montalvo B.M, Katz BH, Eismont FJ, Green BA. Spinal infection evaluation with MR imaging and intraoperative US. Radiology 1998; 169: 765–771.
66. Boden SD, Davis DO, Dina TS, Sunner JL, Wiesel SW. Postoperative discitis: Distinguishing early MR imaging findings from normal postoperative disc space changes. Neuroradiology 1992; 184: 765–771.
67. Modic MT, Feighn DH, Piraino DW, et al. Vertebral osteomyelitis: Assessment using MR. Radiology 1985; 157: 157–166.
68. Smith AS, Blaser SI. Infectious and inflammatory processes of the spine. Radiol Clin North Am 1991; 29: 809–827.
69. Thrush A, Enzmann D. MR imaging of infectious spondylitis. Am J Neuroradiol 1990; 11: 1171–1180.
70. Sharif HS. Role of MR imaging in the management of spinal infections. Am J Roentgenol 1992; 8: 1333–1345.
71. Dagirmanjian A, Schils D, McHenry M, Modic MT. MR imaging of vertebral osteomyelitis revisited. Am J Roengenol 1996; 167:1539–1543.
72. Kapeller P, Fazekas F, Krametter D, et al. Pyogenic infectious spondylitis: Clinical laboratory and MRI features. Eur Neurol 1997; 38: 94–98.
73. Gellin BG, Weingarten K, Gamache FW Jr, Hartman BJ. Epidural abscess. In *Infection of the central nervous system,* WM Scheld, RJ Whitley, DT Durack, eds. New York: Lippincott–Raven, 1997; 923–944.
74. Nussbaum ES, Rigamonti D, Standiford H, et al. Spinal epidural abscesses: A report of 40 cases and review. Surg Neurol 1992; 38: 225–231.
75. Baker AS, Ojemann RG, Swartz MN, et al. Spinal epidural abscess. N Engl J Med 1975; 293: 463–468.
76. Kuker W, Mull M, Mayfrank L, Topper R, Thron A. Epidural spinal infection: Variability of clinical and magnetic resonance imaging findings. Spine 1997; 22: 544–551.
77. Chow GH, Gebhard JS, Brown CW. Multifocal metachronous epidural abscesses of the spine: A case report. Spine 1996; 21: 1094–1097.
78. Tacconi L, Johnston FG, Symon L. Spinal epidural abscess —review of 10 cases. Acta Neurochirugica 1996; 138: 520–523.
79. Borum SE, McLeskey CH, Williamson JB, Harris FS, Knight AB. Epidural abscess after obstetric epidural analgesia. Anesthesiology 1995; 82: 1523–1526.
80. Wang JS, Fellows DC, Vakharia S, Rosenbaum AE, Thomas PS. Epidural abscess — early magnetic resonance imaging detection and conservative therapy. Anesth Analg 1996; 82: 1069 –1071.

81. Obrador GT, Levenson DJ. Spinal epidural abscess in hemodialysis patients: Report of three cases and review of the literature. Am J Kidney Dis 1996; 27: 75–83.
82. Friedman DP, Hills JR. Cervical epidural spinal infection: MR imaging characteristics. Am J Roentgenol 1994; 163: 699–702.
83. Lang IM, Hughes DC, Jenkins JPR, Forbes WSC, McKenna F. MR imaging appearances of cervical epidural abscess. Clin Radiol 1995; 50: 466–471.
84. Teman AJ. Spinal epidural abscess. Early detection with gadolinium magnetic resonance imaging. Arch Neurol 1992; 49: 743–746.
85. Sandhu FS, Dillon WP. Spinal epidural abscess: Evaluation with contrast-enhanced MR imaging. Am J Neuroradiol 1991; 12: 1087–1093.
86. Sadato N, Numaguchi Y, Rigamonti D, et al. Spinal epidural abscess with gadolinium enhanced MRI: Serial follow-up studies and clinical correlation. Neuroradiology 1994; 36: 44–48.
87. Krcun K, Shoemaker El, Choranes Cl, et al. Epidural abscess of the cervical spine: MR findings in five cases. Am J Roentgenol 1992; 158: 1145–1149.
88. Sathi S, Schwartz M, Cortez S, Rossitch E. Spinal subdural abscess: Successful treatment with limited drainage and antibiotics in a patient with AIDS. Surg Neurol 1994; 42: 424–427.
89. Dacey RG, Winn HR, Jane JA, Butler AB. Spinal subdural empyema: Report of two cases. Neurosurgery 1978; 3: 400–405.
90. Fraser RAR, Ratzan K, Wolpett SM, Weinstein L. Spinal subdural empyema. Arch Neurol 1973; 28: 235–238.
91. Levy ML, Wieder BH, Schneider J, Zee CS, Weiss MH. Subdural empyema of the cervical spine: Clinicopathological correlates and magnetic resonance imaging. J Neurosurg 1993; 79: 929–935.
92. Lownie SP, Ferguson GG. Spinal subdural empyema complication of cervical discography. Spine 1989; 14: 1415–1417.
93. Friess MM, Wasenko JJ. MR of staphylococcal myelitis of the cervical spinal cord. Am J Neuroradiol 1997; 18: 455–458.
94. Coker SB, Muraskas JK Thomas C. Myelopathy secondary to neonatal bacterial meningitis. Pediatr Neurol 1994; 10: 259–261.
95. Murphy KJ, Brunberg JA, Quint DJ, Kazanjian PH. Spinal cord infection: Myelitis and abscess formation. Am J Neuroradiol 1998; 19: 341–348.
96. Dev P, Husain M, Gupta A, Gupta RK. MR of multiple intraspinal abscesses associated with congenital dermal sinus. Am J Neuroradiol 1997; 18: 742–743.
97. Gurbani SG, Cho CT, Lee KR. Staphylococcus epidermidis meningitis and an intraspinal abscess associated with a midthoracic dermal sinus tract. Clin Infect Dis 1994; 19:1138–1140.

Chapter 3

TUBERCULOSIS AND OTHER NON TUBERCULOUS BACTERIAL GRANULOMATOUS INFECTIONS

Rakesh K. Gupta

Tuberculosis kills more people worldwide than tropical diseases and AIDS combined. About one-third of the world's population is infected with *Mycobacterium tuberculosis* and presently there are about 9 million new cases of tuberculosis with 3 million deaths worldwide. Despite worldwide infection control programs and powerful antibiotics, tuberculosis of the central nervous system (CNS) is still a prominent cause of morbidity and mortality. In developing countries, it remains a prominent cause of sickness and death; in developed countries; pockets of indigenous poverty and large-scale immigration from endemic countries have contributed to its overall persistence. In India alone, active tuberculosis has been reported in about 200 million and accounts for about one-sixth of the 3 million deaths globally occurring due to tuberculosis. In the United States, recent data confirm its place among infections of the CNS. The increasing prominence of extrapulmonary infection, the prevalence of drug-resistant mycobacteria, and the recent association of the AIDS epidemic with tuberculosis all suggest that CNS tuberculosis will remain a significant clinical entity in the coming years.

Rakesh K. Gupta • Department of Radiodiagnosis, Sanjay Gandhi Post-Graduate Institute of Medical Sciences, Lucknow 226014, India.

MR Imaging and Spectroscopy of Central Nervous System Infection, edited by Gupta and Lufkin. Kluwer Academic / Plenum Publishers, New York, 2001.

M. tuberculosis is responsible for almost all cases of the tuberculous infections of the CNS. Infections caused by *M. bovis* from the ingestion of contaminated milk and other nontuberculous mycobacterial pathogens are now quite rare. *M. tuberculosis* is nonmotile obligate aerobic bacillus with humans as a natural reservoir. The organism grows slowly with a generation time of 15-20 hours. This is responsible for the long wait for the growth of the colonies after inoculation (1). The thick cell wall, demonstrated on Zeihl-Neelsen or fluorochrome stain, possesses complex antigenic structures that include polysaccharides, proteins, peptides, lipids, and glycolipids. These molecules determine the characteristic immune response to tuberculous infection and its resultant pathology (1).

In most cases of CNS tuberculosis, the disease process is believed to originate via hematogenous dissemination from a primary source outside the CNS, such as the lung or gastrointestinal tract. The Rich focus may rupture during primary infection, resulting in coincidental meningeal and pulmonary or meningeal and miliary infections. Rarely, the source of infection is an adjacent focus of extraneural infection in the vertebrae, inner ear or mastoid sinuses (2,3). Most often, the Rich focus, an intracerebral tubercle formed during the initial silent hematogenous dissemination of the organism, gets reactivated with immuno-compromising conditions such as chemotherapy, alcoholism, and diabetes mellitus and may result in secondary hematogenous spread to the CNS.

The specific CNS syndromes associated with infection are a function of the original location of the infecting tubercles. Depending on the location of the infecting tubercle, it may take a variety of forms including meningitis, tuberculomas, abscess formation, and focal cerebritis. We will describe the disease under headings of (1) cranial tuberculosis and (2) spinal tuberculosis.

1. CRANIAL TUBERCULOSIS

1.1. Meningitis

Tuberculous meningitis is the most common cause of chronic meningitis. It is correctly characterized as a meningoencephalitis because it affects not only the meninges but also the parenchyma and the vasculature of the brain. The primary pathologic event is the formation of a thick tuberculous exudate within the subarachnoid space. The exudates centers around the inter-peduncular fossa, enveloping the optic nerves at the chiasma and extending over the pons and cerebellum, often into the Sylvian fissures and rarely up along the cerebral hemispheres. Within the lateral ventricles, a similar exudate often covers the choroid plexus. Macroscopically, the exudate is gelatinous and frequently nodular. Microscopically, it consists of polymorphonuclear leukocytes, red blood cells, macrophages, and lymphocytes with the fibrin network. As the disease progresses, lymphocytes come to predominate; later, fibroblasts and elements of connective tissues are present as well. Typical tubercles may develop within the exudate; occasionally large zones of caseation necrosis are formed. The number of mycobacteria within the exudate is variable (absent to enormous numbers) (4–6).

Clinical presentation of the tuberculous meningitis is quite variable with substantial difference among patients of different ages. Among the children, most commonly reported symptoms on presentation include nausea, vomiting, and behavioral changes. Seizures are an infrequent presenting complaint reported in 10–25% of children before hospitalization (7). Among adults, the disease typically presents in a somewhat more indolent fashion. Median duration of illness is 2–3 weeks but may go to several months before seeking medical attention (8–9). A history of prior clinical tuberculosis is extremely infrequent among adult patients (less than 20%). Headache is the prominent symptom usually occurring in 50–75% of patients. Behavioral changes consisting of apathy, confusion, or bizarre behavior are reported in a large series with increased frequency (9). Focal neurologic signs most frequently consist of unilateral or, less frequently bilateral cranial nerve palsy. Most frequently affected cranial nerve is the 6th followed by the 3rd and 4th, the 7th and less commonly the 2nd, 8th, 10th, 11th and 12th. An equally common neurologic deficit in some series is hemiparesis. Other less common manifestations are a wide variety of movement disorders more frequent in children than adults. It is important to diagnose the disease early before the development of the sequelae (10–20).

Routine laboratory studies are often not helpful. The likelihood of abnormalities on chest radiograph depends on the patient's age. Among adults, abnormalities consistent with active or healed tuberculosis are seen in 25–50% of patients whereas among children these changes are considerably more common (9,21,22). The incidence of miliary disease with meningitis has decreased in the post antimicrobial therapy era, but these still frequently coexist (16). Cerebrospinal fluid (CSF) abnormalities traditionally accompany tuberculous meningitis although these changes can be quite nonspecific. Moderate decreases in CSF glucose and elevation in protein; moderate pleocytosis with lymphocytic predominance are characteristic accompaniments of tuberculous meningitis. Demonstration of *M. tuberculosis* in CSF on culture and/or on direct smear is a definitive confirmatory test for diagnosis but it is positive in 25–86% in different series (10–20). This deficiency has resulted in the development of a number of clinically useful alternatives. Most commonly used tests are detection of mycobacterial antigen, antibody, and polymerase chain reaction with sensitivity and specificity of 79–94 and, 95–100%; 27–100 and 94–100 and 83–100and 80–100%, respectively (23–29).

MRI helps in demonstrating the meningitis and its potential complications, and following the patient on treatment.

The MRI examination usually shows little or no abnormality on conventional spin-echo imaging without contrast enhancement in patients with early disease. Later, distension of the affected subarachnoid spaces occurs with associated mild shortening of T1 and T2 relaxation times as compared with normal CSF (Fig.1). PostgadoliniumT1-weighted images show abnormal meningeal enhancement, more pronounced in the basal cisterns. The interpeduncular fossa, pontine cistern, perimesencephalic and suprasellar cisterns are the sites commonly involved. Some degree of involvement of sulci over the convexities and in the Sylvian fissures is also seen (30–36). Tentorium and abnormal meningeal enhancement in the cerebellar sulci is rarely seen.

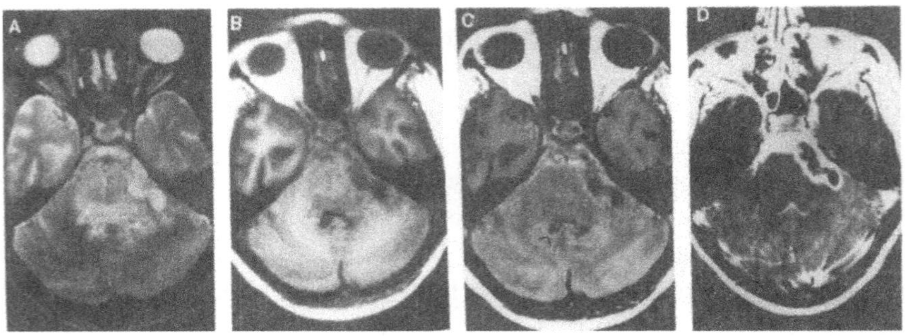

Figure 1. Tuberculous meningitis with CSF loculation. T2-weighted image (a) through the lower pons shows mild dilatation of the right temporal horn of the lateral ventricles along with diffuse hyperintensity in the pons. Note the mixed-intensity loculation in the left cerebellopontine angle. T1-weighted image (b) shows the loculation to be slightly hyperintense than CSF along with mass effect on the fourth ventricle. Magnetization transfer (MT)T1-weighted image (c) shows the margins of the loculation as hyperintense and extending to the opposite side. Note the hyperintensity of the roof of the fourth ventricle (open arrow). Postcontrast MTT1-weighted image (d) shows intense enhancement of the margins of the loculation, thickened meninges, and roof of the fourth ventricle suggesting meningitis with ventriculitis. The thickened meninges and abnormal ependymal lining of the fourth ventricle were clearly visible on precontrast MT imaging that subsequently enhanced on postcontrast MTT1-weighted imaging.

Abnormal meningeal enhancement is usually seen as thick sheathlike enhancement with irregular margins because of thick exudates, especially involving the suprasellar and perimesencephalic cisterns (Fig.2). This abnormal enhancement may continue to be seen on imaging even after a full course of antitubercular chemotherapy. Except in truly caseous regions, these exudates show increased vasculature on microscopic examination (37). The early inflammatory neovessels responsible for this hypervascularity presumably are the cause of the abnormal leptomeningeal enhancement observed. MT spin-echo (MT-SE) imaging following contrast injection is considered superior to postcontrast conventionalT1-

weighted imaging in demonstration of meningeal inflammation. The inflamed meninges are more conspicuously visualized with a smaller amount of contrast administration than conventional SE imaging. Recently, thickened and inflamed meninges in tuberculous meningitis have been clearly shown in early as well as in advanced disease (38) even on precontrast MT-SE images that enhanced subsequently on post-contrast MT-SE images (Fig.2). This is considered to be due to the difference in MT ratio of the inflamed meninges and surrounding parenchyma. Initial experience with quantitation of MT ratios in meningitis of different etiologic agents suggests significantly lower MT ratio in tuberculous meningitis than pyogenic and fungal meningitis and significantly higher MT ratio than viral meningitis. Quantitative MT ratio may be of value in the differentiation of tuberculous meningitis from other chronic meningitis (38). Tuberculous meningitis may precede, follow, or be associated with parenchymal tuberculomas even when the patient is on specific chemotherapy. Fast fluid attenuated inversion recovery (FLAIR) sequence has been shown to be another useful technique in demonstrating meningeal thickening superior to conventional SE imaging in tuberculous meningitis. However, postcontrastT1-weighted imaging was found to be better than FLAIR in the demonstration of CNS infection (39). The characterization of tuberculomas with MRI will be discussed in detail below.

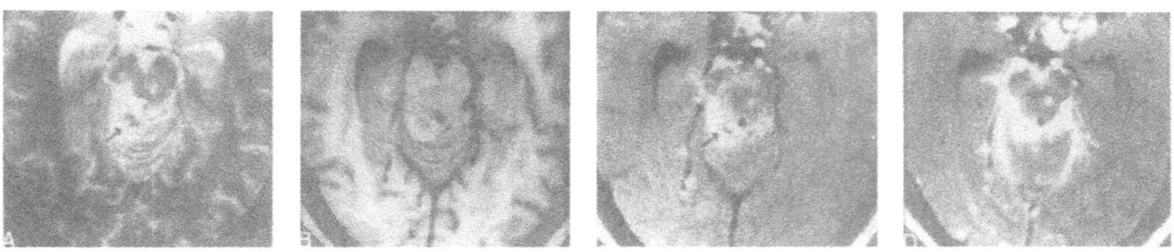

Figure 2. Tuberculous meningitis with tuberculoma.T2-weighted image (a) through the mesencephalon shows a hyperintense lesion on the left side along with a hyperintense region with central hypointense specks (arrow) in the right mesencephalic cistern, merging with the lateral margin of the midbrain. The cisternal lesion appears isointense onT1-weighted image (b). MT-SET1-weighted image (c) shows the right cisternal and left midbrain lesion as hyperintense with central hypointensity extending to the remaining margins of the midbrain (arrow). Postcontrast MT-SET1-weighted image (d) shows marked enhancement of these lesions and meninges.

1.1.1. Sequelae

The complications may develop as the disease progresses or may develop while patient is on specific chemotherapy.

1.1.1.1. Hydrocephalus

This is a common complication of tuberculous meningitis and is well documented on MRI. The hydrocephalus encountered in tuberculosis is usually of the communicating type, secondary to the blockage of the basal cisterns by the inflammatory exudate but occasionally may be the obstructive type secondary to a focal parenchymal lesion with mass effect (Figs.1,2) or due to entrapment of the ventricle by granulomatous ependymitis (37,40). The dilated ventricular system may show a normal pressure or high pressure. The presence of periventricular hyperintensity on proton density- and T2-weighted images usually suggests hydrocephalus under pressure. The hyperintensity on T2- and proton density-weighted images is due to the seepage of CSF across the white matter fibers. Hydrocephalus under pressure is an indication for placement of a shunt for decompression of the ventricular system. Chronic hydrocephalus that is untreated may result in atrophy of gray and white matter.

The choroid plexus is an important CNS structure that may serve as a portal of entry for the pathogens into the CNS. Sometimes, choroid plexus involvement is the only sign of meningitis on imaging (Fig 3); however, it is usually associated with ependymitis,

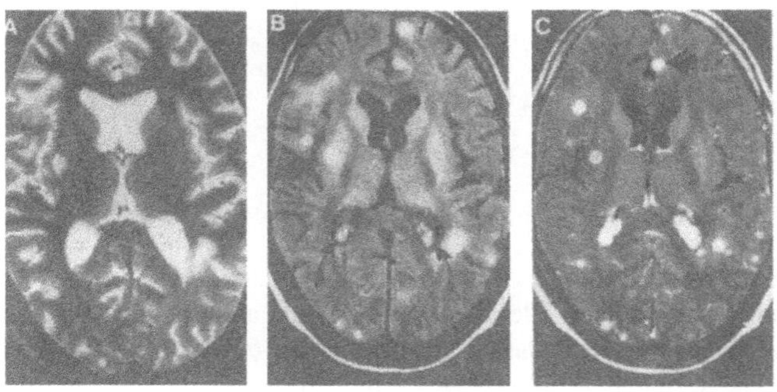

Figure 3. Tuberculous choroid plexitis with multiple tuberculomas. T2-weighted axial image (a) at the level of thalami shows mild dilatation of lateral ventricles with multiple focal lesions involving both cerebral hemispheres. Some of the lesions are hyperintense while others are hypointense with peripheral hyperintensity and are associated with variable perifocal edema. MTT1-weighted image (b) shows thickened choroid plexus with hyperintense foci within (arrows). Note the multiple tuberculomas, some of which are invisible (isointense) onT2-weighted image. These tuberculomas show hypointense center with peripheral hyperintensity or hyperintense foci. Postcontrast T1 MT image (c) shows intense enhancement of the choroid plexus and rim and nodular enhancement of the tuberculomas. Some of the tuberculomas invisible on T2 and visible on pre- and postcontrast T1 MT images are arrow marked (c).

ventriculitis, and meningitis. Contrast-enhanced MRI usually shows prominent enhancement of the choroid plexus and is described as choroid plexitis (Fig 3). It is usually associated with intense enhancement of the ventricular lining (ventriculitis) and meningitis (33,41).

1.1.1.2. Vasculitis

Small and medium-sized vessels are most frequently involved although capillaries and veins may also be affected. It is the second most common complication frequently seen at autopsy in cranial (tuberculous) TB meningitis (37,40). The adventitial layer of the vessels develops changes identical to those of the adjacent tuberculous exudate: Similar cells, tubercles, caseation necrosis are found and are occasionally associated with clumps of tubercle bacilli as well. The intima of the vessels may eventually be similarly affected or it may be eroded by a fibrinoid-hyaline degeneration. A reactive subendothelial cellular proliferation follows that may completely occlude the lumen of the vessel (37). Ischemic cerebral infarction resulting from the vascular occlusion is thus a common sequela of tuberculous arteritis, most often in the distribution of the middle cerebral artery and the lenticulostriate arteries; these are affected as they penetrate the base of the brain (33,37,42). The conventional angiographic features of cranial TB meningitis consist of the triad of a hydrocephalic pattern, narrowing of the arteries at the base of the brain, and narrowed or occluded small or medium-sized arteries with early draining veins (43,44). Because of the predominant basal distribution of exudate, a majority of the infarcts are in the basal ganglia and internal capsule region owing to the involvement of the lenticulostriate arteries (33,45). Brainstem and major territory infarcts are uncommon (45). Dastur et al in an autopsy series of 100 patients found cerebral infarcts in 41% of cases (37). The reported incidence of infarcts detected by CT varies from 20.5 to 38%; however, MR detects more infarcts as well as hemorrhages in the infarcts than CT (30).

MR angiography is a useful technique in the initial diagnosis and evaluation on follow-up of patients with vasculitis secondary to TB meningitis (Fig.4). The amount of vascular occlusion demonstrated (number of vessels and size of the occlusion) on MR angiography does not always correlate with the size of infarction or the distribution of infarction (33). Major infarction of the vascular territory is not so common (Fig.4). It is comparable to the large autopsy series reporting the vasculitis and the site of infarction (33,37). Vascular complications can also occur once the patient is put on antituberculous treatment due to fibrosis of the exudates around the small and medium-sized vessels (33). Demonstration of progression of narrowing of the involved vessel in a patient with TB meningitis while on treatment may alert the clinician for a possible evolution of infarction. MR angiography may also help in detection of another rare vascular complication—pseudoaneurysm (Fig.5)—in a patient with TB meningitis (33).

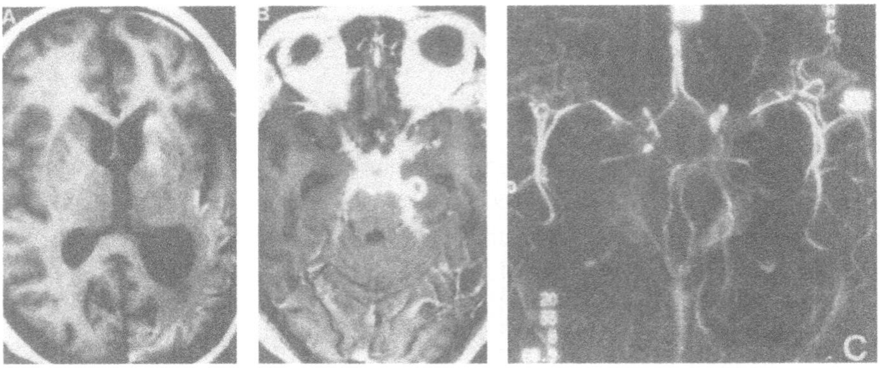

Figure 4. Tuberculous meningitis with vasculitis and hemorrhagic infarction. PrecontrastT1-weighted axial image (a) at the level of third ventricle shows hemorrhagic infarction of the left temporo-occipital region with ipsilateral focal dilatation of the lateral ventricle. Postcontrast T1-weighted image (b) through the level of midbrain shows intensely enhancing thickened meninges of basal cisterns extending into the optic chiasma and optic nerves bilaterally. Postcontrast cranial MR angiogram (c) shows segmental narrowing of both middle cerebral arteries and thin attenuated posterior cerebral arteries suggestive of vasculitis.

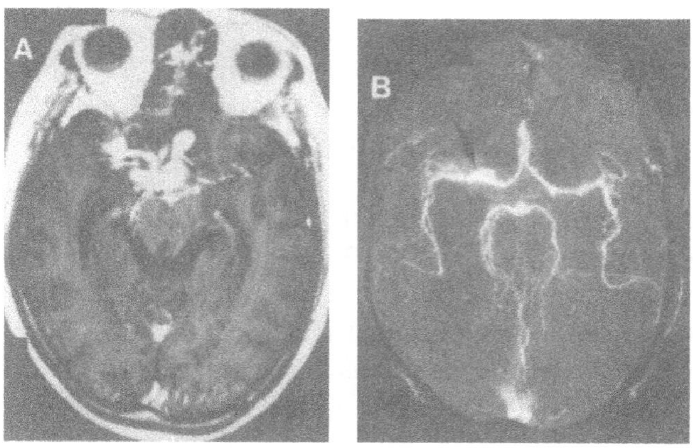

Figure 5. Tubercular meningitis with pseudoaneurysm in a patient who had received 2 years of antituberculous treatment. Postcontrast T1-weighted image (a) through the level of midbrain shows thick nodular enhancing basal exudates in the interpeduncular fossa, extending into the right Sylvian fissure. Post-contrast MR angiogram (b) shows a pseudoaneurysm arising from the right middle cerebral artery (arrow).

1.1.1.3. Cranial Neuropathies

Clinical involvement of the cranial nerves is seen in 17.4–40% of patients with TB meningitis. Cranial neuropathies are thought to be partly due to vascular compromise resulting in ischemia of the nerve (Fig.6). Alternatively, it may be due to entrapment of the nerves by the exudate (Fig. 6a–d). The cranial nerves may appear thick on both T1- and T2-weighted images and may show hyperintensity on T2-weighted image (Fig.6e). On contrast-enhanced MRI, the proximal portion of the root at the nerve entry/exit zone is mostly affected and may be associated with enhancement (33). Cranial nerve involvement has also been shown by the mass effect on the nerve in the subarachnoid space as well as by nuclear and supranuclear involvement of the cranial nerve nuclei in the brainstem due to infarct, focal cerebritis, or tuberculoma (33). MR localization of the cause of neuropathy may help in predicting the recovery of the cranial nerve function in patients undergoing treatment with specific chemotherapy. Patients with brainstem infarction are unlikely to show functional improvement, whereas patients with tuberculous perineuritis/neuritis and tuberculomas are likely to improve with therapy (33).

In some cases, meningitic process may evolve into a true mass-forming granulomatous disease with or without caseation. Parenchymal tuberculomas may also be associated with TB meningitis in 20–30% of cases or these may develop while the patient is receiving specific chemotherapy.

Differential diagnosis. MR features of TB meningitis are not very specific and differential diagnosis includes meningitis due to other infectious agents such as nontuberculous bacteria, viruses, fungi, and parasites; noninfectious inflammatory conditions such as sarcoidosis, and rheumatoid disease; and primary or secondary neoplastic involvement of meningeal surface. Primary CNS tumors that most frequently metastasize via the subarachnoid space include medulloblastoma, pinealoblastoma, ependymoma, and glioblastoma multiforme. In both systemic lymphoma and acute leukaemia, leptomeningeal infiltration by the tumor cells occurs first and then spreads along the Virchow–Robin spaces to invade the brain parenchyma. In all of these conditions, the appearance on postcontrast MRI is nonspecific and clinical features and other laboratory investigations are required to differentiate these entities. However, MT MRI is a promising technique that may be able to differentiate some of the chronic meningitides from TB meningitis.

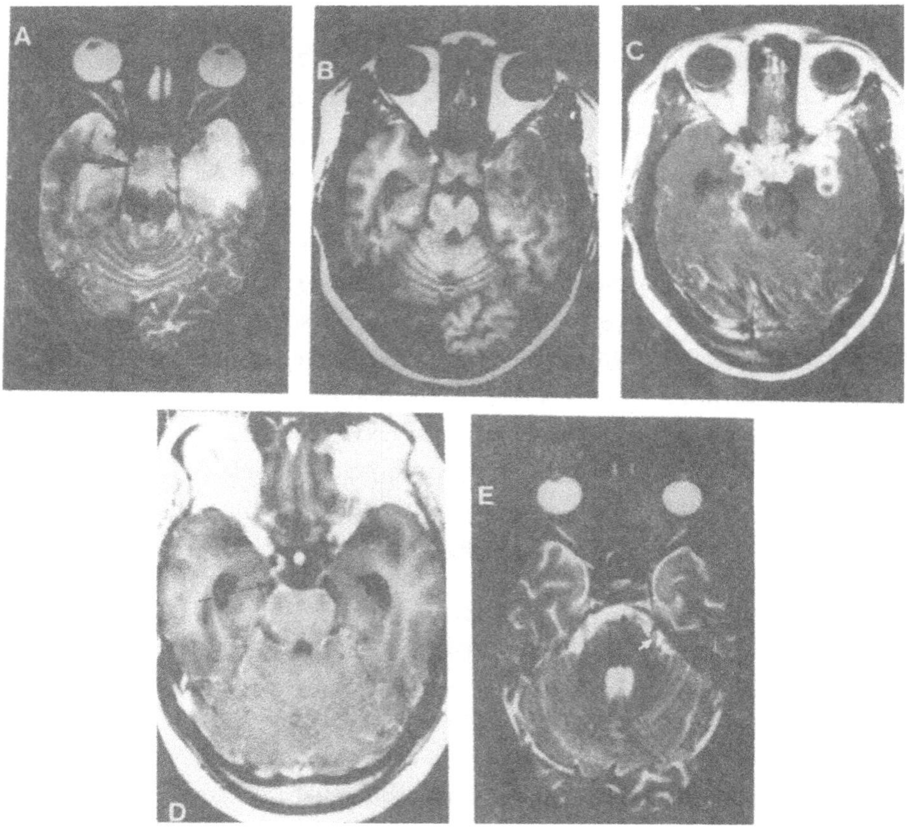

Figure 6. Cranial neuropathies. T2-weighted axial image (a) at the level of optic chiasma shows thickening and heterogeneous signal intensity of the chiasma (arrow). Thickened optic chiasma is seen well onT1-weighted image (b). PostcontrastT1-weighted image (c) shows enhancement of the chiasma along with basal meninges and multiple tuberculomas. In another patient, postcontrastT1-weighted axial image at the level of midbrain shows thickening (black arrow) of the right third nerve (d) along with meningeal enhancement. T2-weighted image (e) through the lower pons shows focal hyperintensity (white arrow) in the center of the left fifth nerve consistent with neuritis in another patient with TB meningitis.

High-resolution MR spectroscopy of CSF. MR spectroscopy of CSF has been performed in various conditions of the CNS with an aim to obtain disease-specific information. As CSF culture is not always positive in TB meningitis, it is possible to look for the mycobacterial specific metabolites using this technique. Five-hundred microlitres of the CSF fluid is obtained from the patient with suspected meningitis before starting any treatment and lyophilized. It is reconstituted in D_2O to make it 500 μl and a one-dimensional proton NMR spectrum is obtained by using two different sequences: a single pulse sequence with frequency-selective presaturation and SE sequence for separating the long T2 components from short T2 components (TE = 80 ms). The typical parameters used are 32,768 data points in sequential mode and a spectral width 3.142 KHz; 64 scans are averaged with a recycle delay of 3 seconds.

In TB meningitis, CSF showed cyclopropyl rings at –0.5 to +0.5 ppm and phenolic glycolipids at 7.1 and 7.4 ppm (Fig.7) this pattern has not been seen in pyogenic or viral meningitis CSF. These metabolites have been shown in all CSF samples culture positive for *M. tuberculosis* and clinically highly suspicious cases of TB meningitis (46). This may be a simple and quick method of confirmation of diagnosis in patients with negative CSF culture for *M. tuberculosis* and may help in making an early diagnosis and treatment.

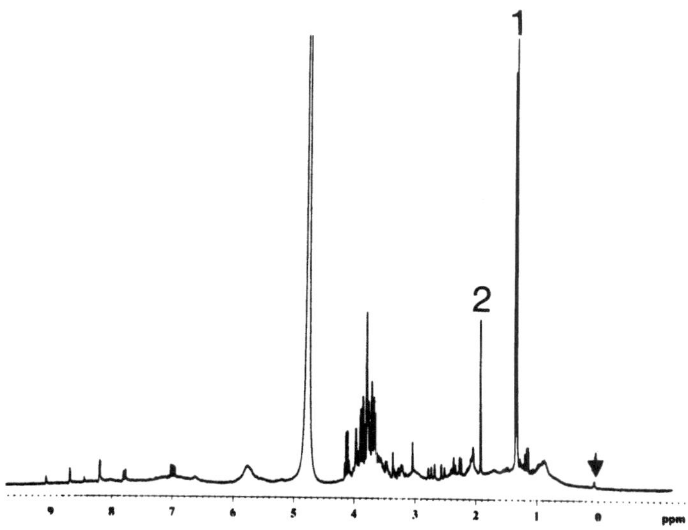

Figure 7. High-resolution proton MR spectroscopy of the CSF from a patient with proven tuberculous meningitis. It shows large resonances from lactate (1), acetate (2), and sugars. Note the resonances at 0.1 ppm (arrowhead) assigned to cyclopropyl rings characteristic of tuberculous meningitis and at 7.1 ppm assigned to phenolic glycolipids. These resonances are characteristically observed in CSF obtained from patients with tuberculous meningitis.

1.2. Focal or Diffuse Pachymeningitis

Isolated involvement of the dura mater is considered as an unusual presentation of CNS tuberculosis. Pachymeningeal tuberculosis is an entity distinct from the inflammation of the dura mater adjacent to an intraparenchymal tuberculoma that has been shown on histopathology to be present in almost 50% of all tuberculomas (47–51). Pachymeningeal tuberculosis consists of either isolated dural involvement or a predominantly dural-based lesion with secondary pial or parenchymal involvement. TB pachymeningitis may also result from the hematogenous spread of the bacilli as in the case of TB meningitis. Pachymeningitis may exist as focal or diffuse involvement of the dura mater (47–51). The focal involvement has also been described as tuberculoma en plaque (47,49). The lesions may be limited to the dura or may secondarily involve the arachnoid or pia mater or the brain parenchyma. Focal pachymeningitis appears as isointense on T1 and iso- to hypointense to the brain parenchyma on T2-weighted images. Postcontrast study usually reveals enhancement (Fig. 8). Diffuse involvement either may appear similar to focal lesions on spin-echo images or may be isointense on T1 and hyperintense on T2-weighted images. Postcontrast study reveals diffuse enhancement of the dura mater (Fig. 9).

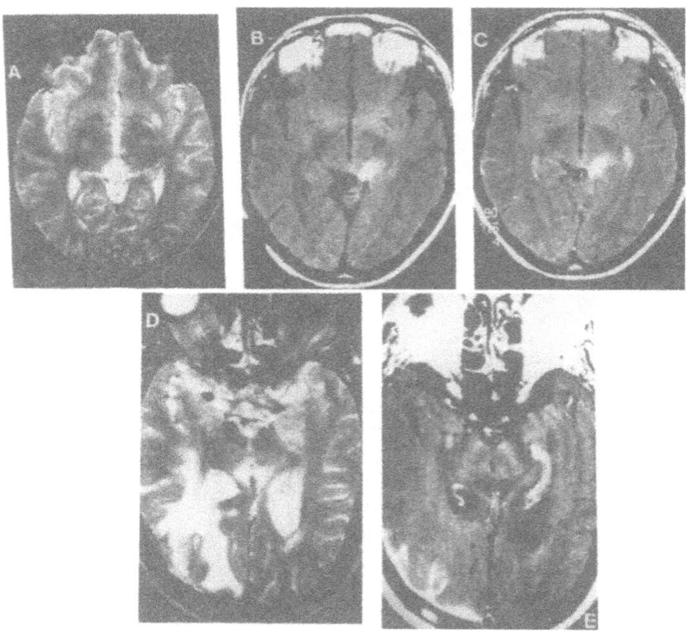

Figure 8. Focal pachymeningitis in two patients. T2-weighted axial image (a) at the level of thalami shows no obvious abnormality. MT T1-weighted image (b) reveals a hyperintensity in the left ambient cistern which shows enhancement in postcontrast MT T1-weighted image (c) suggestive of focal meningitis. In another patient, T2-weighted axial image (d) shows focal hypointensity of the meninges in the right occipital region with edema. Postcontrast MT T1-weighted image (e) shows enhancement of the meninges in the right occipital region, consistent with focal pachymeningitis.

MRI appearance of focal and diffuse pachymeningitis is nonspecific and may be found in a large number of inflammatory and noninflammatory conditions (47–51). Focal lesions may simulate meningioma, lymphoma, sarcoidosis, syphilis, meningeal carcinomatosis, and fibromatosis. Diffuse involvement may also be seen in idiopathic hypertrophic pachymeningitis.

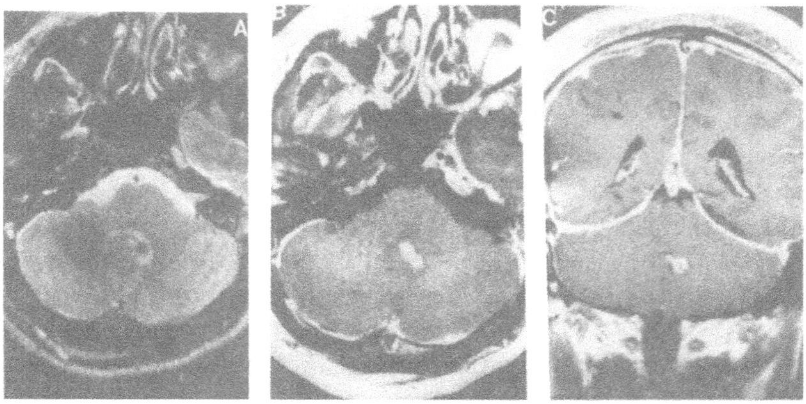

Figure 9. Diffuse pachymeningitis.T2-weighted axial image (a) at the level of fourth ventricle shows a focal hypointense lesion in the fourth ventricle. PostcontrastT1-weighted axial (b) and coronal (c) images show enhancement of the intraventricular lesion and diffuse enhancement of the dura mater.

1.3. Intracranial Tuberculoma

Parenchymal disease may occur with or without meningitis and usually presents as a solitary or multiple tuberculomas. The pathogenesis of CNS tuberculomas is identical to that of TB meningitis. Organisms lodge in the CNS during the silent hematogenous dissemination of primary infection; when cell-mediated immunity develops, small tubercles are formed. However, instead of rupturing into the subarachnoid space, initial tubercles continue to grow, walled off from the brain parenchyma and the meninges by a dense fibrous capsule (52).

In the developing world, tuberculomas account for 15–50% of the intracranial tumors seen. In the developed world, tuberculomas have become relatively uncommon but nonetheless a persistent entity accounting for less than 0.2% of all biopsied brain tumors seen. The clinical sequelae of CNS tuberculomas are those of single or multiple intracranial mass lesions. In large series, the number of identified lesions per patient has ranged from 1to12 (or more) with size varying from 1 mm to 8 cm. Tuberculomas can be seen throughout the cerebral hemispheres, the basal ganglia, the cerebellum, and the brainstem (53). Rarely these are located in the ventricular system (Figs.3, 9). Children are said to develop infratentorial tuberculomas more commonly than adults do, although no precise patterns of localization according to race, age, or sex have been possible. Symptoms are often limited to seizures and correlates of intracranial pressure. Papilledema is seen in most cases, accompanied by neurologic deficits reflecting the location of the lesion. These lesions originate as a conglomerate of microgranulomata in an area of TB cerebritis that join to form a noncaseating tuberculoma. In most cases, subsequent central caseous necrosis develops that is initially solid but that in some instances may eventually liquefy.

CT and MRI are the imaging techniques usually used in the anatomic localization and characterization of such lesions. MRI is considered superior to CT for better localization and characterization of intracranial tuberculomas. The MR features of the individual tuberculoma depend on whether the granuloma is noncaseating or caseating with a solid center, or caseating with a liquid center (30,34,36,45,54–59). The noncaseating granuloma is usually iso/hypointense on T1- and hyperintense on T2-weighted images. These granulomas show homogeneous enhancement after injection of paramagnetic contrast agent. The caseating solid granulomas appear relatively iso/hypointense on T1-weighted images with iso/hyperintense rim and isointense to hypointense on T2-weighted images. The rim may be inseparable from edema when present on T2-weighted images or appears hyperintense in the absence of edema. These lesions show rim enhancement on postcontrast T1-weighted imaging. The lesions which are not visible (isointense on both T1- and T2-weighted images) on conventional SE images, and either enhance after contrast administration or become visible on MT pre/post-contrast images (Figs.3, 11) are referred to as conventional SE invisible lesions (38,60,61). These lesions may show nodular or rim enhancement on a postcontrast study. The degree of hypointensity of the solid caseating granuloma on T2-weighted images depends on the complex relationship between the solid caseation, associated regional fibrosi/ gliosis, macrophage infiltration and macrophage by-products like the release of free radicals, and perilesional cellular infiltrate (58). Kim et al have described three layers in a tuberculoma on MRI and correlated with three layers as observed on histopathology (59). Histologically, the mature tuberculomas are composed of a necrotic caseous center surrounded by a capsule that contains fibroblasts, epitheloid cells, Langhans giant cells, and lymphocytes. They observed that on T1-weighted images, the central isointensity or mixed isointensity and hyperintensity corresponded to caseation necrosis with adjacent inflammatory cellular infiltrates. Surrounding slight hyperintense and hypointense rims corresponded to a layer of collagenous fibrosis and a layer of outer cellular infiltrates respectively. We feel that pathologically and on MRI, there are only two layers that are better visualized on MT-SET1-weighted images. Physically, the caseous material is "cheesy" in texture and is relatively high in lipid content. The granulomas are typically associated with variable degree of surrounding edema that is relatively hypointense on T1-weighted images and hyperintense on T2-weighted images; the wall of the caseating tuberculous granuloma often shows a striking hyperintense rim on T2-weighted images. This rim usually merges with edema and if the granuloma is not associated with edema is clearly visible on T2-weighted images as hyperintense rim. On MT-SE imaging, this rim appears as hyperintense with central hypointensity even in the presence of edema. Gadolinium-enhanced T1-weighted images show rim enhancement in such caseating lesions (Fig.10).

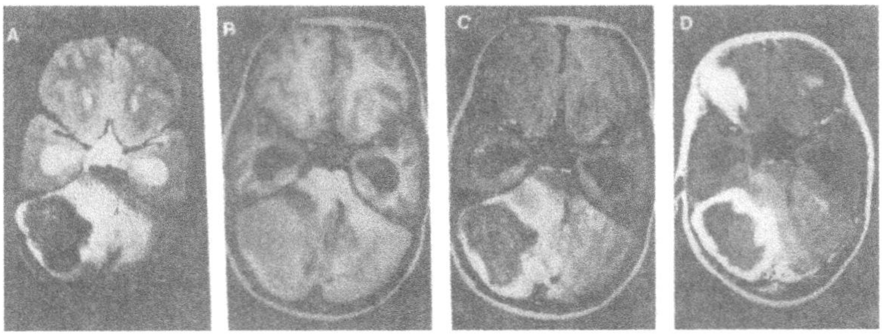

Figure 10. Giant cerebellar tuberculoma with obstructive hydrocephalus. T2-weighted axial image (a) through cerebellum shows a large well defined hypointense mass in right cerebellar hemisphere with perifocal edema and dilated bilateral temporal horns. The mass appears iso- to hypointense onT1-weighted image (b) and is causing mass effect on fourth ventricle. MTT1-weighted image (c) shows the hypointense mass with peripheral hyperintense rim that reveals enhancement on postcontrast MT T1 image (d).

The granulomas with central liquefaction of the caseous material appear centrally hypointense on T1- and hyperintense on T2-weighted images with a variable peripheral hypointense rim on T2-weighted images depending on the amount of liquefaction of the solid caseation with perifocal edema. In the absence of edema, the most peripheral part appears hyperintense and represents the cellular and fibrous component of the granuloma. The images acquired after intravenous gadolinium administration in these cases also show rim enhancement (45). These latter lesions may be indistinguishable on MR from true tuberculous or pyogenic abscess formation. Pathologically, the caseating tubercle contains only rare tubercular organisms, but the tuberculous abscess has a semiliquid center teeming with tubercular bacilli. Parenchymal tuberculous granulomas or abscesses can be seen in the brain, spinal cord and/or in association with the leptomeninges and may be solitary, multiple, or miliary in distribution (45).

The capsules of caseating tuberculous lesions and the whole of noncaseating lesions show evidence of microscopic hypervascularity as determined by sensitive imaging techniques (dynamic bolus contrast-enhanced CT). Therefore, inflammatory neovascularity seems to be responsible both for this hypervascularity and for part of the enhancement observed after contrast agent administration (i.e., blood–brain barrier disruption caused by leaky neovessels) (45).

Miliary brain tuberculosis usually is associated with TB meningitis, although it may be asymptomatic with pulmonary miliary tuberculosis or may precede the meningitis (60,61). Miliary tubercles are less than 2 mm and are either not visible on conventional SE MR images or show tiny foci of hyperintensity on T2-weighted acquisitions. T1-weighted images after gadolinium administration show numerous round, small, homogeneous enhancing lesions. The SE invisible lesions, which may or may not enhance after intravenous injection of gadolinium are clearly visible on MT-SE T1-weighted imaging. MT-SE imaging helps in defining the true disease load (Fig. 11) (38).

It may be useful here to discuss the types of changes in appearance that have been observed within tuberculomas while the patient was receiving antituberculous therapy. In those receiving antituberculous treatment, there is a progressive decrease in the amount of edema associated with a reduction in the overall size of the lesion. If treatment is successful, the lesion may disappear altogether as judged by conventional MR. However, the residual lesion or gliosis, which may not be visible on conventional SE imaging, may still show on MT-SE images. Nonenhancing residual encephalomalacia with or without calcification is also a sign of healing. Calcium deposits, when present, typically appear hypointense on T1-, T2- and T2*-weighted images. Only rarely has a paradoxical response been observed in which they increased in size or even first appeared while the patient was on antituberculous therapy. Nevertheless, these latter tuberculomas were seen to regress gradually with continued medical treatment (45).

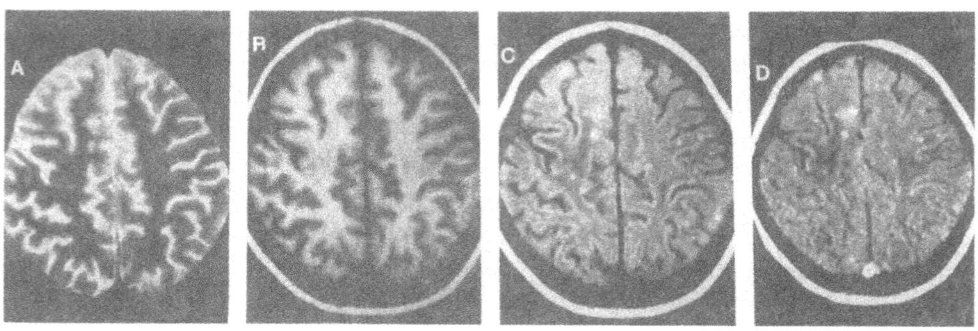

Figure 11. Miliary brain tuberculomas. T2- (a) and T1-weighted (b) axial images at supraventricular level show no definite lesion. Precontrast MTT1 image (c) shows multiple hyperintense lesions in both cerebral hemispheres and reveals enhancement following contrast administration (d). Patient had miliary pulmonary tuberculosis.

As a general rule, different phenomena have been demonstrated concerning tuberculous lesions: (1) lesions in a single patient may be in a different stage of evolution from one another at the time of initial diagnosis; (2) different lesions in the same patient may respond to medical therapy differently (i.e., resolve at different rates); (3) similar appearing lesions in different patients may respond to medical treatment differently; (4) larger centrally caseous lesions in any location in general take longer to resolve than smaller noncaseous lesions; (5) the larger the lesion initially, the greater is the likelihood of permanent sequelae (e.g., calcification, encephalomalacia) (45).

Differential diagnosis. Healing stage of neurocysticercosis, fungal granulomas, chronic pyogenic brain abscess, lymphomas, some gliomas and metastases may simulate tuberculomas and should be considered in its differential diagnosis. Quantitative MT ratio and *in vivo* proton MR spectroscopy may help in differentiation of these conditions from tuberculomas (38, 62–66). However, at times even large tuberculomas may appear as predominantly hyperintense on T2- and of mixed intensity on T1-weighted images and may show heterogeneous enhancement on postcontrast study. These lesions mimic neoplastic lesions on MRI. MR spectroscopy may not be very useful in such patients, as these lesions are cellular and show raised choline and lipid, a feature characteristically described in neoplastic lesions (Fig.12).

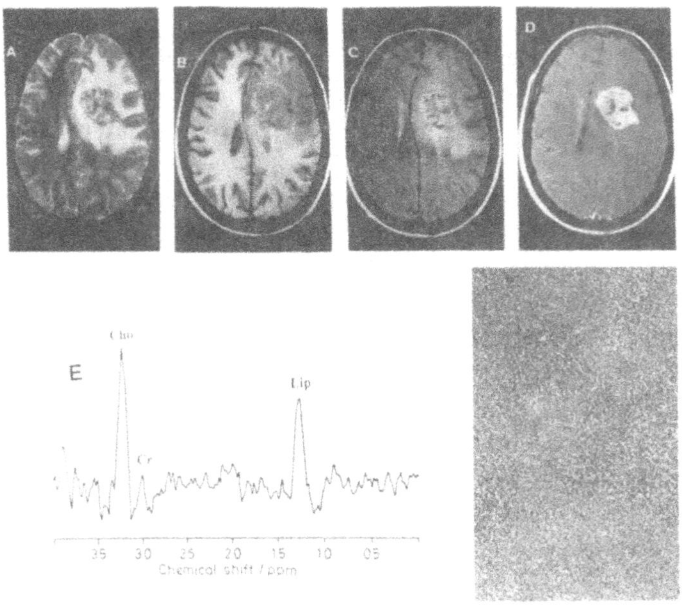

Figure 12. Large tuberculoma simulating neoplasm. T2-weighted axial image (a) shows a large left periventricular mass with perifocal edema and mass effect. It appears predominantly hyperintense with small hypointense foci. The mass is predominantly hypointense with well-defined margin on T1-weighted image (b). The areas of T2 hyperintensity show peripheral rim-like subtle hyperintensity on T1-weighted image and demonstrate its pronouncement on MTT1-weighted image (c). Gradient echo image showed no evidence of susceptibility effect ruling out the presence of calcification and hemorrhage. Postcontrast MTT1-weighted image (d) shows heterogeneous enhancement of the mass. MR spectroscopy (e) shows a high choline and lipid resonances suggestive of a neoplastic lesion. Histopathology shows presence of giant cells, endothelial cells with small areas of caseation (f). Ziehl –Neelsen stain showed acid-fast bacilli consistent with *Mycobacterium tuberculosis* within the lesion.

Proton MR spectroscopy. In vivo proton MR spectroscopy is a novel, noninvasive, and powerful technique that can provide biochemical information of the pathophysiologic process of the tissue in question. The technique has been used in differentiation of neoplastic from inflammatory intracranial mass lesions (62). The lesions, which may have similar imaging features with different pathologic process, can be distinguished by *in vivo* MR spectroscopy. *In vivo* spectra are found to be specific for intracranial tuberculomas when combined with imaging and demonstrate the biochemical fingerprints of the tubercle bacilli in a granuloma (62–66).

Resonances observed on in vivo MR spectroscopy in a tuberculoma. Resonances have been observed at 0.9, 1.3, 2.0, 2.8, and 3.7 ppm on STEAM 20-ms sequence (Fig.13). On SE 135 ms, resonances at 1.3 and 0.9 ppm show marked reduction in signal intensity, while other resonances are barely visible. The resonances at 0.9, 1.3, 2.0, 2.8 and 3.7 ppm have been assigned to terminal methyl (CH_3), acyl chain $(CH_2)_n$, CH_2= CH in fatty acyl chain, = CH-CH_2CH= in fatty acyl chain, and phosphoserine, respectively. Neoplastic lesions show high choline, low to very low NAA and lactate and/or lipid and are clearly distinct from infective lesions. The spectral patterns in all cases of tuberculomas have been very consistent, so far studied by *in vivo* MR spectroscopy (64). One of the close differential diagnoses of T2 black lesion is healing cysticercus granuloma. Unfortunately, these lesions are usually less than 1 cm and cannot be cleanly studied with the current technique available. There is no literature clearly showing the spectra in such lesions backed up by *ex vivo* and *in vitro* study. Brain abscess shows acetate, succinate, leucine, isoleucine, valine, alanine, lipid, lactate, glycine, and other amino acids that have not been described in other infective lesions and neoplasm on *in vivo* study (64).

Ex vivo and in vitro study. These techniques have been used for the purpose of confirming the assignment of the resonances and to assess the utility of these techniques for better diagnosis and management of such lesions. The spectral resolution of *in vivo* MR spectroscopy is relatively poor due to the low magnetic field used for clinical studies and low homogeneity.

Resonances observed on ex vivo and in vitro study. Ex vivo single-pulse spectrum (Fig.13) shows resonances at 0.9, 1.3, 1.58, 2.02, 2.24, 2.80, 3.22, 3.79, 4.1, 4.29, and 5.33 ppm assigned to terminal methyl (CH_3), acyl chain $(CH_2)_n$, OC-CH_2-CH_2 in fatty acyl chain, CH_2 = CH in fatty acyl chain, $COCH_2$ in fatty acyl chain, =CH-CH_2CH= in fatty acyl chain, -$N(CH_3)_3$ of choline, phosphoserine, glycerol backbone of phospholipids (4.1 and 4.29 ppm), and olefinic group of lipids, respectively.

SE Fourier-transformed (SEFT) spectrum shows the phase reversal of resonances at 1.3 and 3.7 to 3.8 ppm assigned to lactate and serine (serine-containing compounds, namely, serine, phosphoserine, and phosphatidylserine), respectively. Sharp resonance at 2.29 ppm was assigned to acetoacetate. Terminal methyl at 0.9 ppm, acyl chain at 1.3 ppm, and olefinic group at 5.3 ppm show marked reduction in signal. Other resonances seen in the single-pulse spectrum were not visible in the SEFT spectrum, confirming that those resonances were due to the different lipids (64).

Ex vivo, 2D correlated spectroscopy (COSY) has confirmed all of the resonances seen on 1D single-pulse and SEFT spectra. Alanine (1.48 and 3.76 ppm) and threonine (1.34 and 4.25 ppm) are also observed (Fig.13).

PCA and lipid extracts of granuloma and M. tuberculosis. The PCA and lipid extracts of tuberculoma and *M. tuberculosis* have shown similarity of spectra.

Lipid extracts: Besides lipids, which are usually seen in any tissue, the unusual resonances seen in granuloma at 0.5 and 0.1 ppm are assigned to cyclopropane rings and at 7.1 and 7.4 ppm assigned to phenolic lipids. These resonances have also been observed in *M. tuberculosis* confirming that the granuloma is of mycobacterial origin.

PCA extracts: PCA extracts of tuberculoma show leucine/valine (0.95 ppm), beta-hydroxybutyrate (1.2 ppm), lactate (1.33, 4.1 ppm), acetate (1.92 ppm), glutamate (2.19 ppm), acetoacetate (2.29 ppm), lysine (3.46 ppm), serine (3.7–3.9 ppm), adenine (8.1, 8.4 ppm), and guanine (8.2 ppm). The PCA extract of *M. tuberculosis* has shown a small amount of serine confirmed on SEFT and 2D COSY study (Fig.13).

A number of *ex vivo* studies of different intracranial neoplastic lesions have shown high choline, lipids, lactate, and some amino acids (67–69). These are dependent on the consistency of the mass lesions from which the spectrum is obtained. Presence of lipids suggests hypoxia along with micro/macronecrosis (67–69). Granulomas show caseation necrosis and lipids are visible on *ex vivo* study.

However, the presence of serine on *ex vivo/in vitro* study is distinct in tuberculomas and has not been observed so far in any other intracranial mass lesion. One of the characteristic features of mycobacterium is the presence of a lipid-rich cell wall dominated by a wide variety of species and type-specific glycolipids. Phenolic glycolipids are present in virulent and avirulent strains of different species of mycobacterium. Demonstration of phenolic glycolipids on *ex vivo/in vitro* study in tuberculomas suggests that phenolic lipids represent the constituents of cell wall of the living/dying or dead mycobacteria in a granuloma (64).

Tissue destruction leading to caseation is a particular feature of tuberculosis and is rarely found in other immunologic and nonimmunologic granulomas. Caseation is a cheeselike necrosis and is composed of necrosed tissue, inflammatory cells, and dead/fragmented/lysed mycobacteria. Presence of phenolic lipids represents the

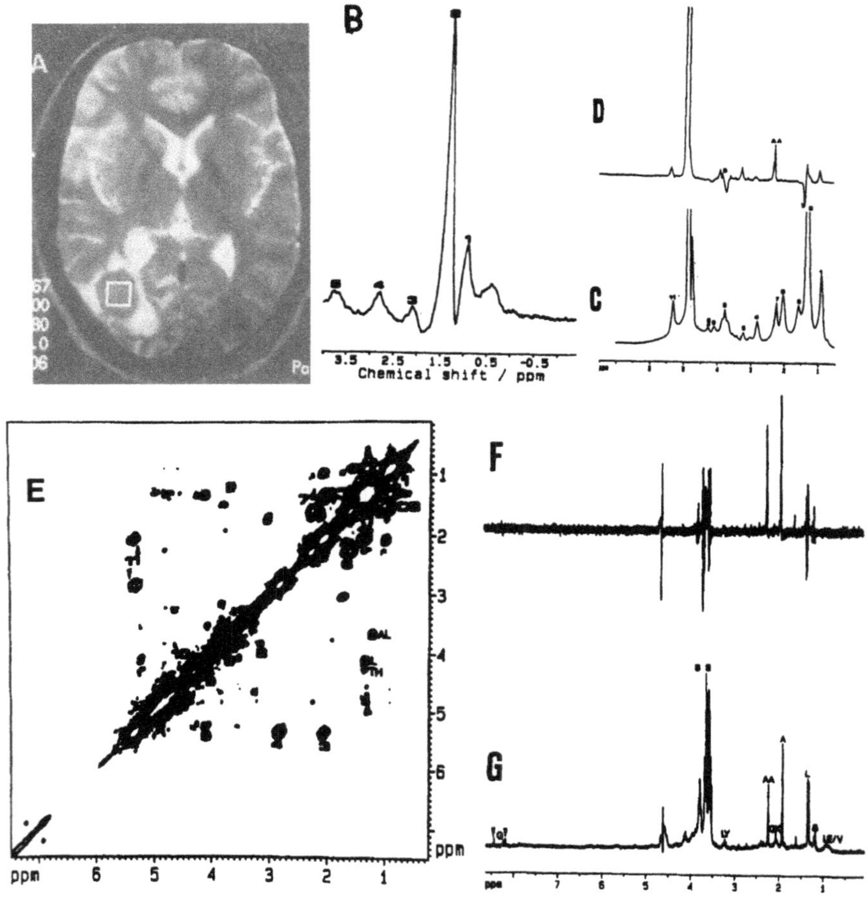

Figure 13. *In vivo*, *ex vivo*, and *in vitro* spectroscopy of an intracranial tuberculoma. T2-weighted axial image (A) of the brain shows hypointense well-defined mass in the right occipital region with hyperintense perifocal edema. Inset shows the volume of interest where the *in vivo* proton MR spectrum was obtained. (B) *In vivo* proton MR spectrum using STEAM sequence (TE = 20) from the tuberculoma shows different lipid resonances [1, -CH_3; 2, $(CH_2)_n$; 3, $H_2C{=}CH$; 4, CH-$CH_2CH{=}$; 5, phosphoserine]. (C) *Ex vivo* single-pulse spectrum shows lipid resonances (6, OC-CH_2-CH_2; 7, OC-CH_2; 8, choline; 9, phospholipids; 10, triglyceride backbone; 11, olefinic group). (D) SEFT spectrum (TE = 80 ms) shows phase reversal of lactate (L) and serine (S) with acetoacetate (AA) at 2.29 ppm. Note the marked reduction or disappearance of various lipid resonances seen in (C). 2D COSY (E) spectrum (Al, alanine; TH, threonine; CS, cholesterol). Single-pulse (F) and spin-echo spectra (G) of perchloric acid extract of tuberculous granuloma (LE/V, leucine/valine; B, beta-hydroxybutyrate; A, acetate; Gx, glutamate; LY, lysine; G, guanine; arrowheads, adenine).

biochemical fingerprints of *M. tuberculosis* in a granuloma. Isolation of *M. tuberculosis* from the tissue on either smear or culture is possible in a very small number of cases and diagnosis rests on demonstration of caseating granuloma on histology. In a significant number of cases, even caseation is not seen and it becomes difficult to differentiate tuberculous granuloma from nontuberculous granulomas. Besides culture having a low yield, it takes about 6–8 weeks for the results; and treatment cannot be withheld for such a long time due to prognostic reasons. Lipid extraction for the purpose of MR spectroscopy is a relatively simple procedure and can be performed routinely in most well-equipped laboratories. It may help as an additional tool in the fingerprinting of *M. tuberculosis* cell wall biochemicals in the granuloma and may help in facilitating the diagnosis, where routine histology and culture are not conclusive (64).

1.4. Tuberculous Brain Abscess

Tuberculous brain abscesses are extremely rare. In two series of cases from developing countries, tuberculous abscess was seen in 4–7% of patients with CNS tuberculosis without HIV infection. Tuberculous abscess is a condition distinct from CNS tuberculomas. The histopathologic diagnosis of tuberculous brain abscess depends on the following criteria: macroscopic evidence of pus in the abscess cavity, microscopic evidence of acute inflammatory changes in the abscess wall, and isolation of *M. tuberculosis* (70,71). Tuberculous brain abscess formation, characterized by this pyogenic reaction, is considered distinct from the more common granulomatous response seen in the development of cerebral tuberculomas. Abscesses tend to be solitary and larger and progress more rapidly than tuberculomas (70,71). Tuberculous abscesses on MRI are generally larger, solitary (Fig.14), and frequently multiloculated ring-enhancing lesions with surrounding edema and mass effect (70,71). Bacterial and fungal infections may simulate tuberculous brain abscess on MRI and should be considered in the differential diagnosis.

Paradoxical response to treatment in patients with intracranial tuberculosis. The paradoxical progression of intracranial tuberculomas during the treatment of CNS tuberculosis has been recognized (fig.15) as a rare response to tuberculostatic chemotherapy (45,57,70). Nearly 34 cases have been reported in the literature worldwide (71,72). Most of these cases had deeply located intracranial tuberculomas developing a few weeks or months after the start of chemotherapy. Progression of the lesions on antituberculous therapy usually suggests the alternative etiology of the lesion. Hejazi and Hassler have described a case of multiple tuberculomas that continued to enlarge on antituberculous therapy for 7 months and one of the larger lesions was subsequently excised and confirmed as tuberculoma on histology and microbial culture (73). The phenomenon of symptomatic cerebral tuberculomas developing and enlarging during adequate antituberculous chemotharpy can not be explained by poor drug compliance as most reports have ensured

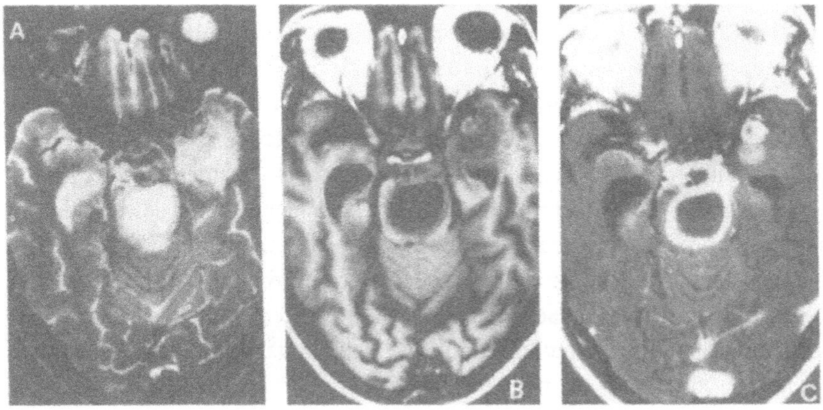

Figure 14. Tuberculous brain abscess. T2-weighted axial image (a) through the level of pons shows a large well-defined area in the pons with a thin slightly less hyperintense rim compared to the center. Note the T2 hypointense tuberculoma in the left anterior temporal lobe. OnT1-weighted image, the pontine lesion appears as hypointense with isointense rim (b) that shows thick intense enhancement in postcontrast T1-weighted image (c). Note the enhancement of basal meninges and the tuberculomas in the left temporal lobe.

drug compliance. The mechanisms of this entity are still controversial. Possibly, the expansion of the intracranial tuberculomas has an immunologic basis (70,71). The infected host develops hypersensitivity to an array of mycobacterial proteins. The tuberculostatic drugs cause destruction of the mycobacterial structures and liberation of bacilli proteins. These provoke a delayed hypersensitivity immune reaction. The CNS tuberculomas grow slowly and get encapsulated after a latent period, resulting in paradoxical progression of the existing lesions, supported by an immunologic phenomenon, namely, the local perilesional secondary granulomatous vasculitis associated with intimal proliferation and degeneration of the vessel wall and occlusion of the lumen that worsen the penetration of the drug into the lesion. It is important to know that in a patient with imaging and spectroscopic characteristic features of a tuberculoma, the patient may still not improve due to paradoxical response to the drug treatment.

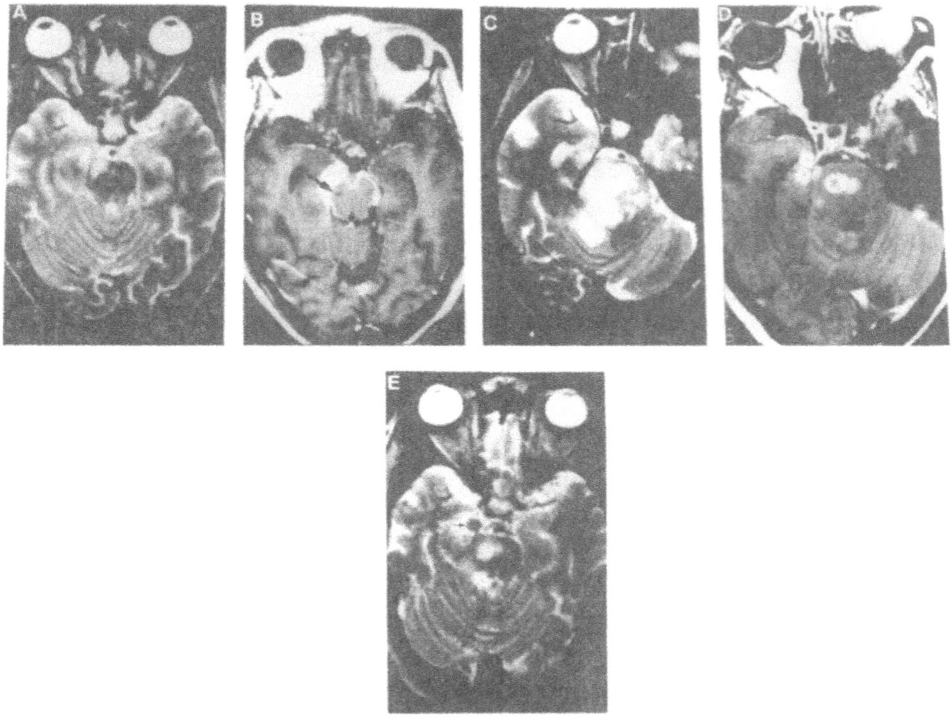

Figure 15. Paradoxical response to antituberculous therapy in a case of tuberculous meningitis. T2-weighted axial image (a) through the pons shows hyperintensity in the right upper part of the pons. Postcontrast T1-weighed image (b) shows thick meningeal enhancement in the perimesencephalic cisterns. Note the focal hypointensity in the right upper pons (arrow) consistent with an infarct. Following 3 months of specific antituberculous treatment, the patient showed clinical deterioration and repeat study (c) showed an increase in the size of the lesion on T2-weighted images. Postcontrast study (d) showed reduction in the thickened enhanced meninges along with appearance of ring-enhancing lesions in the pons and right medial temporal lobe suggesting appearance of tuberculomas while patient was on specific treatment. Patient was continued with the same treatment and a repeat study after another 4 months revealed marked reduction in the T2 abnormality (e) with regression of tuberculomas. The tuberculomas in the right temporal lobe changed from T2 hyperintense to T2 hypointense lesions (small arrow). Patient was followed up for 15 months and showed small residual lesions at the end of the treatment.

1.5. Miscellaneous Cranial and Intracranial Tuberculous Lesions

1.5.1. Tuberculous Encephalopathy

This disease is reported to be apparently independent of other pathology. It consists of cerebral edema, occasionally with perivascular demyelination or hemorrhagic leukoencephalopathy. These changes occur deep in the white matter at a distance from vascular abnormalities or exudate (74). Although systemic immunopathologic studies of this entity are lacking, some observers hypothesize that it is a purely allergic phenomenon, mediated by tissue hypersensitivity to bacillary antigen or possibly to a myelin-related antigen of brain tissue itself. No definite CT or MRI findings are available for this condition. Recently, we have observed that patients of TB meningitis presenting with encephalopathy show a diffuse alteration of MT ratio in the white matter that reverts to normal once the patient recovers clinically. The white matter appeared normal on T2-weighted images in the MT abnormal regions. This was not associated with the change in meningitis status on imaging. It appears that MT may help in understanding this poorly understood presentation of CNS tuberculosis.

1.5.2. Focal Tuberculous Cerebritis

Jinkins described this condition as specific clinical, radiologic, and histologic manifestations of CNS tuberculosis based on a retrospective analysis of five patients (75). The histology in these patients showed extensive inflammatory exudate, Langhans giant cells; reactive parenchymal changes and diffuse caseating and noncaseating microgranulomata in the cortex. CT shows intense focal gyral enhancement and on digital subtraction angiogram, gyral blush is noted. The MRI appearance of this condition differs from what has been described by Jinkins on CT. Focal cerebritis appear as hyperintense on T2- and hypointense on T1-weighted images. Postcontrast T1-weighted images either show no enhancement or show small areas of patchy enhancement (33). This is considered as a precursor of the tuberculoma formation. If the entity is recognized early, it may result in cure.

1.5.3. Hypophyseal Tuberculoma

These are rare clinical entities and may clinically simulate pituitary adenoma with or without manifestations of hypopituitarism. MRI shows an isointense mass on T1 and hypointense to hyperintense on T2-weighted images. Postcontrast study may show dense enhancement along with thickened stalk. However, this is a nonspecific finding and may be seen in other granulomatous conditions such as sarcoidosis, lymphocytic hypophysitis, and idiopathic granulomatous hypophysitis (76).

1.5.4. Osteomyelitis

Skull tuberculosis is seen more commonly in the frontal and parietal bones than in the occipital and temporal bones. The route of infection is usually hematogenous. Clinically, it usually results in a soft tissue, swelling in relation to the bony lesion, but when the continuity is established between the extradural space and the soft tissue swelling becomes less prominent. MRI features in a case of skull tuberculosis with tuberculous abscess involving the extradural space and extending to the intrasellar and suprasellar region have been described. The abscess is hyperintense with foci of hypointensity on T2-weighted images and isointense to the gray matter on T1-weighted images (Fig.16) (77). This appearance is nonspecific and may be seen in abscess secondary to pyogenic and fungal abscesses.

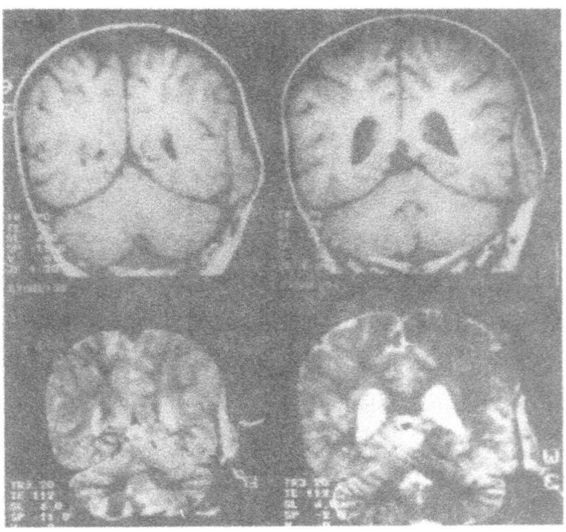

Figure 16. Tuberculous osteomyelitis of the skull. T2-weighted coronal images in the lower row show extradural hyperintense collection extending below the skin along with destruction of the skull bone. The collection appears hyperintense to the CSF on T1-weighted images in the upper row. The pus culture from the lesion was positive for *M. tuberculosis*.

2. SPINAL TUBERCULOSIS

2.1. Intraspinal Tuberculosis

As in intracranial disease, the lesions described fall into two categories: (1) space-occupying lesions of the cord or spinal meninges and (2) more diffuse inflammatory arachnoiditis analogous to typical intracranial meningitis. The inflammatory spinal disease caused by, *M. tuberculosis* has been variously called spinal meningitis or spinal arachnoiditis (78–80). Spinal pathology could complicate the course of treatment of tuberculous meningitis, could precede it, or could occur alone in the absence of overt intracranial disease. The pathophysiology of the spinal meningitis is analogous to that of tuberculous meningitis: A submeningeal tubercle formed during primary infection ruptures into the subarachnoid space and elicits all the mediators of delayed hypersensitivity (78–80). As in intracranial disease, a thick tuberculous exudate fills the subarachnoid space, frequently extending into the subdural space as well. The exudate characteristically extends over many segments of the cord. Generally, the exudate is circumferential but occasionally noted to be particularly prominent posteriorly, probably due to the gravitational pull noted in patients lying supine for a long time. As in intracranial disease, its microscopic appearance consists of granulomatous inflammation, areas of caseation, and tubercles with development of fibrous tissue in chronic or treated cases. Acid-fast organisms are almost invariably absent and the diagnosis is generally based on the characteristic tuberculous histology pattern alone (78–80).

The extensive spinal exudates are accompanied by vasculitic changes, as well secondary effects on the cord parenchyma. As in tuberculous meningitis, vessels traversing the exudate are subject to medial and subintimal changes, with fibrosis and thrombotic occlusion (78–80). The vasculitis may be limited or extensive; it has been reported to be less prominent in arteries than in veins. In contrast to intracranial disease, sudden arterial occlusion with spinal cord infarction, although reported, is uncommon. Its rarity is thought to reflect the extensive anastomotic network present in the spinal circulation. The gross pathologic changes include an edematous sponginess of the white matter probably secondary to venous congestion, zones of demyelination with degeneration of axons and honeycombing of the parenchyma, circumscribed central necrosis secondary to an ischemic myelomalacia, and withering away of the cord tissue with glial cell replacement (78–80). Nerve roots show gross compression but no infiltration by the exudate.

The MR features include CSF loculation and obliteration of the spinal subarachnoid space with loss of outline of the spinal cord in the cervicothoracic spine and matting of the nerve roots in the lumbar region (Fig.17). Gadolinium-enhanced MR images reveal nodular, thick, linear intradural enhancement, often completely filling the subarachnoid space; this may occur when the unenhanced MR image appears entirely normal (Fig.18) (81–85). When chronic, the gadolinium-enhanced images may not show any enhancement even when unenhanced images show signs of arachnoiditis (45,83). The spinal cord may be

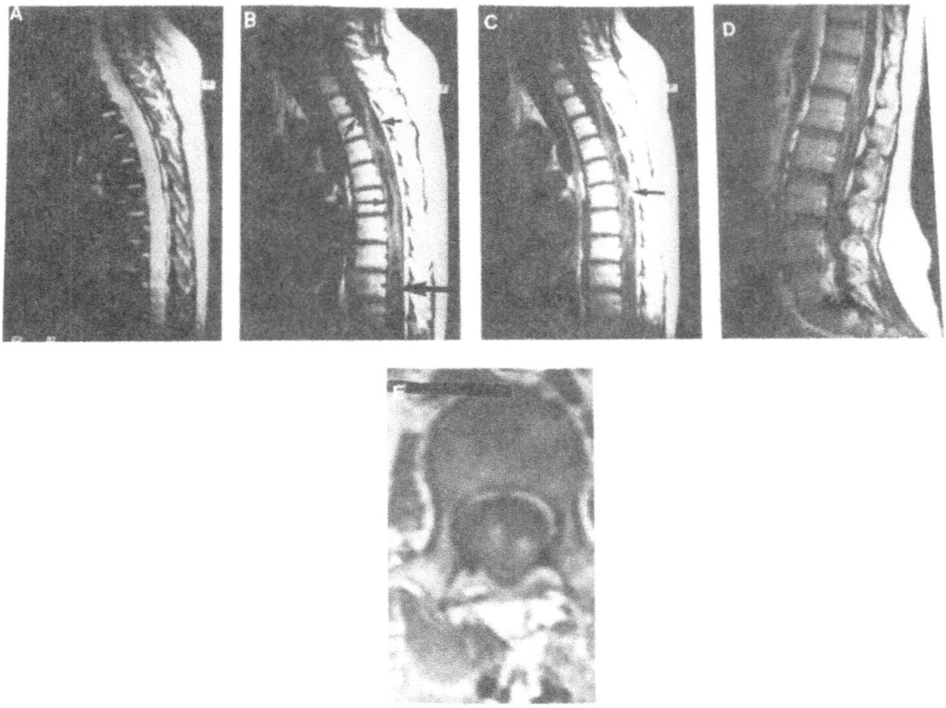

Figure 17. Arachnoiditis involving cervicothoracic and lumbar region. T2-weighted sagittal section (a) of cervicothoracic region shows predominantly hyperintense signal with loss of details of the spinal canal. Note the CSF pulsations causing loss of signal in the lower part of the cord.T1-weighted image (b) shows loculation of CSF anterior and posterior to the cord (arrowheads) in the cervicothoracic region, myelomalacia (small arrow) with a syrinx in the lower part of the thoracic cord (large arrow). Postcontrast T1-weighted image (c) shows a focal enhancing area (arrow). Postcontrast T1-weighted image (d) of the lumbosacral region of the same patient shows thickened and clumped nerve roots with enhancement. Axial image at second lumbar level (e) shows clumping of roots with enhancement

involved either as a complication of arachnoiditis that includes infarction and syringomyelia or it may be complicated with parenchymal TB myelitis and tuberculoma formation (45,83). Syringomyelia is seen as cord cavitation that typically demonstrates CSF intensity on both

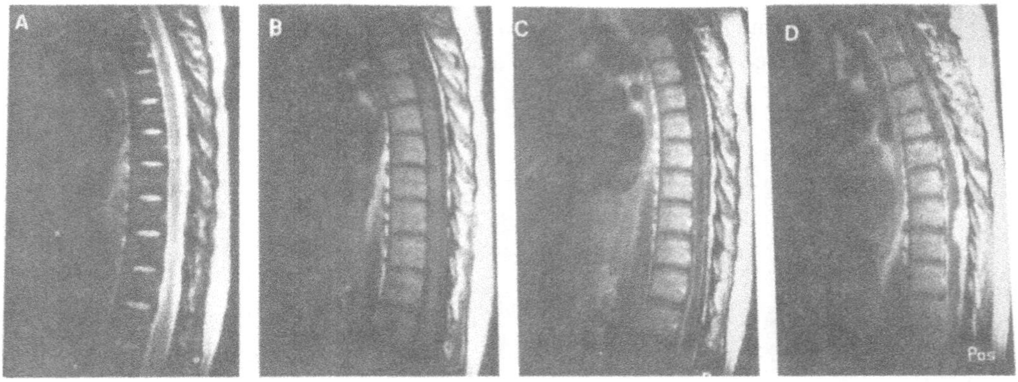

Figure 18. A case of arachnoiditis worsening during treatment. T2-weighted sagittal section (a) of the thoracolumbar spine shows only an area of focal hyperintensity in the lower cord suggesting myelitis. Sagittal T1-weighted image (b) shows no obvious abnormality. Postcontrast T1-weighted image (c) shows diffuse leptomeningeal enhancement. After 8 weeks of antituberculous treatment, patient showed clinical deterioration and a repeat postcontrast study shows focally enhancing dural mass (d) indenting the cord with enhancement of dura above and below the mass. On surgery, the mass was arising from the dura and adherent to the posterior surface of the cord. Histopathology of the mass was consistent with tuberculous granuloma.

T1- and T2-weighted images and does not enhance after intravenous gadolinium administration (45,83). The imaging features of intraspinal tuberculoma formation (Figs.19,20) are the same as described for cranial parenchymal tuberculoma. Usually tuberculomas are associated with a small area of perifocal edema but extensive cord edema involving the whole of the cord (Figs. 19,20) is also seen and it may simulate an intramedullary tumor (45,83,86).

2.1.1. TB Myelitis

MRI features of TB myelitis are nonspecific and are similar to cerebritis (Fig. 21). It has been observed that 1 week after the initiation of treatment the region of myelitis becomes less diffusely hyperintense on T2-weighted images, with more clearly defined marginal enhancement on postcontrast T1-weighted images (83,87). The surrounding edema continues to be more extensive than the margins of enhancement. This phase has been referred to as the late stage of myelitis (87). This finding is thought to represent the beginning of an intramedullary abscess formation. The central cavitary portions of the intraaxial necrotic areas are seen on T1-weighted noncontrast images as areas of low signal intensity and as hyperintense foci on T2-weighted images (87). The T2 abnormalities are known to subside in several weeks, while foci of contrast enhancement may persist for several months (87). In the absence of meningeal and root enhancement and arachnoiditis, cord swelling and edema may be associated with compromised cord perfusion and relative ischemia (87). The differential diagnoses usually include cord ischemia, acute disseminated myelitis, intramedullary tumor, multiple sclerosis, and other bacterial and parasitic infections.

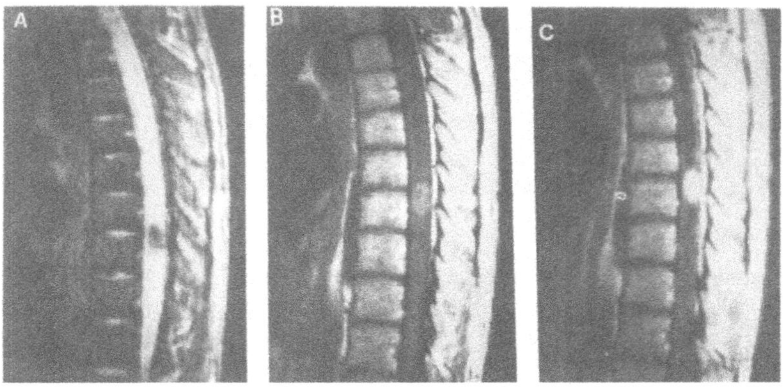

Figure 19. Intramedullary T2 hypointense tuberculoma. T2-weighted sagittal image (a) of the thoracic spine shows a hypointense mass in the cord with extensive perilesional hyperintensity of the cord suggestive of edema. On T1-weighted image (b), the lesion appears mildly hyperintense, and postcontrast T1-weighted image (c) shows intense enhancement of the lesion. Note also the mild enhancement of the pia-arachnoid above and below the lesion.

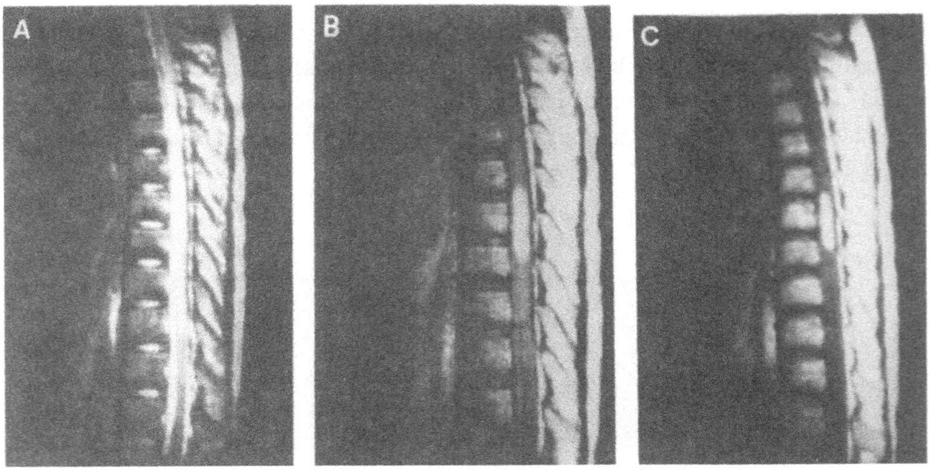

Figure 20. Intramedullary T2 hyperintense tuberculoma. T2-weighted sagittal image (a) shows a hyperintense mass that appears hyperintense on proton density image (b) and diffusely enhances on postcontrast T1-weighted image (c). The mass was excised and histopathology was consistent with tuberculoma.

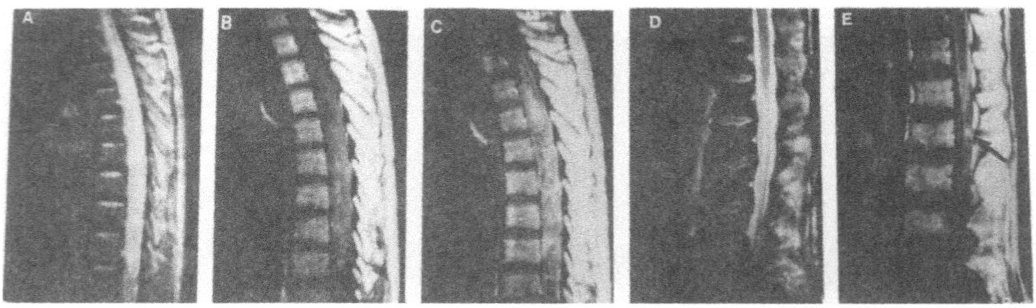

Figure 21. Myelitis with intradural tuberculoma. T2-weighted sagittal image (a) of the thoracic spine shows a diffusely thickened and heterogeneously hyperintense thoracic cord. T1-weighted image (b) also shows heterogeneous signal intensity of the cord. Postcontrast T1-weighted image (c) shows heterogeneous enhancement of the cord with enhancing leptomeninges. T2-weighted sagittal image (d) of the lumbar spine of the same patient shows hyperintensity in the lower part of the cord. Postcontrast T1-weighted image (e) shows enhancing meninges along with an enhancing tuberculoma (arrow).

2.1.2. Dural and Subdural Pathology

The formation of tuberculous pus may be noted between the dura and the leptomeninges and may appear as loculation. It is seen as hyperintense on T2 and iso- to hypointense on T1-weighted images (Fig.22). The dural granulomas may be seen as developing during the course of treatment or otherwise may appear as hypo- to isointense on T2- and isointense on T1-weighted images (Fig.21,23). Rim enhancement is noted after injection of gadolinium (83). We have noted that during the course of treatment some of the patients worsen with more focal compression due to fibrosis and better definition of the granulomas (Fig. 18) (45,83).

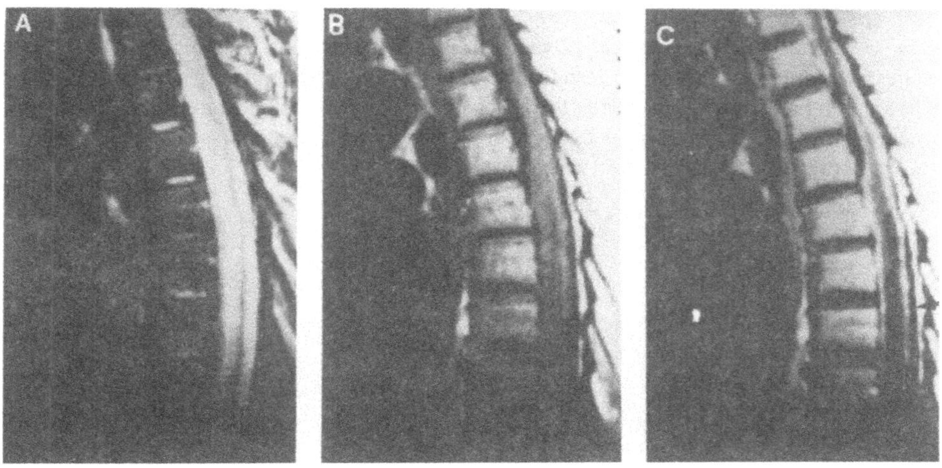

Figure 22. Arachnoiditis with subdural collection. T2-weighted sagittal image (a) of the thoracic spine shows diffusely hyperintense cord with posterior margin visualized. On T1-weighted image (b) the center of the cord appears hypointense as compared with the periphery and the cord above. Posterior to the cord, there is a thick linear hypointense area. Postcontrast image (c) shows the hypointense area (arrows) to lie between the enhancing leptomeninges and the posterior dura consistent with subdural collection.

Most extradural tuberculous reactions are believed to be secondary to extension from adjacent vertebral or diskal disease. It is important to note that it is possible to have extradural tuberculosis in conjunction with normal signal characteristics within the vertebral bodies and intervertebral disks, which suggests that the route of infection is most likely to be hematogenous. Epidural TB lesions generally appear to be isointense to the spinal cord on T1-weighted images and have mixed intensity on T2-weighted images (Figs. 23,24). Enhancement after gadolinium will be uniform if the TB inflammatory

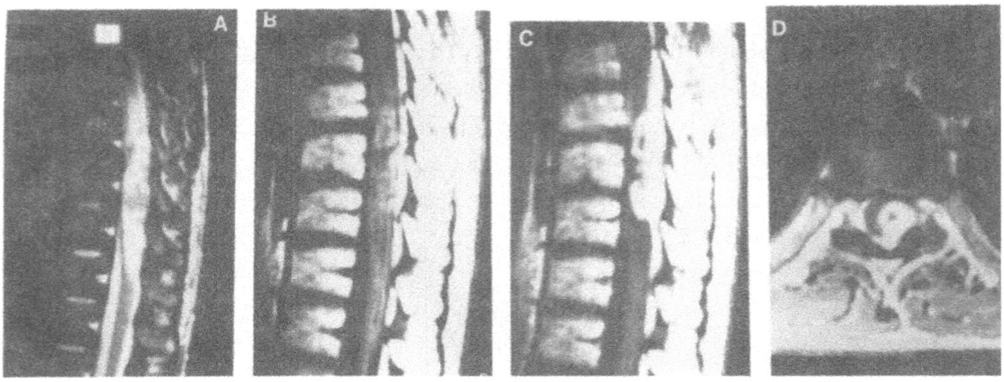

Figure 23. Dural-based tuberculoma. T2-weighted sagittal image (a) of the thoracic spine shows an irregular mixed intensity area on the dorsal aspect of the cord with hyperintensity in the cord rostral to the lesion. On T1-weighted image (b) the lesion is of mixed intensity. Postcontrast T1-weighted sagittal (c) and axial (d) images show thick nodular enhancement of the lesion along with the adjacent dura. The cord is displaced to the right side and anteriorly.

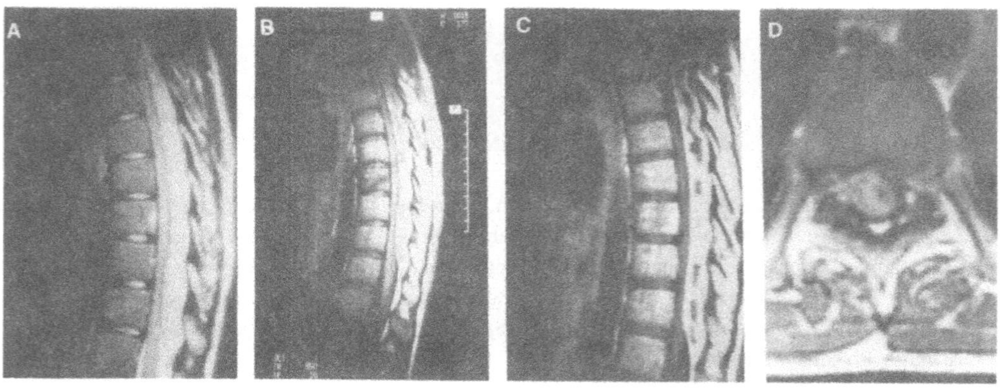

Figure 24. Epidural collection. T2-weighted sagittal image (a) of the thoracic spine shows hyperintense area posterior to the cord and displacing it anteriorly. On T1-weighted image (b) the collection appears heterogeneously hyperintense to the adjacent cord. Postcontrast T1-weighted sagittal (c) and axial (d) images show heterogeneous enhancement of the lesion. The cord is displaced to the left and anteriorly by the collection.

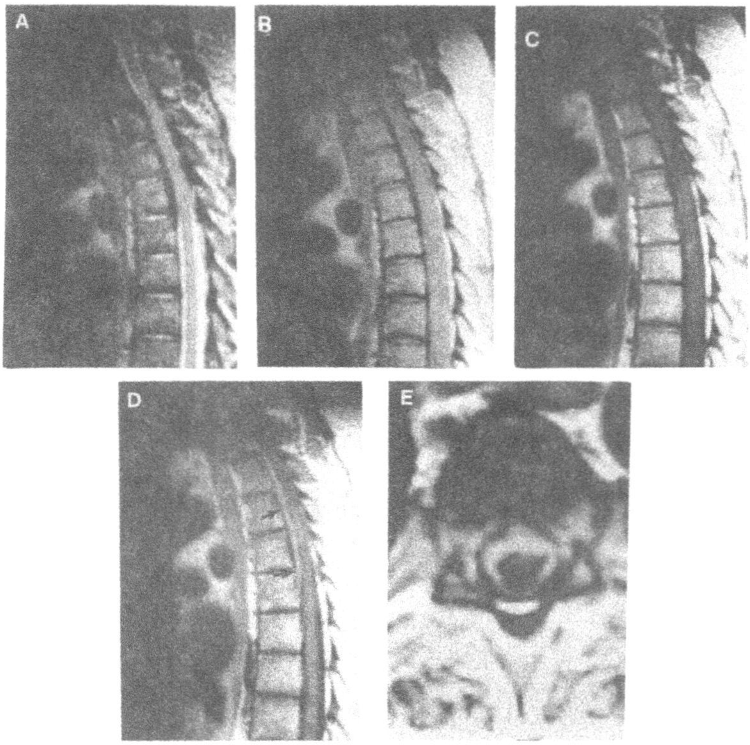

Figure 25. Focal hypertrophic pachymeningitis. T2-weighted (a) and proton density (b) sagittal images of the thoracic spine show a linear hypointense area lying in anterior relation to the cord. The lesion is hypointense on T1-weighted sagittal image (c) and shows enhancement (arrows) in the postcontrast sagittal (d) and axial (e) images suggestive of focal meningitis.

process is phlegmonous in nature or peripherally enhancing if true epidural abscess formation or caseation has developed (45,83). Epidural tuberculous abscess may be seen as primary or in association with arachnoiditis, myelitis, spondylitis, and intramedullary and dural tuberculomas (45,83). Occasionally, focal or diffuse pachymeningitis is observed in spinal tuberculosis and imaging is similar to cranial pachymeningitis as described earlier (Fig.25).

2.2. Tuberculous Spondylitis

Tuberculous spondylitis continues to be an important cause of spinal disease in developing countries and the AIDS epidemic has resulted in a worldwide resurgence of tuberculosis. Early diagnosis and prompt treatment are essential to avoid permanent neurologic damage or spinal deformity.

Tuberculous spondylitis is an infection that involves one or more of the extradural components of the spine. The vertebral bodies are most frequently affected, but the posterior osseous elements, intervertebral disks, epidural space, and paraspinal soft tissue can be primarily or secondarily involved (Fig.26) (88). The vertebral column is affected in 25–60% of the cases of skeletal tuberculosis (88). The dorsal and lumbar spine are mainly involved, especially the thoracolumbar junction. However, in a series of 60 cases, the cervicothoracic region was most frequently affected by tuberculosis (89). The sacrum and cervical spine are less common sites with the C2 to C7 region being involved in 3–5% of cases and the atlantoaxial articulation in less than 1% of cases (Fig.27). Solitary vertebral lesions can occur, but it is usual to find more than one vertebral level affected. The skip lesions are reported to occur in up to 4% of cases. It is advisable to screen the whole spine to detect the clinically silent skip lesions. Overall, the posterior arch is reported to be involved in 2–10% of the cases of spinal tuberculosis; however, involvement of the posterior arch with sparing of the vertebral body has been seen in less than 2% of the spinal tuberculosis cases (Fig.28), with the pedicle being most often affected (88).

Symptoms and signs of tuberculous spondylitis vary considerably. Patients with musculoskeletal involvement may be afebrile and free of systemic complaints until the late stage of the disease. In spinal infection, the presenting symptom is often persistent spinal pain and local tenderness, with limitation of spinal mobility evident on clinical examination. A large number of patients present with neurologic deficit when the dura is compressed by the extradural collection, granulation tissue, and/or destroyed vertebrae. Sometimes clinical signs of arachnoiditis or intramedullary granulomas are the primary clinical presentation (88).

The anterior aspect of the vertebral body is the initial site of infection more frequently than the posterior, and occasionally the posterior osseous arch may be involved in isolation. It is generally believed that tuberculous spondylitis results from hematogenous spread. Although controversy has existed regarding the primary vascular pathology (arterial versus venous), most workers believe that the veins (especially Batson plexus) are more important in spread of tuberculosis to the spine. In most cases, the initial focus of infection is the anterior part of the vertebral body adjacent to the subchondral bone plate (Fig.29). As the disease progresses, infection may spread to the intervertebral disk through the subchondral end plate or beneath the anterior or posterior ligaments. Subsequently, the neighboring vertebral bodies and intervertebral disks may be involved by either of the above-mentioned mechanisms; subligamentous spread can lead to erosion of the anterior vertebral surface. The mechanism of isolated involvement of the posterior arch is unclear; the posterior external venous plexus, lying on the posterior surface of the laminae and surrounding the spinus, transverse and articular processes, anastomoses freely with other vertebral venous plexuses and may constitute the final pathway by which the infection reaches the neural arch (88–91).

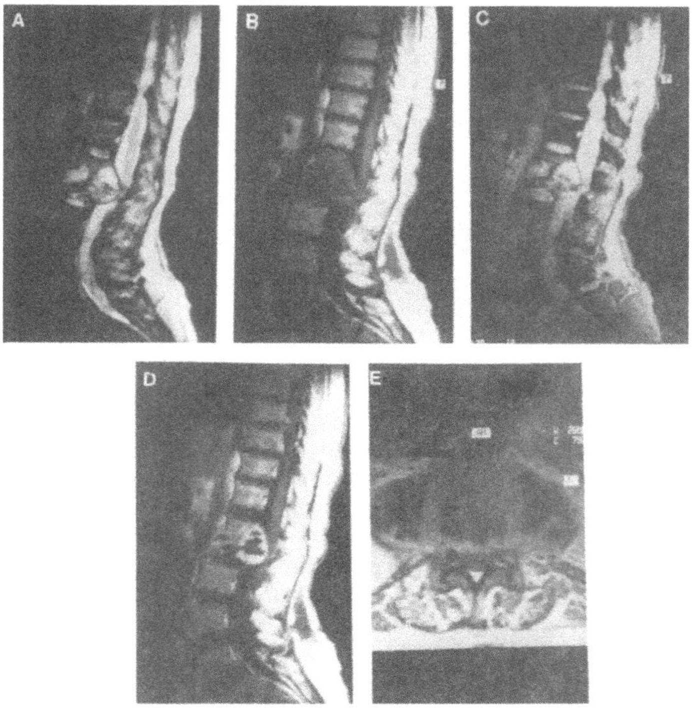

Figure 26. Tuberculous spondylitis. T2-weighted sagittal image (a) shows hyperintense destructive process involving the first and second lumbar vertebrae and the intervening disk. There is evidence of a hyperintense collection with areas of hypointensity, extending into the spinal canal and causing extradural compression. The vertebrae and collection appear hypointense on T1-weighted image (b) and the rim of the collection is slightly hyperintense. Some of the hypointense areas visible on T2-weighted image show susceptibility effect on T2*-weighted (GRE) image (c) consistent with bone fragments. Postcontrast T1-weighted sagittal (d) and axial images (e) show peripheral enhancement of the lesion along with rim enhancement of the pre- and paravertebral collections.

With progression of the disease, necrotic tissue, debris, and caseous material accumulate to form an abscess. The tuberculous abscess may remain localized to the site of infection, but more often bursts through or passes the anterior or posterior longitudinal ligament and may track long distances from the site of infection (Fig.29). However, in patients who present with early neurologic symptoms, the abscesses may not be large and may be localized to the site of infection. Abscess formation in tuberculosis is considered characteristic.

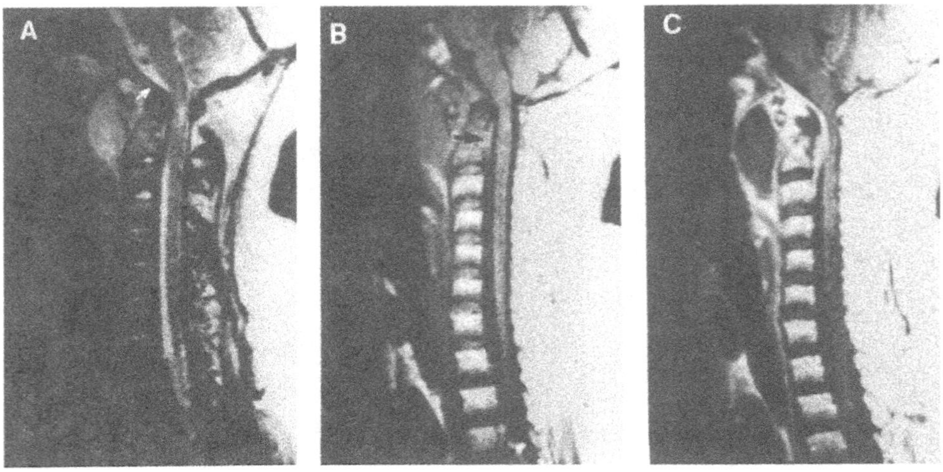

Figure 27. Skeletal tuberculosis involving craniovertebral (CV) junction. T2-weighted sagittal image (a) of the CV junction shows a hyperintense collection anterior to and around the anterior arch of atlas. Note the destruction of the clivus (white arrow), atlantoaxial dislocation with focal hyperintensity at cervicomedullary junction. Corresponding T1-weighted image (b) shows cortical discontinuity of posterior aspect of dens (black arrow). Postcontrast T1-weighted image (c) shows peripheral rim enhancement of the collection along with enhancement of the C2 vertebra, the granulation tissue around the anterior arch, and part of the destroyed clivus.

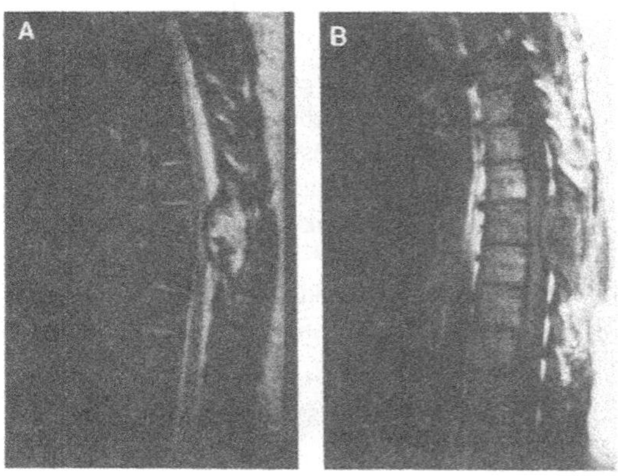

Figure 28. Isolated involvement of the posterior elements of the vertebra by tuberculous process. T2-weighted sagittal image (a) through the dorsal vertebrae shows localized hyperintensity of the posterior elements of the D6 vertebra with hypointense rim anteriorly. Note the extradural compression of the spinal cord at that level. The lesion shows central hypointensity on T1-weighted image (b). Note the normal signal of the vertebral bodies.

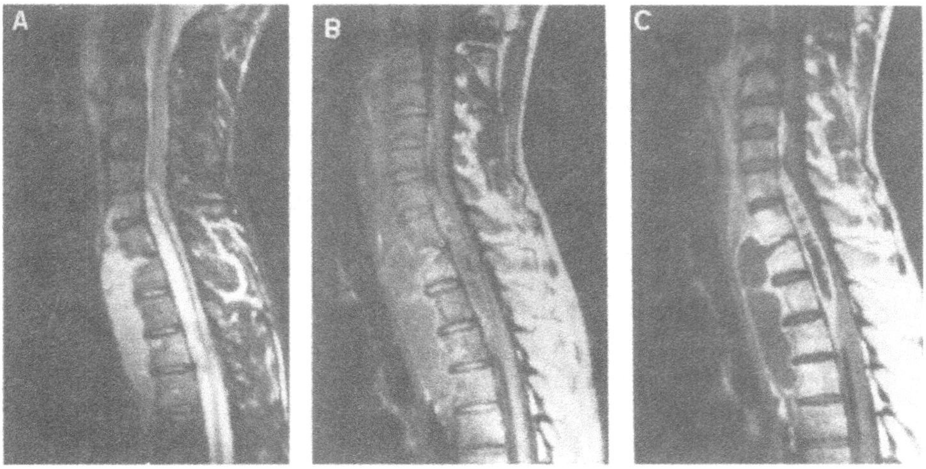

Figure 29. Anterior vertebral spondylitis with epidural abscess. T2-weighted image (a) shows hyperintense-localized collection anterior to the vertebral bodies, extending from C7 to D5 vertebrae. The collection is extending into the D1/D2 disk space. Note the large extradural collection anterior to the cord extending from C3 to D4 vertebral level causing extradural cord compression. Precontrast T1-weighted image (b) shows the collection to be slightly hyperintense to CSF. Postcontrast study (c) shows rim enhancement of the collection.

MRI is sensitive in the early detection of tuberculous spondylitis even in patients with normal radiographs, in part because of its ability to detect marrow abnormalities before any bony destruction. It appears as hyperintense on T2- and hypointense on T1-weighted images showing the vertebral body involvement in a majority of cases (88,89,92–95). As the disease progresses, classical diskovertebral involvement is noted (Fig.26). It appears as hyperintense on T2 with destruction and reduction of the disk space and hypointense on T1-weighted images. Vertebral intraosseous abscesses, paraspinal abscesses, diskitis, skip lesions, and spinal canal encroachment are all readily seen on routine spin-echo imaging. T1-weighted images usually show decreased signal within the affected vertebral marrow, as normal fat content is replaced by edematous inflammatory involvement. The disk height is reduced in the later stages of the infection, and morphologic alteration of the paraspinal soft tissue becomes apparent. The latter is indicated by a loss of the uniform psoas muscle signal intensity with asymmetric enlargement of the affected muscle belly. Epidural extension is demonstrated by thecal sac displacement and spinal cord compression (87,89,92–95). On T2-weighted images, a relative increase in signal intensity is noted within involved vertebral bodies, disks and soft tissues, thereby reflecting the increased water content of inflamed, edematous structures. The internuclear cleft within the disk is a normal finding in individuals over 30 years of age and its loss on imaging in combination with a prolonged T2 value can be a sign of inflammation (88). Enhanced MR studies are particularly useful for characterizing tuberculous spondylitis. Rim enhancement around intraosseous and paraspinal soft tissue abscesses, on postcontrast T1-weighted images, is characteristic of tuberculous spondylitis and is only occasionally demonstrated in other spinal infections (88). The rim enhancement classically suggests abscess formation, although it may also be seen in solid caseating tuberculomas similar to the solid caseating intracranial tuberculoma (89). Meningeal and epidural inflammation is better defined on a postcontrast study and similar to that seen for epidural abscess and arachnoiditis (88,89,92–95). While the patient is treated with specific chemotherapy, progressive sequential increases in signal intensity on T1-weighted images in previously affected vertebrae suggest fatty marrow replacement and indicate healing (88). This has been found to correlate well with resolving clinical signs and symptoms.

CT demonstration of bone fragments in the intra- and/or extraspinal soft tissue has been described as characteristic of tuberculous spondylitis (96). This finding has not been recorded to date in patients with other spinal infections and neoplasm (96). This feature is attributable to the lack of proteolytic enzymes required to lyse the bone in the tuberculous inflammatory exudate. CT is considered superior to MRI for demonstration of these small bone fragments. T2*-weighted images have been shown (Fig.26) to demonstrate calcification better than spin-echo images by accentuating the diamagnetic susceptibility properties of calcium salt. The low signal is more prominent on T2*-weighted than on spin-echo images and closely matches the calcification seen on CT. Demonstration of bone fragment on T2*-weighted images is also considered characteristic of tuberculous spondylitis even in the absence of abscess formation (89).

Differential diagnosis. The imaging features that favor the diagnosis of tuberculous spondylitis are involvement of one or more segments of the spine, destruction of the intervertebral disk, abscess formation, and demonstration of small bone fragments. Differentiation of tuberculous spondylitis from pyogenic vertebral osteomyelitis can sometimes be difficult (97–102). With similar imaging features, clinical data in such cases may help in suggesting the diagnosis. Other differential diagnoses include degenerative disk disease, metastatic disease, lymphoma, multiple myeloma, fungal infections, and echinococcosis.

3. CNS TUBERCULOSIS AND AIDS

CNS tuberculosis represents a disease that complicates AIDS with an increasing incidence in endemic regions as well as in developed countries in those patients who have some risk factors such as intravenous drug abuse. Although tuberculous infection of the CNS in AIDS patients may follow a rapidly progressive course, the imaging findings are similar to those of the nonimmunocompromised population as described above. Meningeal enhancement, hydrocephalus, parenchymal granulomas, and infarcts are seen frequently and often observed in combination with one another (32,36,103). However, the differential diagnosis in AIDS patients must include other opportunistic infections and primary or metastatic lymphoma of the CNS.

Spinal tuberculosis is usually seen in the form of an epidural abscess secondary to tuberculous spondylitis, although it may be seen in isolation of the spinal column involvement. The differential diagnosis includes spinal lymphoma or pyogenic abscess formation. Radiculomyelitis or isolated intramedullary tuberculomas are much less frequently observed and can be suspected on imaging only if there is concomitant evidence of intracranial tuberculosis (103).

4. SYPHILITIC INFECTIONS

The term "neurosyphilis" describes a spectrum of manifestations of CNS involvement in individuals with *T. pallidum* infection and, as such, tends to contribute to confusion in discussing the disease. There is a sharp rise in the number of cases of neurosyphilis following the emergence of AIDS epidemic (104,105). Neurosyphilis is most often asymptomatic. CNS involvement may occur at any stage of the systemic infection. Neurosyphilis may occur weeks to decades after the initial infection and occurs in one third of patients who progress to the late stages of syphilis. Symptomatic cases of neurosyphilis can be divided into four types based on predominant clinical findings: meningeal, vascular, general paresis, and tabes dorsalis. Although they may be seen in various combinations, neurosyphilis most commonly takes the meningeal or vascular form.

4.1. Meningeal Syphilis with Gumma Formation

Meningeal syphilis generally has the clinical features of any acute meningitis, including cranial neuritis, hydrocephalus, and formation of focal leptomeningeal granulomas or gummas of various sizes (104,105). There is a widespread thickening of the meninges, meningeal lymphocytic infiltrates, and perivascular lymphocytic infiltrates. Cranial nerve involvement is most commonly with second and eighth cranial nerves. Syphilitic gummas are the circumscribed masses of granulation tissue surrounded by mononuclear epithelial and fibroblastic cells with occasional giant cells and perivasculitis. Gummas are created by an intense localized leptomeningeal inflammatory reaction early in the meningeal phase of the disease. Gummas originate from the meningeal connective tissue and blood vessels with spread to the adjacent parenchyma. These are usually seen overlying the cerebral convexities, adherent to both dura and parenchyma. The lesions vary in size from 1 mm to 4 cm, mostly solitary but may be multiple. Patients with leptomeningitis on MRI show focal or diffuse enhancement of the leptomeninges with or without hydrocephalus (104–106).

Sometimes meninges may be thick and hypertrophied and may not be possible to differentiate from other causes of hypertrophic pachymeningitis. The syphilitic gummas appear iso- to hypointense on T1-weighted images and hyperintense on T2-weighted images (Fig.30). Postcontrast T1-weighted images show diffuse or rim enhancement on the surface of the brain. Adjacent meningeal enhancement may also be present and may give a dural tail appearance on MRI (107). Agrones et al. described a gumma with a rim of hyperintensity on T1-weighted images and a rim of hypointensity on T2-weighted images (108). Differential diagnoses of a syphilitic gumma on MRI include pyogenic, tuberculous, or fungal lesions as well as neoplasm, usually lymphoma and metastasis.

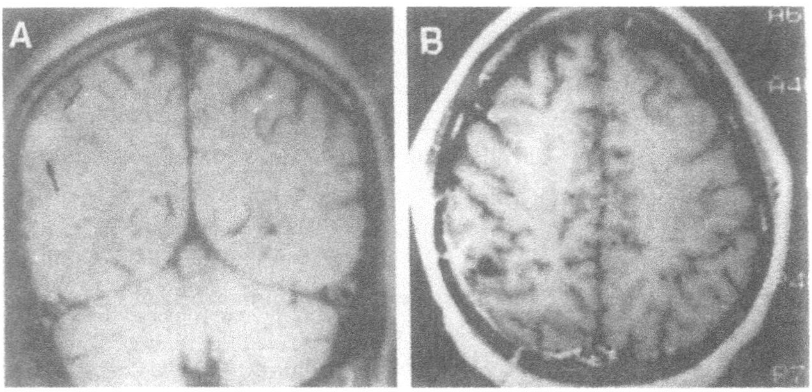

Figure 30. Cerebral gumma. Coronal T1-weighted image (a) after contrast administration reveals an ill-defined enhancing mass in the right parietal cortex with surrounding edema (arrows). Slightly more posteriorly, there was a faint tag of meningeal enhancement (not shown). Axial enhanced T1-weighted image (b) at the level of the lesion after approximately 3 months of high-dose penicillin therapy shows resolution of the mass and edema. (Reprinted with permission from Brightbill TC, Ihmeidan IH, Post MJD, Berger JR, Katz DA. Neurosyphilis in HIV-positive and HIV-negative patients: Neuroimaging findings. Am J neuroradiol 1995; 16: 703–711.)

4.2. General Paresis

General paresis is a reflection of diffuse parenchymal damage and may result in alteration of personality in addition to illusions, delusions, hallucinations, loss of memory and judgment and insight associated with speech disturbances, hyperactive reflexes, and Argyll Robertson pupil. General paresis is associated with chronic meningoencephalitis. Neuropathologic examination reveals cortical atrophy and ependymitis. Microscopy reveals degenerative neuronal changes with gliosis and microglia. Spirochetes may be found in the cortical gray matter. Frontal and temporal lobe cortical atrophy is a common finding in neurosyphilis, although it is nonspecific. Zifko et al described MRI features in four patients with general paresis and showed frontal and temporal lobe atrophy, subcortical gliosis in all patients and increased ferritin in the basal ganglia in one patient. Follow-up MRI in these patients after treatment with penicillin showed progression of atrophy and degeneration along with clinical deterioration (109).

Diagnosis of neurosyphilis is based on a positive serum fluorescent treponal antibody (FTA) absorption test, CSF pleocytosis, elevated CSF protein levels, and a positive VDRL test. A positive VDRL test is highly specific for active neurosyphilis but it is negative in about 50% of these patients. Specific criteria for the diagnosis of neurosyphilis with negative VDRL test are elevated cells and protein in CSF and positive serum FTA. Meningovascular syphilis responds promptly to treatment with penicillin G while treatment of general paresis and tabes dorsalis is far less successful.

4.3. Vascular Neurosyphilis

Vascular neurosyphilis is usually characterized by headache and transient hemiparesis followed by an acute or subacute vascular event and abnormal CSF findings. Two types of vascular involvement have been described: Heubner's endarteritis and Nissel's endarteritis.

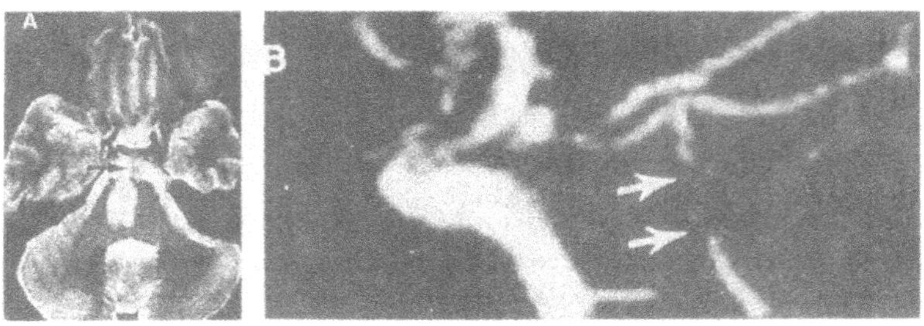

Figure 31. Brainstem infarction with basilar arteritis in a case of neurosyphilis. T2-weighted axial image (a) shows a large area of increased signal in the pons. No normal flow void is present in the basilar artery. The area of low intensity anterior to the pons (arrow) represents pulsatile CSF flow in the basilar cisterns in this patient with meningovascular syphilis. The contrast-enhanced T1-weighted images (not shown) revealed mild pontine enhancement. MR angiogram (b) shows narrowing of the basilar artery (arrows). (Reprinted with permission from Brightbill TC, Ihmeidan IH, Post MJD, Berger JR, and Katz DA. Neurosyphilis in HIV-positive and HIV-negative patients: Neuroimaging findings. Am J Neuroradiol 1995; 16: 703–711.)

Heubner's endarteritis is more commonly encountered; affecting large and medium-sized arteries with resultant luminal narrowing and ectasia. Nissel's endarteritis is less frequent and involves the small arteries in which the luminal narrowing occurs because of intense proliferation of the endothelial and adventitial cells. MRI features of neurosyphilis are varied and non-specific of the disease. In patients with vascular syphilis, small foci of infarcts are visible secondary to vasculitis as multiple focal hyperintensities on T2-weighted images involving both gray and white matter in cortical and subcortical locations. Multiple arterial distributions are affected involving both supra- and infratentorial structures with predilection for the basal ganglia. Postgadolinium images may reveal enhancement in areas of subacute infarction (104–107). MR angiogram demonstrates narrowing and/or beading of the major intracranial vessels such as basilar artery, middle cerebral artery, and carotid artery suggesting vasculitis (Fig.31,32) (110). Since this is a noninvasive technique and can be performed repeatedly, it may be useful in assessing the effect of treatment on vasculitis. As the infarction may occur in any region of the brain, the patient may present with a specific symptom pertaining to the infarcted region and some time simulates the known syndromes (111, 112).

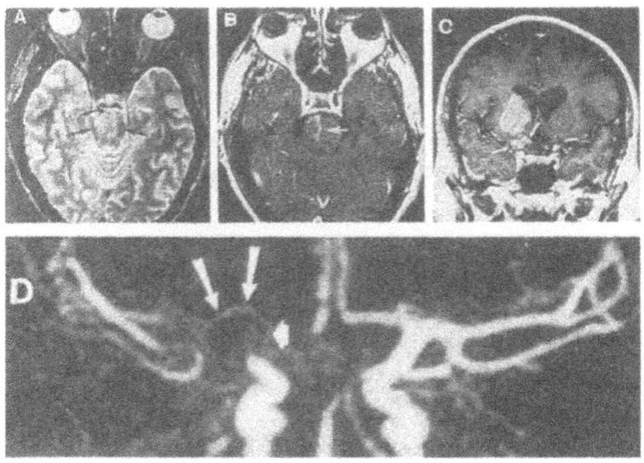

Figure 32. Neurosyphilis with meningitis and vasculitis. Axial T2-weighted image (a) reveals poorly defined regions of increased signal in the pons (straight arrows) corresponding to areas of brainstem infarction. A flow void in the basilar artery is seen (curved arrow). Axial enhanced T1-weighted image (b) demonstrates a linear band enhancement within the area of pontine infarction (arrow). Coronal enhanced T1-weighted image (c) reveals extra-axial enhancement around the supraclinoid segment of the right internal carotid artery corresponding to localized meningitis. There is also a large enhancing infarction in the right gangliocapsular structures producing mass effect on the ventricular system. MR angiogram (d) shows severe stenosis of the right supraclinoid internal carotid artery (short, thick arrow), with decreased caliber of the ipsilateral proximal middle cerebral artery (long, thin arrow) (Reprinted with permission from Brightbill TC, Ihmeidan IH, Post MJD, Berger JR, Katz DA. Neurosyphilis in HIV-positive and HIV-negative patients: Neuroimaging findings. Am J Neuroradiol 1995; 16: 703–711.)

4.4. Syphilitic Cranial Neuropathies

In the preantibiotic era, syphilis was a major cause of sensorineural hearing loss. The disease could be acquired perinatally, resulting in congenital hearing loss, or in adult life.

The pathology includes a meningoneurolabyrinthitis in the early stage of congenital syphilis and in the acute meningitides of the secondary or tertiary stage, and temporal bone osteitis in the late congenital form and in tertiary syphilis. The chronic lesions are identical regardless of the acquisition mode and are characterized by endolymphatic hydrops and degeneration of the sensory and neural structures. Contrast-enhanced MRI is useful in demonstrating the labyrinthitis by showing enhancement of the cochlea. This finding is nonspecific and can be seen in labyrinthitis due to any cause. Optic nerve may also be affected and in such cases, fat-suppressed postcontrast T1-weighted image shows enhancement of the nerve suggesting neuritis. This is a nonspecific finding on MRI and may be seen in conditions other than syphilis (113).

4.5. Syphilitic Myelitis

Syphilitic transverse myelitis and polyradiculitis are rare manifestations of secondary syphilis. The disease is characterized pathologically by a plasmocytic infiltrate surrounding the vessels and affects primarily the meninges and the pia. The clinical features are indistinguishable from transverse myelitis from any other cause. MRI appearance of patients with syphilitic myelitis has been reported. The precontrast images demonstrated a nonspecific increase in signal intensity of the cord on T2-weighted images. After contrast, there was enhancement of the pia and of the spinal cord nerve roots. The lesion in one case resolved on follow up imaging after adequate therapy with penicillin (114). The high-intensity area in the cord on T2-weighted images may reflect ischemic changes secondary to the severe vasculitis, which may lead to cord infarction. MRI features are nonspecific of syphilitic myelitis and may be seen in myelitis due to other infective conditions.

5. BRUCELLOSIS

Brucellosis is a zoonosis of worldwide distribution that is caused by small, gram-negative, nonencapsulated coccobacilli of the genus *Brucella*. Of the four species implicated in human infection, *Brucella melitensis* is the most common, the most invasive, and the one most frequently associated with infective spondylitis. Ingestion of unpasteurized contaminated milk and milkproducts is the most common form of transmission in countries in which the disease is endemic. In the United States and Western Europe, brucellosis is mostly an occupational hazard among animal handlers, slaughterhouse workers, and vegetarians. The diagnosis is established by detecting definite elevation of antibrucella antibodies and the presence of characteristic imaging features. Isolation of the organism from blood culture and/or from tissue removed from the site of involvement is extremely rare and histopathology shows nonspecific inflammatory changes in the presence or absence of small granulomas. Patients respond promptly and adequately to antibrucella chemotherapy; surgical intervention is rarely needed (85,88).

Brucella spondylitis is unique among different forms of infective spondylitis because it can manifest as either focal or diffuse. In the focal form, the osteomyelitis is localized to the anterior aspect of an end plate (classically fourth lumbar vertebra) at the diskovertebral junction. In the diffuse form, the infection initially involves an osseous end plate and ultimately extends to involve two consecutive vertebrae and the intervening disk. The lower lumbar spine is the area of predilection, but other areas of the spine can be affected. The focal form of the disease is self-limiting, and in the great majority of cases is seen as an incidental finding on lumbar spine radiographs (85,88).

The MR features of the diffuse form of brucella spondylitis are similar (Fig.33) but less severe than those of pyogenic spine infections. Vertebral body morphology and cortical margins are intact despite evidence of osteomyelitis; no associated spinal deformity is present; moderate granulation tissue is produced, which may herniate into the spinal canal; and the disc is reduced in size and shows an increase in signal intensity on heavily T2-weighted images in the absence of a nuclear cleft. The presence of paraspinal abscesses is exceedingly rare. Enhanced images do not seem to provide information additional to the T2-weighted images except in demonstrating epidural extension and meningeal involvement, which are infrequent (85,88,97).

As with other spinal infections, posttherapy follow-up demonstrates increased signal intensity on T1-weighted images from previously affected vertebrae. This is a reflection of partial destruction of the cellular marrow with predominance of the signal from the fatty component, and has been noted to correlate well with healing of the lesion.

Other CNS manifestations of brucellosis are rare and include acute or chronic meningoencephalitis, polyradiculomyelitis, brain abscess, and subdural empyema (115,116). The MRI features in a case of subdural empyema have been described and are nonspecific (116).

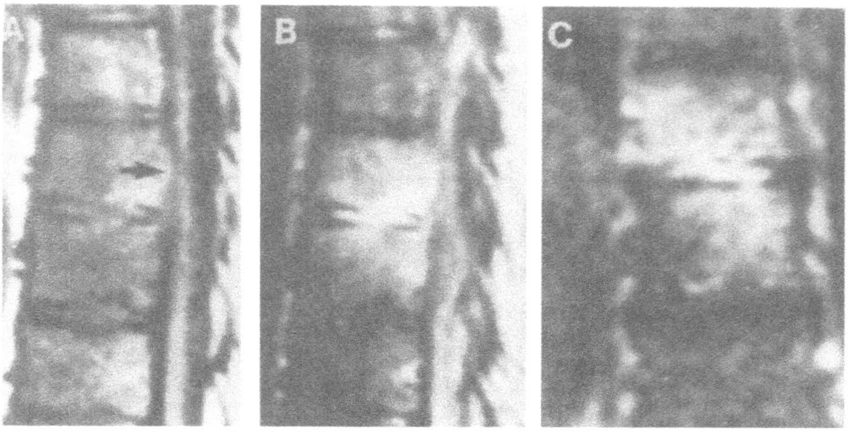

Figure 33. Brucellar spondylitis in lower thoracic spine of a 66-year-old Saudi man. Midsagittal T1-weighted image (a) of the spine shows minimal loss of signal in T9 and T10 vertebrae with a small area of intermediate signal intensity in the posterior part of T9 with epidural extension (arrow). T2-weighted image (b) shows diffuse high signal intensity in the vertebrae, epidural space, and the intervening disk. Coronal T2-weighted image (c) shows no evidence of paraspinal soft tissue abnormality. (Reprinted with permission from Sharif HS, Aideyan OA, Clark DC, et al. Brucellar and tubercular spondylitis: comparative-imaging features. Radiology 1989; 171: 419–425.)

REFERENCES

1. Des Prez RM, Goodwin RA. Mycobacterium tuberculosis. In *Principles and practice of infectious diseases,* GL Mandell, RG Douglas, JE Bennett, eds. New York: Wiley, 1985; 1383–1406.
2. Skolnik PR, Nadol JB, Baker AS. Tuberculosis of the middle ear: Review of literature with an instructive case report. Rev Infect Dis 1986; 8: 403–410.
3. Samuel J, Fernandes CM. Tuberculous mastoiditis. Ann Otol Rhinol Laryngol 1986; 95: 264–266.
4. Dastur DK, Lalitha VS. The many facets of neurotuberculosis: An epitome of neuropathology. In *Progress in neuropathology,* HM Zimmerman, ed. New York: Grune & Stratton, 1973; 351–408.
5. Rich AR, McCordock HA. The pathogenesis of tuberculous meningitis. Bull Johns Hopkins Hosp 1933; 52: 5–37.
6. Auerbach O. Tuberculous meningitis: Correlation of therapeutic results with the pathogenesis and pathologic changes. Part II: Pathologic changes in untreated and treated cases. Am Rev Tuberc 1951; 64: 419–429.
7. Smith AL. Tuberculous meningitis in childhood. Med J Aust 1975; 1: 57–60.
8. Bateman DE, Newman PK, Foster JB. A retrospective survey of proven cases of tuberculous meningitis in the northern region, 1970–1980. J R Coll Physicians London 1983; 17: 106–110.

9. Traub M, Colchester ACF, Kingsley DPE, Swash M. Tuberculosis of the central nervous system. Q J Med 1984; 53: 81–100.
10. Barrétt-Connor E. Tuberculous meningitis in adults. South Med J 1967; 1061–1067.
11. Ogawa SK, Smith MA, Brennessel DJ, Lowy FD. Tuberculous meningitis in an urban medical center. Medicine 1987; 66: 317–326.
12. Alvarez S, McCabe WR. Extrapulmonary tuberculosis revisited: A review of experience at Boston City and other hospitals. Medicine 1984; 63: 25–54.
13. Klein NC, Damsker B, Hirschman SZ. Mycobacterial meningitis: Retrospective analysis from 1970–1983. Am J Med 1985; 79: 29–34.
14. Kennedy DH, Fallon RJ. Tuberculous meningitis. JAMA 1979; 241: 264–268.
15. Hinman AR. Tuberculous meningitis at Cleveland Metropolitan General Hospital. Am Rev Respir Dis 1967; 95: 670–673.
16. Idriss ZH, Sinno AA, Kronfol NM. Tuberculous meningitis in childhood. Am J Dis Child 1976; 130: 364–367.
17. Haas EJ, Madhavan T, Quinn EL, Cox F, Fisher E, Burch K. Tuberculous meningitis in an urban general hospital. Arch Intern Med 1977; 137: 1518–1521.
18. Naughten E, Weindling AM, Newton R, Brower BD. Tuberculous meningitis in children. Lancet 1981; 2: 973–975.
19. Kent SJ, Crowe SM, Yung A, Lucas CR, Mijch AM. Tuberculous meningitis: A 30-year review. Clin Infect Dis 1993; 17: 987S994.
20. Hsieh LY, Chia LG, Shen WC. Locations of cerebral infarctions in tuberculous meningitis. Neuroradiology 1992; 34: 197–199.
21. Clark WC, Metcalf JC, Muhlbauer MS, Dohan FC, Robertson JH. Mycobacterium tuberculosis meningitis: A report of twelve cases and a literature review. Neurosurgery 1986; 18: 604–610.
22. Stockstill MT, Kauffman CA. Comparison of cryptococcal and tuberculous meningitis. Arch Neurol 1983; 40: 81–85.
23. Hernandez R, Munoz O, Guiscafre H. Sensitive enzyme immunoassay for early diagnosis of tuberculous meningitis. J Clin Microbiol 1984; 20: 533–535.
24. Kalish SB, Radin RC, Levitz D, Zeiss R, Phair JP. The enzyme-linked immunosorbent assay method for IgG antibody to purified protein derivative in cerebrospinal fluid of patients with tuberculous meningitis. Ann Intern Med 1983; 99: 630–633.
25. Sada E, Ruiz-Palacio GM, Lopez-Vidal Y, Ponce de Leon S. Detection of mycobacterial antigens in cerebrospinal fluid of patients with tuberculous meningitis by enzyme-linked immunosorbent assay. Lancet 1983; 2: 651–652.
26. Krambovitis E, McIllmurray MB, Lock PE, Hendrickse W, Holzel H. Rapid diagnosis of tuberculous meningitis by latex particle agglutination. Lancet 1984; 2:1229–1231.
27. Kadival GV, Samuel AM, Mazarelo TBMS, Chaparas SD. Radioimmunoassay for detecting Mycobacterium tuberculosis antigen in cerebrospinal fluids of patients with tuberculous meningitis. J Infect Dis 1987; 155: 608–611.
28. Kaneko K, Onodera O, Miyatake T, Tsuji S. Rapid diagnosis of tuberculous meningitis by polymerase chain reaction (PCR). Neurology 1990; 40: 1617–1618.

29. Liu PY, Shi Z, Lau Y, Hu B. Rapid diagnosis of tuberculous meningitis by a nested amplification protocol. Neurology 1994; 44: 1161–1164.
30. Chang KH, Han MH, Roh JK, et al. Gd-DTPA enhanced MR imaging in intracranial tuberculosis. Neuroradiology 1990; 32: 19–25.
31. Schoeman J, Hewlett R, Donald P. MR of childhood tuberculous meningitis. Neuroradiology 1988; 30: 473–477.
32. Villoria MF, de la Torre J, Fortea F, Munoz L, Hernandez T, Alarcon JJ. Intracranial tuberculosis in AIDS: CT and MRI findings. Neuroradiology 1992; 34: 11–14.
33. Gupta RK, Gupta S, Singh D, Sharma B, Kohli A, Gujral RB. MR imaging and angiography in tuberculous meningitis. Neuroradiology 1994; 36: 87–92.
34. Kioumehr F, Dadsetan MR, Rooholamini AA. Central nervous system tuberculosis: MRI. Neuroradiology 1994; 36: 93–96.
35. Wilson JD, Castillo M. Magnetic resonance imaging of granulomatous inflammations: Sarcoidosis and tuberculosis. Top Magn Reson Imaging 1994; 6: 32–40.
36. Whiteman M, Espinoza L, Post MJD, Bell MD, Falcone S. Central nervous system tuberculosis in HIV-infected patients: Clinical and radiographic findings. Am J Neuroradiol 1995; 16: 1319–1327.
37. Dastur DK, Lalitha VS, Udani PM, Parekh U. The brain and meninges in tuberculous meningitis. Gross pathology in 100 cases and pathogenesis. Neurology (Bombay) 1970; 18: 86–100.
38. Gupta RK, Kathuria MK, Pradhan S. Magnetization transfer MR imaging in central nervous system tuberculosis. Am J Neuroradiol 1999; 20: 867–875.
39. Tsuchiya K, Inaoka S, Mizutani Y, Hachiya J. Fast fluid attenuated inversion recovery MR of intracranial infections. AJNR 1997; 18: 909–913.
40. Tandon PN, Bhatia R, Bhargava S. Tuberculous meningitis. In *Handbook of clinical neurology*, Vol. 8, PJ Vinken, GW Bruyn, HZ Klawans, eds. Amsterdam: Elsevier, 1988; 196–226.
41. Cho IC, Chang KH, Kim YH, Kim SH, Yu IK, Han MH. MRI features of choroid plexitis. Neuroradiology 1998; 40: 303–307.
42. Leiguarda R, Berthier M, Starkstein S, Nogues M, Lylyk P. Ischaemic infarction in 25 children with tuberculous meningitis. Stroke 1988; 19: 200–204.
43. Lehrer H. The angiographic triad in tuberculous meningitis. A radiographic and clinicopathologic correlation. Radiology 1966; 87: 829–835.
44. Mathew NT, Abraham J, Chandy J. Cerebral angiographic features in tuberculous meningitis. Neurology 1970; 20: 1015–1023.
45. Jinkins JR, Gupta R, Chang KH, Carbajal RJ. MR imaging of central nervous system tuberculosis. Radiol Clin North Am 1995; 33: 771–786.
46. Roy R, Gupta RK, Kishore J, Taparia S, Poptani H, Bhakuni V. High resolution proton MR spectroscopy of the cerebrospinal fluid from children with tuberculous meningitis. Proceedings of annual meeting of ISMRM 1996; p 1157.
47. Glass JD, Becker PS, Moses H, Troncoso JC. Dural scrofula. Neurology 1989; 39: 1123-1124.
48. Callebout J, Dormont D, Dubois B, Chiras J, Borries J. Contrast enhanced MR imaging of tuberculous pachymeningitis cranialis hypertrophica: Case report. Am J Neuroradiol 1990; 11: 821–822.
49. Linder A, Schneider C, Hofmann E, Soerensen N, Toyka KV. Isolated meningeal tuberculoma mimicking meningioma: Case report. Surg Neurol 1995; 43: 81–84.

50. Brismar T, Hugosson C, Larsson SG, Lundstedt C, Nyman R. Tuberculosis as a mimicker of brain tumor. Acta Radiol 1996; 37: 496–505.
51. Goyal M, Sharma A, Mishra NK, Gaikwad SB, Sharma MC. Imaging appearance of pachymeningeal tuberculosis. Am J Roentgenol 1997; 169: 1421–1424.
52. Zuger A, Lowy FD. Tuberculosis. In *Infections of the central nervous system,* WM Scheld, RJ Whitley, DT Durack, eds. Philadelphia: Lippincott–Raven, 1997; 417–443.
53. Dastur HM, Desai AD. A comparative study of brain tuberculomas and gliomas based upon 107 case records of each. Brain 1965; 88:375–386.
54. Just M, Higer HP., Betting O, Bockenheimer S, Pfannenstiel P. MRI in cranial tuberculosis. Eur J Radiol 1987; 7: 276–278.
55. Gupta RK, Jena A, Sharma A, Guha DK, Khushu S, Gupta AK. MR imaging of intracranial tuberculomas. J Comput Assist Tomogr 1998; 12:280–285.
56. Salgado P, Del Brutto OH, Talamas O, Zenteno MA, Carbajal JR. Intracranial tuberculomas: MR imaging. Neuroradiology 1989; 31:299–302.
57. Gupta RK, Jena A, Singh AK, Sharma A, Puri V, Gupta M. Role of magnetic resonance (MR) in the diagnosis and management of intracranial tuberculomas. Clin Radiol 1990; 41:120–127.
58. Gupta RK, Pandey R, Khan EM, Mittal P, Gujral RB, Chhabra DK. Intracranial tuberculomas: MRI signal intensity correlation with histopathology and localized proton spectroscopy. Magn Reson Imaging 1993; 11: 443–449.
59. Kim TK, Chang KH, Chong JK, Goo JM, Kook MC, Han MH. Intracranial tuberculoma: Comparison of MR with pathologic findings. Am J Neuroradiol 1995; 16: 1903–1908.
60. Gee GT, Bazan III C, Jinkins JR. Miliary tuberculosis involving the brain: MR findings. Am J Roentgenol 1992; 159: 1075–1076.
61. Gupta RK, Kohli A, Gaur V, Lal JH, Kishore J. MRI of the brain in patients with miliary pulmonary tuberculosis without symptoms or signs of central nervous system involvement. Neuroradiology 1997; 39: 699–704.
62. Gupta RK, Poptani H, Kohli A, Chhabra DK, Sharma B, Gujral RB. In vivo localized proton magnetic resonance spectroscopy of intracranial tuberculomas. Indian J Med Res 1995; 101: 19–24.
63. Poptani H, Gupta RK, Roy R, Pandey R, Jain VK, Chhabra DK. Characterization of intracranial mass lesions with in vivo proton MR spectroscopy. Am J Neuroradiol 1995; 16: 1593–1603.
64. Gupta RK, Roy R, Dev R, et al. Finger printing of Mycobacterium tuberculosis in patients with intracranial tuberculomas by using in vivo, ex vivo, and in vitro magnetic resonance spectroscopy. Magn Reson Med 1996; 36: 829–833.
65. Gupta RK, Roy R. MR imaging and spectroscopy of intracranial tuberculoma. Curr Sci 1999; 76: 783–788.
66. Poptani H, Kaartinen J, Gupta RK, Niemitz M, Hiltunen Y, Kauppinen RA. Diagnosis of focal brain disorders by in vivo proton magnetic resonance spectroscopy and artificial neural network. J Cancer Res Clin Oncol 1999; 125: 343–349.
67. Usenius JR, Kauppinen RA, Vainio PA, et al. Quantitative metabolic patterns of human brain tumors: Detection of 1H NMR spectroscopy in vivo and in vitro. J Comput Assist Tomogr 1994; 18: 705–713.
68. Kussel AC, Donnelly SM, Halliday W, Sutherland GR, Smith ICP. Mobile lipid and metabolic heterogeneity of brain tumors as detectable by ex vivo 1H NMR spectroscopy. NMR Biomed 1994; 7: 172–180.

69. Peeling J, Sutherland G. High-resolution 1H NMR spectroscopy studies of extracts of human cerebral neoplasms. Magn Reson Med 1992; 24: 123–126.
70. Velasco-Martinez JJ, Guerrero-Espeijo A, Gomez-Mampasco E, et al. Tuberculous brain abscess should be considered in HIV/AIDS patients. AIDS 1995; 9: 1197–1199.
71. Farrar DJ, Flanigan TP, Gordon NM, Gold RL, Rich JD. Tuberculous brain abscess in a patient with HIV infection: Case report and review. Am J Med 1997; 102: 297—301.
72. Teoh R, Humphries MJ, O'Mahony SG. Symptomatic intracranial tuberculoma developing during treatment of tuberculosis: A report of 10 patients and review of the literature. Q J Med 1987; 241: 449–460.
73. Hejazi N, Hassler W. Multiple intracranial tuberculomas with atypical response to tuberculostatic chemotherapy. Acta Neurochir (Wien) 1997; 139: 194—202.
74. Udani PM. Pathology and pathogenesis of tuberculous encephalopathy. Acta Neuropathol 1966; 6: 311–326.
75. Jinkins JR. Focal tuberculous cerebritis. Am J Neuroradiol 1988; 9:121–124.
76. Sinha S, Singh AK, Tatke M, Singh D. Hypophyseal tuberculoma: Direct radiosurgery is contraindicated for a lesion with a thickened pituitary stalk: case report. Neurosurgery 2000; 46: 735–739.
77. Gupta RK, Jena A, Sharma A. Sellar abscess secondary to tuberculous osteomyelitis of the skull: its demonstration on MR. Am J Neuroradiol 1989; 10: 448.
78. Brooks WDW, Fletcher AP, Wilson RR. Spinal cord complications of tuberculous meningitis. Q J Med 1954; 23: 275–290.
79. Dastur DK, Wadia NH. Spinal meningitides with radiculomyelopathy: Part 2. Pathology and pathogenesis. J Neurol Sci 1969; 8: 261–293.
80. Wadia NH, Dastur DK. Spinal meningitides with radiculomyelopathy: Part 1. J Neurol Sci 1969; 8: 239–260.
81. Chang KH, Han MH, Choi YW, Kim IO, Han MC, Kim CW. Tuberculous arachnoiditis of the spine: Finding on myelography, CT, MR imaging. Am J Neuroradiol 1989; 10: 1255–1262.
82. Kumar A, Montanera W, Willinsky R, TerBrugge KG, Aggarwal S. MR features of tuberculous arachnoiditis. J Comput Assist Tomogr 1993; 17: 127–130.
83. Gupta RK, Gupta S, Kumar S, Kohli A, Misra UK, Gujral RB. MRI in intraspinal tuberculosis. Neuroradiology 1994; 36: 39–43.
84. Gero B, Sze G, Sharif H. MR imaging of intradural inflammatory disease of the spine. Am J Neuroradiol 1991; 12: 1009–1019.
85. Sharif HS. Role of MR imaging in the management of spinal infections. Am J Roentgenol 1992; 158: 1333–1345.
86. Jena A, Banerji AK, Tripathi RP, et al. Demonstration of intramedullary tuberculomas by magnetic resonance imaging: A report of two cases. Br J Radiol 1991; 64: 555–557.
87. Murphy KJ, Brunberg JA, Quint DJ, Kazanjian PH. Spinal cord infection: Myelitis and abscess formation. Am J Neuroradiol 1998; 19: 341—348.
88. Sharif HS, Aabed MY, Haddad MC. Magnetic resonance imaging and computed tomography of infectious spondylitis. In *MRI and CT of the musculoskeletal system: A text atlas,* JL Bloem, DJ Satoris, eds. Baltimore: Williams & Wilkins, 1992; 580–602.
89. Gupta RK, Agarwal P, Rastogi H, Kumar S, Phadke RV, Krishnani N. Problems in differentiating spinal tuberculosis from neoplasm on MRI. Neuroradiology 1996; 38: S97–S104.

90. Smith AS, Weinstein MA, Mizushima A, et al. MR imaging characteristics of tuberculous spondylitis vs vertebral osteomyelitis. Am J Neuroradiol 1989; 10: 619–625.
91. Resnick D, Niwayama G. Osteomyelitis, septic arthritis and soft tissue infection of the axial skeleton. In *Diagnosis of bone and joint disorders,* D Resnick, G Niwwayama, eds. Philadelphia: Saunders, 1988; 2619–2646.
92. Modic MT, Feiglin DH, Piraino DW, et al. Vertebral osteomyelitis: Assessment using MR. Radiology 1985; 157:157-166.
93. Post MJD, Sze G, Quencer RM, Eismont FJ, Green BA, Gahbauer H. Gadolinium-enhanced MR in spinal infection. J Comput Assist Tomogr 1990; 14:721–729.
94. Thrush A, Enzmann D. MR imaging of infectious spondylitis. Am J Neuroradiol, 1990; 11: 1171–1180.
95. Hoffman EB, Crosier JH, Cremin BJ. Imaging in children with spinal tuberculosis. A comparison of radiography, computed tomography and magnetic resonance imaging. J Bone Joint Surg Br Vol, 1993; 75:233–239.
96. Jain R, Sawhney S, Berry M. Computed tomography of vertebral tuberculosis. Clin Radiol 1993; 47:196–199.
97. Sharif HS, Aideyan OA, Clark DC, et al. Brucellar and tubercular spondylitis: Comparative-imaging features. Radiology 1989; 171:419–425.
98. Van Lom KJ, Kellerhouse LE, Pathria MN, et al. Infection vs. tumor in spine: Criteria for distinction with CT. Radiology 1988; 166:851–855.
99. Ragland RL, Abdewahab IF, Braffman B, Moss DS. Posterior spinal tuberculosis: A case report. Am J Neuroradiol, 1990; 11:612–613.
100. Lin-Greenberg A, Cholankeril J. Vertebral arch destruction in tuberculosis: CT features. J Comput Assist Tomogr 1990; 14:300–302.
101. Shanley DJ. Tuberculosis of the spine: imaging features. Am J Roentgenol 1995; 164: 659–664.
102. Fu WK, Wu WC, Ip FK. Concomitant tuberculosis and pyogenic infection of the cervical spine. Spine 1998; 23: 139–143.
103. Villoria MF, Fortea F, Moreno S, Munoz L, Manero M, Benito C. MR imaging and CT of central nervous tuberculosis in patients with AIDS. Radiol Clin North Am 1995; 33: 805–820.
104. Berger JR, Waskin H, Pall L, Hensley G, Ihmedian I, Post MJD. Syphilitic cerebral gumma with HIV infection. Neurology 1992; 42: 1282–1287.
105. Tien RD, Gean-Marton AD, Mark AS. Neurosyphilis in HIV carrier: MR findings in six patients. Am J Roentgenol 1992; 158: 1325–1328.
106. Brightbill TC, Ihmeidan IH, Post MJD, Berger JR, Katz DA. Neurosyphilis in HIV-positive and HIV-negative patients: Neuroimaging findings. Am J Neuroradiol 1995; 16: 703–711.
107. Bourekas EC, Wildenhain P, Lewin JS, et al. The dural tail sign revisited. Am J Neuroradiol 1995; 16: 1514–1516.
108. Agrones GA, Han SS, Husson MA, Simeone F. MR imaging of cerebral gumma. Am J Neuroradiol 1991; 12: 80–81.
109. Zifko U, Wimberger D, Lindner K, Zier G, Grisold W, Schindler E. MRI in patients with general paresis. Neuroradiology 1996; 38: 120–123.
110. Gallego J, Soriano G, Zubieta JL, Delgado G, Villanueva JA. Magnetic resonance angiography in meningovascular syphilis. Neuroradiology 1994; 36: 208–209.

111. Tyler KL, Sandberg E, Baum KF. Medial medullary syndrome and meningovascular syphilis: A case report in an HIV- infected man and a review of literature. Neurology 1994; 44: 2231–2235.
112. Stepper F, Schroth G, Sturezenegger M. Neurosyphilis mimicking Miller- Fisher syndrome: A case report and MRI findings. Neurology 1998; 51: 269–271.
113. Frohman L, Wolansky L. Magnetic resonance imaging of syphilitic optic neuritis/perineuritis. J Neuro-ophthalmol 1997; 17: 57–59.
114. Nabatame H, Nakamura K, Matuda M, Fujimoto N, Dodo Y, Imura T. MRI of syphilitic myelitis. Neuroradiology 1992; 34: 105–106.
115. Shakir RA, Al-Din AS, Araj GF, Lulu AR, Mousa AR, Saadah MA. Clinical categories of neurobrucellosis: A report on 19 cases. Brain 1987; 110: 213–223.
116. Pera M, Pera M, Moreno A. Chronic subdural empyema: A new presentation of neurobrucellosis. Clin Infect Dis 1996; 23: 400–401.

Chapter 4

VIRAL INFECTIONS

Rakesh K. Gupta and Robert B. Lufkin

Involvement of the CNS due to viral infection has tremendous importance because of the potential for death and neurologic damage. These viruses display tissue tropism and cause illness with a characteristic temporal course. Viral infections of the CNS may present as meningitis, encephalitis, meningoencephalitis, or meningoencephalomyelitis or myelitis with or without radiculitis. Depending on the duration of the illness, it is labeled as acute (days) or chronic when it evolves over weeks. Another way to classify viral infection of the CNS is based on the pathogenesis. The encephalitis that results from direct viral entry into the CNS is referred to as primary encephalitis. However, subsequent damage may occur as a consequence of host immune response, but invasion by the pathogen initiates the CNS damage. An encephalitis temporally associated with a systemic viral infection without evidence of direct viral invasion in the CNS is referred to as postinfectious or parainfectious encephalitis (1). Primary encephalitis will be discussed in the present chapter and parainfectious encephalitis and HIV encephalitis will be discussed separately in later chapters.

Rakesh K. Gupta • Department of Radiodiagnosis, Sanjay Gandhi Post-Graduate of Medical Sciences, Lucknow 226014, India. Robert B. Lufkin • Department of Radiological Sciences, University of California, Los Angeles Medical Center Los Angeles, California 90095-1721.

MR Imaging and Spectroscopy of Central Nervous System Infection, edited by Gupta and Lufkin. Kluwer Academic / Plenum Publishers, New York, 2001.

Acute viral meningitis and meningoencephalitis represent the majority of the viral brain infections and frequently occur in epidemics. Enteroviruses and arboviruses constitute the majority of the infections seen as epidemic and endemic while herpes simplex viral infections occur throughout the year without seasonal variation, affecting all ages with fatality (2). Spread of the virus to the CNS occurs via either hematogenous or neuronal route.

We will discuss pathology, clinical, and imaging features of the common viral infections that infect the CNS, namely, the herpes group of viruses, arthropod-borne viruses, measles, mumps, and Creutzfeldt–Jakob disease. The HIV virus will be covered in a later chapter.

1. HERPES GROUP OF VIRUSES

There are eight common double-stranded DNA viruses that can cause infection of the CNS: herpes simplex virus 1 (HSV-1), herpes simplex virus 2 (HSV-2), varicella zoster virus (VZV), human herpesvirus 6 (HHV-6), cytomegalovirus (CMV), Epstein–Barr virus (EBV), human herpesvirus -7 (HHV-7), and human herpesvirus-8 (HHV-8). These viruses share similar molecular and biologic characteristics. They are double-stranded DNA viruses and have the unique ability to establish latency and reactivate. These agents, among the most common encountered by humans, are frequent causes of CNS infection (2).

1.1. Herpes Simplex Encephalitis

The gross appearance of the brain in adults with herpes simplex encephalitis (HSE) initially reveals acute inflammation, congestion and/or hemorrhage, and softening, most prominently in the temporal lobes and usually disposed asymmetrically. Adjacent limbic areas show involvement as well. The meninges overlying the temporal lobes may appear clouded or congested. After about 2 weeks, these changes proceed to frank necrosis and liquefaction. Microscopically, involvement extends beyond areas that appear grossly abnormal (3). At the earliest stage, the histologic changes are not dramatic and may be nonspecific. Congestion of capillaries and other small vessels in the cortex and subcortical white matter is evident along with the presence of petechiae. Perivascular cuffing and hemorrhagic necrosis become prominent after second and third weeks of infection. Intranuclear inclusions are present at an early stage in about 50% of the cases and support the diagnosis of viral infection (3,4). The application of *in situ* polymerase chain reaction (PCR) shows a broader distribution of viral antigen than suggested by histopathology (5).

Clinically, patients with HSE typically present with fever headache, seizures, focal neurologic signs, and impaired consciousness. Rapid diagnosis and early treatment with acyclovir can decrease the high mortality associated with HSE. The detection of HSV DNA in cerebrospinal fluid has been shown to be a very sensitive and specific method for the diagnosis of HSE (6). ELISA test for antibody detection in serum and CSF for HSE is considered to be not very useful as it may be positive in non-HSE patients and has not shown any correlation with biopsy-positive cases (6).

MRI features have been well described in the literature. The characteristic MRI findings reflect the necrotizing encephalitis involving the orbital surface of the frontal lobes, inferomedial temporal lobes including part of the limbic system, insular cortex, and external capsule, and sometimes the cingulate gyrus. Basal ganglia, thalamus, internal capsule, frontal, parietal, and occipital lobes are less frequently involved. Some authors (7–10) have also demonstrated rhomboencephalitis. MRI findings appear as high-signal-intensity lesions on T2-weighted and low intensity on T1-weighted images (Fig.1). Occasionally, MR imaging can also show hemorrhage (Fig.2) in these lesions. FLAIR imaging is found to be superior to conventional spin-echo imaging in patients with encephalitis. This is due to the involvement of cortical gray matter in encephalitis which is seen better on FLAIR images due to suppression of CSF resulting in elimination of artifacts due to CSF motion and CSF volume averaging with the brain parenchyma (11). Blood–brain barrier disruption, as demonstrated by contrast enhancement, is usually detected after the first week of disease. The leptomeningeal and gyral enhancement associated with HSE typically occur along the temporal lobes, insular cortex, and subfrontal area and cingulate gyrus. Recently, postcontrast MT spin-echo MRI has been found to be better for detection of leptomeningeal involvement in patients with HSE (Fig.3) (12,13). MRI may demonstrate involvement of the limbic system in HSE before there are any CT abnormalities; however, the sensitivity and specificity of MRI for the diagnosis of HSE remain to be determined (13). Rarely, HSE may manifest as diffuse brainstem encephalitis without altering the level of consciousness (14). Either disseminated cortical, subcortical, or isolated cortical involvement of occipital or frontal lobe has also been reported (15,16). The correlation has been shown between the CSF PCR and involvement of temporal lobe as demonstrated on MRI in 7/8 patients (13). MRI is also found to be normal in patients with HSE even when PCR is positive for HSV DNA (13). Demonstration of normal MRI does not rule out HSE when there is a strong clinical suspicion. However, MRI helps in separating other causes of temporal lobe involvement, which simulate HSE on clinical grounds (13). In patients, who get through to the chronic phase, the MRI shows encephalomalacia, cortical atrophy (Fig.4), calcification, and features of old hemorrhage (16, 17). MRI has been used to assess the response to antiviral therapy and lesions are known to disappear completely after 4–6 weeks of treatment (15). The lesions may not show any change or may decrease in size on a follow-up study after specific therapy. It has also been reported that the disease may continue to be active even after acyclovir therapy and patients may develop intractable seizure (17). MRI in such patients may show signs of chronic and destructive focal disease.

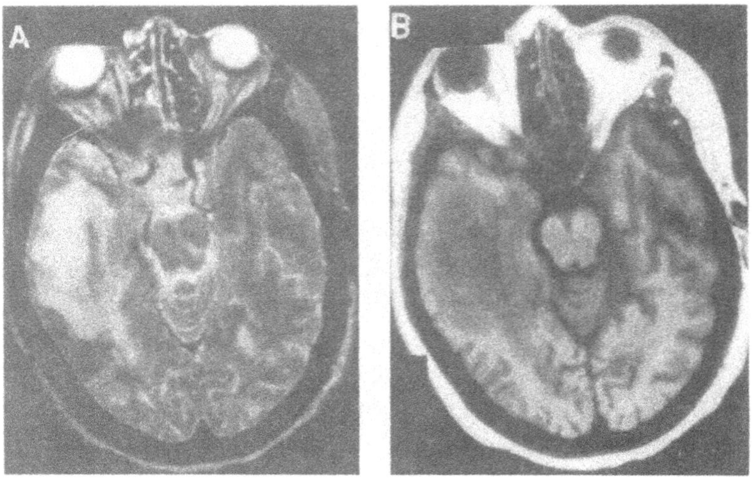

Figure 1. Herpes simplex encephalitis. T2 axial image (a) at the midbrain level shows evidence of hyperintense, swollen right temporal lobe with effacement of sulci. Note the focal hyperintensity in the left midbrain. On T1-weighted image (b) the involved region appears hypointense.

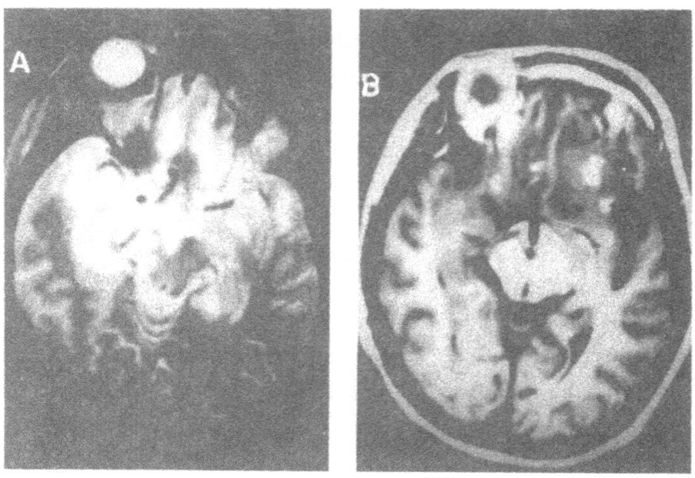

Figure 2. Herpes simplex encephalitis with hemorrhage. T2-weighted axial image (a) shows evidence of hyperintensity in the temporal lobes and basal-frontal region. Right midbrain also shows hyperintensity. On T1-weighted image (b) there is evidence of focal hyperintensity in the left frontal region consistent with subacute hemorrhage.

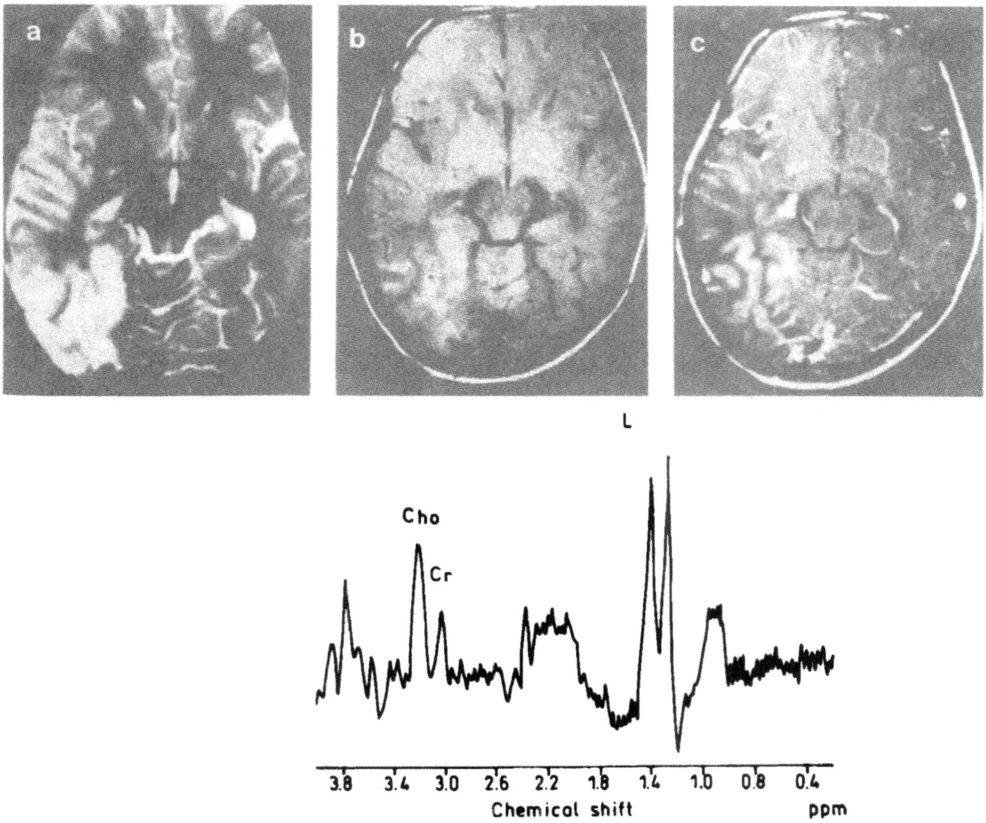

Figure 3. Herpes simplex encephalitis. T2 axial image (a) shows hyperintense right temporal lobe with hyperintensity extending into the right frontal region. MTT1-weighted image (b) shows hyperintensity in the posterior part of the temporal lobe. Postcontrast MTT1-weighted image (c) shows enhancement along the gyri and the meninges on the right side. In vivo proton MR spectroscopy from the right temporal lobe (d) shows presence of large resonance of lactate (L) along with altered choline/creatine (Cho/Cr) ratio.

In vivo proton MR spectroscopy has been used in the evaluation of HSE (18–20). The NAA/Cr ratio has been shown to decrease between 2 and 17 weeks after the onset of the disease and is presumed to represent the loss of structural/functional activity of the neurons (18–20). A longitudinal study has been performed in a case of HSE toward the end of the 7th week using single-voxel technique when the imaging abnormality had cleared. It showed a marked decrease in NAA/Cr ratio compared to the corresponding contralateral region and showed a gradual increase to reach the normal levels over a period of 1 year (20). It has been suggested that neuronal dysfunction improves over a long period while

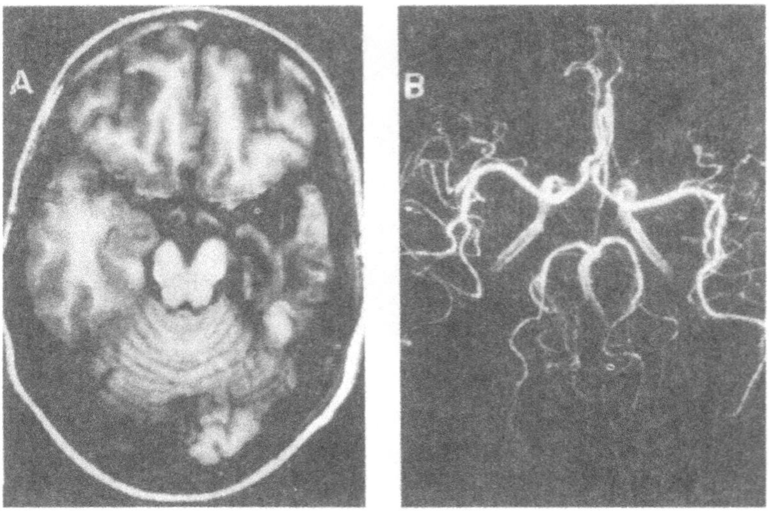

Figure 4. Atrophy following herpes simplex encephalitis. T1-weighted axial image (a) shows evidence of atrophy of left temporal lobe. MR angiogram (b) shows no obvious abnormality.

the imaging abnormalities are due to interstitial edema and regress early. In another study, an increase of Cho/Cr ratio along with progressive decrease in NAA/Cr ratio has been reported in a case of HSE with neurologic sequelae and probably reflects the demyelination along with neuronal dysfunction (18). The spectroscopic results suggest that proton MRS may help in the disease prognosis.

1.2. Herpes Varicella Zoster Encephalitis

Cutaneous VZV is a common, usually self-limited dermatomal infection caused by reactivation of latent VZV. It occurs with greater frequency among elderly and immunocompromised patients. Neurologic complications associated with VZV infection in normal and immunocompromised hosts include vasculopathy, aseptic meningitis, leukoencephalopathy, dorsal root or cranial nerve ganglionitis, polyradiculoneuritis, myelitis, ventriculitis, necrotizing angiitis, and meningoencephalitis (2).

Encephalitis, a rare neurologic complication of VZV infection, usually occurs acutely during the rash but may have a chronic course in an immunocompromised patient. Thus, the clinical presentation and initial diagnostic and laboratory studies may be nonspecific, leading to failure or delay in diagnosis. Highly specific and sensitive tests such as PCR technique or VZV antibody assay on CSF should be considered for the diagnosis of VZV.

More than one episode of cutaneous zoster in an immunocompromised patient before the development of chronic progressive encephalopathy makes VZV encephalitis highly probable, but the absence of such a history does not exclude VZV encephalitis from the diagnostic consideration. The spherical white matter lesions seen on MRI have a unique appearance (Fig.5) and should suggest VZV encephalitis (21,22). They mimic the round foci of infection seen pathologically. Unlike tumor, they lack surrounding edema and are scattered throughout the white matter. They do not follow any arterial distribution and are too discrete for an infarct. They may be confused with other types of infection such as toxoplasmosis. The VZV lesions are initially restricted to white matter and are nonenhancing, whereas toxoplasmosis prefers gray matter and the lesions enhance with associated surrounding edema (21–24). Untreated VZV encephalitis is uniformly fatal unless early treatment with acyclovir is initiated before extensive tissue destruction ensues. The MR appearance in early cases may suggest the diagnosis and help in starting the early therapy with acyclovir (24).

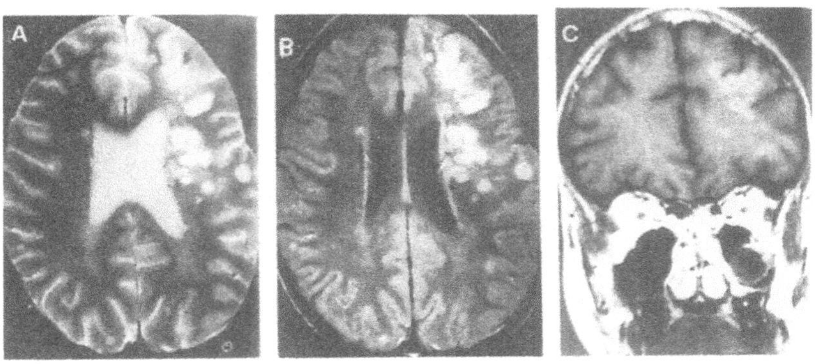

Figure 5. Varicella zoster encephalitis. T2-weighted (a) and proton density (b) axial images show evidence of multiple small rounded hyperintensities in the left periventricular and frontal region with ventricular dilatation. Postcontrast T1-weighted coronal image (c) shows multiple regions of rim enhancement. (Reprinted with permission from Lentz D, Jordan JE, Pike GB, Enzmann DR. MRI in varicella-zoster virus leukoencephalitis in the immunocompromised host. J Comput Assist Tomogr 1993; 17: 313–316.)

1.3. Herpesvirus 6 Encephalitis

HHV-6 infects children early in life causing roseola infantum and febrile illness. Although the infection is usually mild, small proportions of patients develop severe complications including encephalitis (25). Thereafter, the virus persists life-long in the individual, and HHV-6 DNA may be detected in peripheral blood mononuclear cells and saliva. HHV-6 can reactivate in immunocompromised subjects (25–27).

MRI shows both selective and multifocal involvement of the cortical gray matter with a partially confluent asymmetric pattern at the level of frontal, temporoparietal, and occipital lobes with a lack of meningeal enhancement on postcontrast study (25–27). Pathologically, lesions have been demonstrated in the white matter of both hemispheres and cortical gray matter of the temporal lobes. The FLAIR sequence (Fig.6) can be used to better delineate the cortical involvement that helps in excluding the progressive multifocal leukoencephalitis, subacute panencephalitis, and progressive rubella panencephalitis that may present with similar clinical findings (25–27).

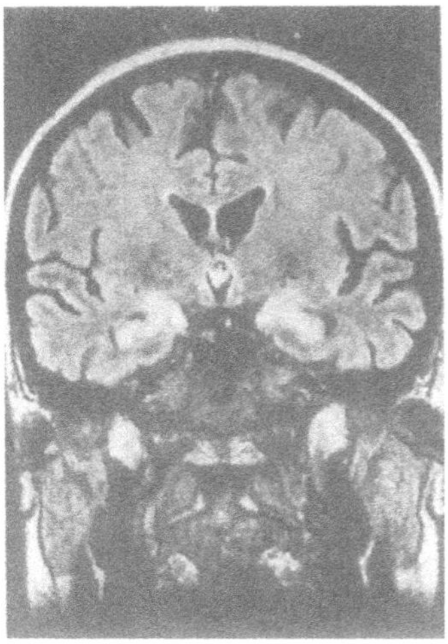

Figure 6. Herpes 6 encephalitis. T2-weighted fluid attenuated inversion recovery (FLAIR) coronal image at the level of hippocampus demonstrating high signal intensity in bilateral hippocampal region. (Reprinted with permission from Bethge W, Beck R, Jahn G, et al. Successful treatment of human herpes virus-6 encephalitis after bone marrow transplantation. Bone Marrow Transplant 1999; 24: 1245–1248.)

1.4. Cytomegalovirus Disease

CMV has the largest genome among the viruses known to infect humans. Individual cases of CMV encephalitis have been described in patients who are immunocompromised but not infected with HIV; such cases are still the exception rather than the rule. The AIDS epidemic has changed this perspective; CMV infection is found so frequently at autopsy that it must be the most opportunistic infection of the CNS in AIDS patients. The discussion here will exclude the perinatal and neonatal CMV infections as these are discussed in a later chapter.

The histologic changes may be classified into one of the following major groups: (1) nodular encephalitis, in which widely disseminated microglial nodules, only a minority of which may contain inclusion-bearing cells; (2) isolated inclusion-bearing cells, which may be astrocytes, neurons, or macrophages; (3) focal parenchymal necrosis characterized by accumulation of macrophages with axonal destruction and occasional inclusion-bearing cells; ventriculoencephalitis, characterized by inflammation and necrosis involving the ependyma and subependymal glia (28).

With the exception of a cauda equina syndrome resulting from myeloradiculitis, CMV infection of the CNS produces no specific clinical features. Diagnosis is often made at postmortem. A clinicopathologic association has been seen between the detection of CMV in the CNS and an antemortem diagnosis of undiagnosed encephalopathy, consistent with, but not proof of, an etiologic role for CMV. Usually, the imaging studies are normal despite autopsy specimen showing florid CMV lesions. In patients with AIDS, the focal ring-enhancing lesions are occasionally noted in the brain parenchyma which are confirmed as CMV encephalitis on histology and closely resemble toxoplasma, brain abscess, lymphoma, and tuberculosis. The ring-enhancing lesions represent necrotic encephalitis (29).

1.5. Epstein–Barr Virus

EBV is a member of the herpesvirus family. Involvement of the nervous system in infectious mononucleosis is more common than is generally appreciated. About 50% of the patients have headache on presentation. Neck stiffness without meningitis is a frequent finding. Abnormal electroencephalograms, which most often reveal diffuse slowing pattern, have also been reported in patients with infectious mononucleosis, but without clinically significant CNS disease (30,31). Significant neurologic complications of acute infectious mononucleosis are rare, occurring in less than 0.5% of the patients with infectious mononucleosis in a review (30). CNS symptoms may occur shortly before, during, or after infectious mononucleosis (31,32).

There are few reports describing detailed neuroimaging infections of the CNS associated with EBV (32–35). Kou et al. reported a Japanese girl who showed clinical signs of encephalitis during acute infection with EBV (33). MRI showed low- and high-intensity lesions of both basal ganglia (predominantly in the putamen) on T1- and T2-weighted images, respectively. Similar findings were subsequently reported by Ono et al. (35). While EBV encephalitis may involve any area of the brain, the cerebellum is frequently involved. Numerous cases of cerebellitis manifested as acute cerebellar ataxia have been reported (30–32). Unusual complications of EBV encephalitis include acute aqueductal stenosis and syndrome of inappropriate antidiuretic hormone secretion. Shouji et al reported a 52-year-old man with cerebellitis associated with EBV seroconversion. He improved with a course of prednisolone but developed recurrent cerebellitis after a year (34). The CNS involvement related to EBV is either due to viral infection or due to immunologic reaction to the viral antigen. Absence of viral genome on PCR from the involved tissue suggests that it is more of an immunologic reaction. MRI abnormalities usually resolve completely on follow-up (34,35). Definite diagnosis is usually based on fourfold rise in EBV-specific IgM antibodies against viral capsid antigen in serum and positive PCR for EBV DNA in CSF (30–35).

2. ARTHROPOD-BORNE ENCEPHALITIDES

Arthropod-borne viruses (arboviruses) are common causes of infections of the CNS, especially encephalitis, worldwide. Well over 20 arboviruses are capable of producing encephalitis. Most common viruses are Eastern equine, Western equine, Venezuelan equine, St. Louis, Japanese, Murray Valley, West Nile, Iiheus, Rocio, Far Eastern, Central European, Kyasanur Forest, and Colorado tick fever. In addition to causing encephalitic syndromes, these viruses also cause a variety of other clinical syndromes, ranging from benign febrile illnesses to severe, life-threatening hemorrhagic fevers. With regard to arbovirus causes of encephalitis, these pathogens are, for the most part, geographically and seasonally restricted (36). Fortunately, only a few arboviruses exist in any one region.

Pathogenesis of these agents is relatively similar from one offending pathogen to another; that is, arboviruses can be inoculated subcutaneously or intravenously from infected secretions of the insect vector (mosquito or tick). After initial replication at either a skin or a muscle site, primary viremia ensues in many instances. Seeding of the reticuloendothelial system or direct invasion of the CNS follows primary viremia. The common resulting pathologic lesion is perivascular cuffing. Spread within the CNS is usually from cell to cell (36). We will discuss the clinical features and imaging characteristics of some of the common arboviruses seen in different parts of the world.

2.1. Eastern Equine Encephalitis

EEE occurs principally along the east and gulf coasts of the United States, Caribbean and South America. Diagnosing EEE is difficult because its symptoms are nonspecific and confirmation requires either specific serologic findings or the demonstration of the virus in CSF or brain tissue (37–42). Since CT and MRI are performed for most patients with encephalitis, the identification of MRI findings suggestive of EEE could hasten its recognition (42). Although no specific treatment is available, rapid diagnosis is important for the institution of public health controls.

Gross inspection of the brains of patients who died early in the course of EEE reveals edema, leptomeningeal vascular congestion, and hemorrhage and encephalomalacia (40). On cut sections, small necrotic foci are widely distributed. Histopathologic manifestations include vascular, neuronal, and glial components. Vascular findings include arteriolar and venular inflammation, endothelial proliferation, small vessel thrombosis, and perivascular cuffing. An inflammatory meningeal infiltrate is variably present. Focal parenchymal infiltration by neutrophils and macrophages causes neuronal dysfunction, neuronophagia, focal necrosis, and spotty demyelination. Glial proliferation results in glial nodules (40). In the brains of the patients who died later in the course of illness, cortical atrophy predominates. Results of autopsy series differ on whether certain areas of the brain are particularly vulnerable to EEE virus. Faber et al. suggested that the basal ganglia and brainstem are the most severely affected (39); Hauser suggested the midbrain (41). Dent studied Hauser's patients but found no predilection for any particular areas (40). Deresiewicz et al. suggested that although focal lesions are variably distributed, involvement of the basal ganglia and thalami is prominent (42). These changes are evident early in the course of EEE and often resolve with recovery. Accordingly, they suggest that the lesions represent ischemia, inflammation, or edema, at least initially, rather than necrosis. It is difficult to explain this anatomic predilection but it has been described in other encephalitides such as Japanese encephalitis, measles, mumps, echovirus encephalitis, and in Creutzfeldt–Jakob disease (42).

MRI is exceedingly sensitive to the CNS manifestations of EEE. MRI of the brain is abnormal very early in the disease and is positive even before the patient goes into coma. Thus, the diagnosis of EEE should be questioned in comatose patients with normal findings on MRI (42). The lesions appear hyperintense on T2- and hypointense to isointense on T1-weighted images and usually involve thalami, basal ganglia, and brainstem. Cortical lesions, meningeal enhancement on postcontrast T1-weighted images, and periventricular white matter changes are less common (37,42). HSE clinically resembles EEE, and typically produces abnormalities in the frontal and temporal lobes. Rarely, HSE involves the basal ganglia. The involvement of basal ganglia is lateral and late in HSE, while it is medial and early in EEE. However, a minority of the patients with EEE have changes that are characteristic of HSE (42).

2.2. Japanese Encephalitis

Epidemics of Japanese encephalitis (JE) occur in all countries throughout Asia. It usually appears in summer and early fall, and is often severe, with rapid progression to coma. The characteristic neurologic findings during the acute stage are extrapyramidal signs, such as tremor, dystonia, and rigidity. Seizures are more common in children than adults. It is important to distinguish JE from other types of encephalitis, particularly HSE, because antiviral therapy for HSE is very effective in the acute stage. Specific antiviral therapy is not available for JE; however, its treatment is supportive (43–49).

Diffuse meningoencephalitis affecting both gray and white matter of the cerebral hemispheres, basal ganglia, brainstem, cerebellum, and thalamus has been reported in the autopsy study. Diffuse inflammatory changes are also found in the spinal cord, particularly in the anterior and lateral column. Histology reveals cerebral edema, congestion, small hemorrhages, and perivascular cuffing (48,49). MRI shows areas of hyperintensity on T2-weighted images involving the thalamus, basal ganglia, brainstem, and cerebellum (Fig.7). Involvement of cortex of the cerebral hemispheres, and deep white matter is also noted in association with involvement of thalami and basal ganglia (43–46). Involvement of thalami and basal ganglia is usually seen with or without involvement of other regions. However, isolated involvement of the substantia nigra has also been described in JE (50). The lesions are uniformly hyperintense on T2- and iso- to hypointense on T1-weighted images. Hemorrhage in the thalami and basal ganglia is reported in a series but is not a usual feature of JE in general (43). Characteristics of hemorrhage depend on the onset of bleed and the time when MRI is performed. Postcontrast study may show meningeal enhancement especially on MT-SE T1-weighted images. The parenchymal lesions show minimal or no enhancement after injection of gadolinium. Diagnosis of JE is confirmed by showing a fourfold rise in the antibody titer in a paired sample (49). However, a patient presenting with encephalitic illness from an endemic zone of JE with demonstration of lesions in the basal ganglia, thalami, and brainstem should help in differentiating JE from atypical HSE. There are reports of follow-up MRI study in JE. The lesions remain well-defined hyperintense on T2- and hypointense on T1-weighted images (Fig.8). The lesions decrease in size and may clear from some of the regions but the residual lesions may persist for a long time. There is no relationship between the imaging and the patient's prognosis (44).

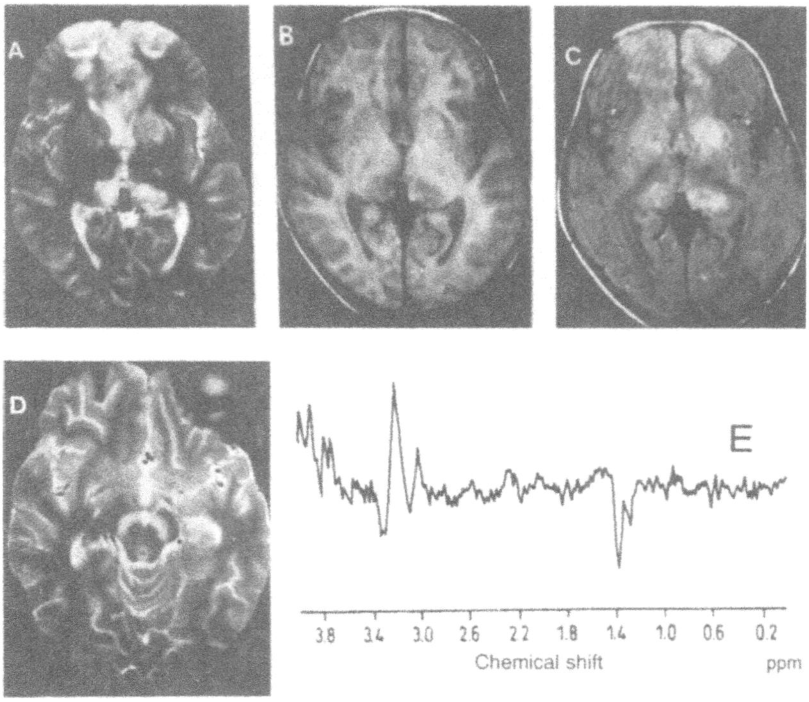

Figure 7. Japanese B encephalitis. T2 axial image (a) shows evidence of hyperintensity in the medial part of bilateral thalami, caudate nuclei, and medial part of bilateral frontal lobes. The lesions appear iso- to hypointense on T1-weighted image (b). On MTT1-weighted image (c) the lesions appear hyperintense and are conspicuous compared to T1-weighted image. Note the hyperintense bilateral substantia nigra in T2 image at a level below (d). Single-voxel *in vivo* proton MR spectroscopy (e) from the frontal region (SE, TR/TE/n = 3000/135/128) shows presence of lactate at 1.33 ppm, choline at 3.22 ppm, and creatine at 3.02 ppm with no visible NAA resonance.

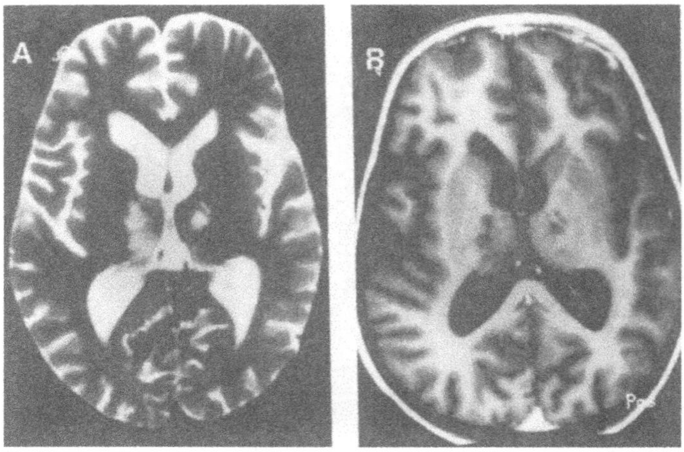

Figure 8. Japanese B encephalitis. Follow-up T2 image (a) in a case of JE shows evidence of hyperintense lesions in the medial part of bilateral thalami that appear hypointense on T1-weighted image (b).

2.3. St. Louis Encephalitis

This disease occurs in both endemic and epidemic form throughout the Americas, being one of the most important and common epidemic arbovirus infections in the United States. Clinical features are similar to other arbovirus infections.

There is only one report describing the MRI features of St. Louis encephalitis. Cerna et al. described imaging abnormalities in 2 out of 10 patients studied with MRI and found isolated involvement of substantia nigra as hyperintense on T2, isointense on T1, and no enhancement after contrast administration. They felt that these findings were characteristic of St. Louis encephalitis and have not been reported in other conditions. However, similar findings have been reported in patients with JE (51).

3. MEASLES ENCEPHALITIS

Measles is an acute, highly contagious infection caused by the rubeola virus, an enveloped RNA virus of the paramyxovirus family. There are three CNS syndromes related to measles infection: acute measles encephalitis (AME), subacute sclerosing panencephalitis (SSPE), and subacute measles encephalitis (SME) (52).

3.1. Acute Measles Encephalitis

AME presents as a recrudescence of fever during convalescence from measles and is probably an autoimmune process. It is characterized by absence of the measles virus and by the presence of demyelination, necrosis, vascular damage, and hemorrhage. Typically, patients recovering from measles present with an abrupt onset of renewed fever and a wide variety of neurologic symptoms that include seizures in approximately 50%, altered mental status, and multifocal neurologic signs. The onset is most often between 2 and 7 days after the appearance of the rash, but occasionally it predates or appears up to 3 weeks after the rash. Often, the patient is recovering from measles when there is a sudden reappearance of fever and depression of level of consciousness. The fatality is variable and neurologic residua in those who survive may be substantial and are present in a majority of the cases. Serum and CSF antibody titers against measles virus are usually high and are helpful in its diagnosis (52,53).

3.2. Subacute Sclerosing Panencephalitis (SSPE)

SSPE is an invariably fatal neurodegenerative disease, developing as a sequel to early childhood measles infection. Following the original measles infection, the virus becomes altered and remains dormant intracellularly, only to manifest as SSPE a decade or so later. Measles vaccination has now all but eradicated the disease in the developed countries, but SSPE is still endemic in many developing countries due to improper vaccination policy. The disease also is of interest as a model for persistent viral infection of the brain (52–55).

SSPE is a dramatic disease, starting as minor disturbances in behavior in a previously entirely healthy child. Myoclonic attacks then develop and become increasingly more frequent, dementia follows, and within months or a few years, the child is comatose, void of higher mental function. Neuronal and glial viral cell inclusions, subacute inflammatory changes, subacute demyelination, and extensive gliosis are characteristic histopathologic features. It is now known that the disease is linked to a measles infection in early childhood (52,55).

The average age of onset of SSPE has been reported to vary from 9 to 13 years. History of preceding measles infection is available in 65–75% of the cases. The age of onset of measles is between 2 and 3 years and the interval from the measles infection to the onset of SSPE symptoms averages from 8 to 11 years. It is still not known how the measles virus manages to survive clinically dormant for many years and why it becomes active again and causes SSPE. Possibly the immature immune system fails to destroy the virus completely, and the partially degraded virus remains in the CNS (55–59). Viral mutation may alter the surface antigen of the virus and thereby make it invisible to the immune response while retaining the ability to reproduce and spread from cell to cell. Perhaps nonproductive, cell-associated forms of the measles virus occur naturally during a measles infection but are kept passive by the hosts' defense mechanisms, such as interferon. If suppression then fails, for instance from development of viral forms less sensitive to interferon or from a too low production of interferon, the virus may reproduce and spread within the CNS, causing SSPE. This hypothesis receives some support from the finding that intraventricular interferon seems to induce a clinical improvement, or at least it temporarily arrests the disease in more than half the cases (55–61).

The initial symptoms of SSPE are usually behavioral changes, such as irritability, impaired school performance, disobedience, inappropriate affection, and withdrawal. These symptoms may be ongoing for many years, and may be recognized only in retrospect. Myoclonic spasms then appear, often seen as drop attacks. These are initially sporadic but may later occur at intervals of only a few seconds, and will eventually prevent the patient from ambulating. The mental deterioration progresses and a characteristic EEG pattern develops with generalized brief, bilaterally synchronous bursts of spike-wave and/or slow-wave complexes. Spasticity and ataxia become prominent and may later be followed by choreoathetosis. Language difficulties progress, patients speak less and have poor verbal comprehension; visual problems may proceed to cortical blindness. Seizures follow; the patient becomes bedridden, and may need tubal feeding. The spasticity progresses to opisthotonus and the patient lapses into a coma. Terminally, the muscular hypertonia decreases, myoclonus disappears, and the patient passes into a neurovegetative state and eventually dies (54,55,58).

Studies of pathology in SSPE show that mild inflammation of the cortical gray matter is prominent in the earlier period of the disease, later progressing to subcortical and deep white matter involvement. Later still, the inflammation subsides, giving way to demyelination, necrosis, and gliosis. Earlier in the course of the disease, the posterior regions are predominantly affected (53,56,62).

MRI has a major advantage in demonstrating the white matter changes. The imaging may be normal in early as well as advanced cases of SSPE. Though MRI is more sensitive than CT in demonstrating the brain involvement, the diagosis is usually based on clinical features, EEG, and CSF serology for measles antibody in SSPE (63–67). The lesions appear as focal hyperintense on T2-weighted images and hypointense on T1-weighted images (Fig.9,10) and involve the posterior parietal, occipital, and temporal regions, basal ganglia, corona radiata, and subcortical and deep white matter. Brainstem and cerebellum may also show T2 hyperintensity. There may be cortical atrophy, mild to severe ventricular dilatation, and atrophy of the brainstem. Basal ganglion involvement was noted in patients who had severe disease and of longer duration (68).

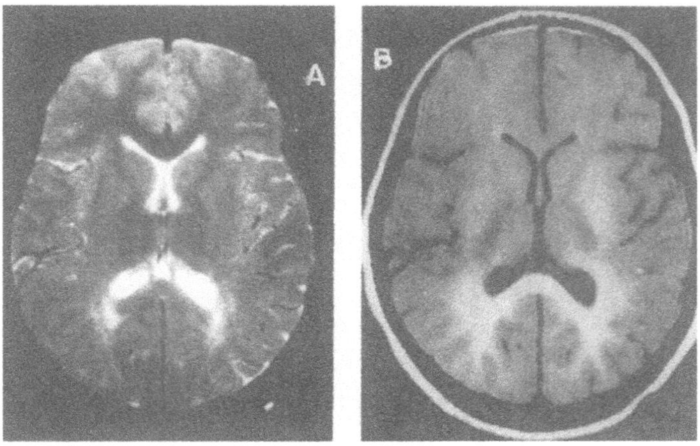

Figure 9. Subacute sclerosing panencephalitis. T2 axial image (a) shows evidence of minimal hyperintensity around the occipital horn bilaterally. MTT1-weighted image (b) shows hyperintensity involving the white matter around the occipital and the splenium not demonstrated in (a). The MT ratio from the lesion was lower than the normal white matter.

The correlation between the clinical stage of SSPE disease and the MR findings is often poor. Thus, patients bedridden with severe disease may still have normal findings at MR examinations. Improvement of MR findings despite progress of the disease has also been reported. Several regimens have been tried for the treatment of SSPE and an arrest of the disease or even an improvement of the clinical status has been reported (69,70). The clinical improvement has been associated with improvement in MRI findings (67,71,72). However, patients have shown improvement while MRI findings remained stationary or even progressed. It is most likely that in order to alter significantly the course of the disease, therapy must be initiated at an early stage. Presently, the diagnosis of SSPE is usually not made until major, permanent brain damage has taken place. An increased awareness of the possibility of SSPE as the cause of behavioral changes in a patient with normal findings on CT or MRI of the brain is probably one of the prerequisites for better therapeutic results (67,71,72).

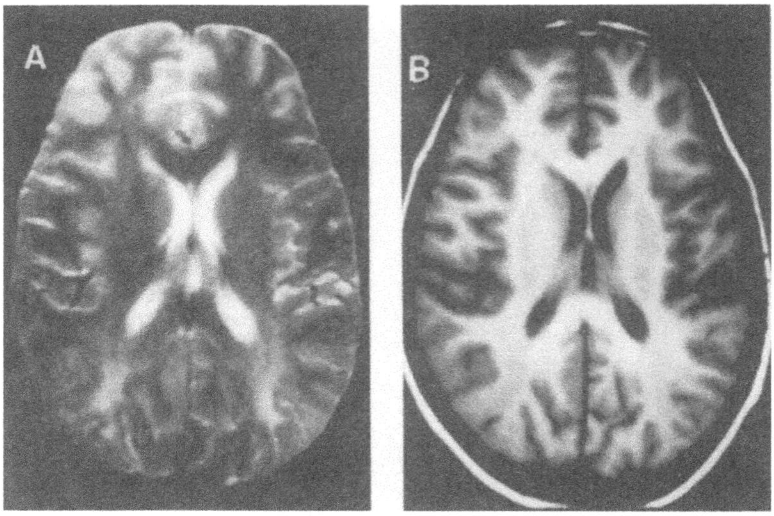

Figure 10. Subacute sclerosing panencephalitis. In another patient the abnormality on T2-weighted image (a) is more widespread compared to Fig.9 involving the occipital and right frontal lobes. However, the T1-weighted image (b) does not show any obvious abnormality.

3.3. Subacute Measles Encephalitis

SME usually occurs as an opportunistic infection in immunodeficient individuals with a latent period of 1 to 7 months. The prominence of cognitive and motor dysfunction is similar to that of SSPE, but it has a subacute onset with more rapid evolution (73,74). The presence of seizures rather than myoclonus is the distinctive feature of SME. Neither CSF nor serum antibody titers against measles are markedly increased as in SSPE and AME. Brain biopsy reveals pathologic changes similar to those in SSPE but with minimal perivascular inflammatory infiltrate. Brain biopsy is the definitive diagnostic tool as histologic, electron microscopic, and immunocytochemical studies reveal typical intranuclear inclusions of paramyxovirus nucleocapsids, and measles virus antigen in neurons and oligodendrocytes (73).

MRI findings are nonspecific and serial MRI clearly documents focal onset and the progression of cerebral damage. The lesions appear hyperintense on T2- and isointense on T1-weighted images involving both occipital cortical and subcortical regions (Fig.11). Postcontrast study shows some enhancement (74).

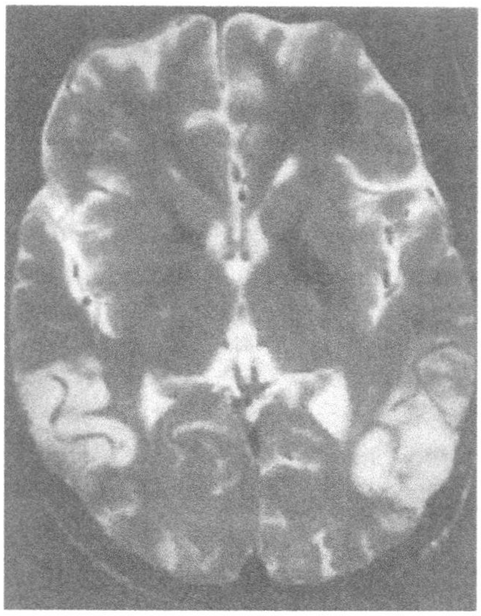

Figure 11. Subacute measles encephalitis. Axial T2-weighted image shows widespread and severe signal abnormalities in the bilateral cerebral cortex. (Reprinted with permission from Gazzola P, Cocito L, Capello E, Roccatagliata L, Canepa M, Mancardi GL. Subacute measles encephalitis in a young man with immunosuppression for ankylosing spondylitis. Neurology 1999; 52:1074–1077.)

4. MUMPS

Mumps virus previously was the single most common agent producing aseptic meningitis and mild encephalitis. Immunization has markedly decreased the incidence and subsequent complications of mumps, although the vaccine is successful in only 75–90% of patients (75). In the CNS, mumps produces meningismus in more than half the cases. Associated encephalitis is rare but also has a significant mortality and morbidity. Pathologically, mumps produces areas of perivascular inflammation and demyelination scattered throughout the CNS (75,76). The diagnosis is usually based on demonstration of a fourfold rise of antibody titer in paired samples. MRI features are similar to postinfectious encephalitis (76).

5. CREUTZFELDT–JAKOB DISEASE

CJD belongs to the group of subacute spongiform encephalopathies found in humans. Other slow virus encephalitides found in humans belonging to this group are kuru and Gerstmann–Sträussler syndrome. CJD is an uncommon cause of rapidly progressive dementia that may be difficult to distinguish clinically from other forms of dementia. CJD is thought to be caused by an unconventional virus known as prion (83). Prions are small proteinaceous infectious particles that differ from standard viruses in that they contain little or no nucleic acid and do not evoke an immune response during infection. Several hypothetical mechanisms for the origin and spread of CJD have been proposed, including exposure to infected meat, iatrogenic transmission from patients, and appearance of spontaneous somatic mutations to yield an infectious protein agent *de novo*. Incidence is estimated at 0.25 to 2 cases per million population per year worldwide (83–85).

Clinical presentation is classically that of a rapidly progressive dementia associated with upper motor neuron dysfunction and myoclonic jerks. The usual age of presentation is between 40 and 80 years. Some patients present with sensory abnormalities, confusion, and/or inappropriate behavior, or with cerebellar ataxia. The prognosis is poor, with a mean survival of less than 1 year from the onset of symptoms and only 10% surviving for more than 2 years. EEG reveals diffuse slowing with superimposed bursts of sharp waves. CSF is usually normal, but may show mild elevation of protein. However, the diagnostic triad of progressive dementia, myoclonic jerks, and periodic sharp-wave EEG activity may be lacking in as many as 25% of patients and the antemortem diagnosis remains a problem (84,85).

Microscopically, CJD is characterized by spongiform changes in the cortical as well as subcortical gray matter, with loss of neurons and replacement gliosis. The spongiform change is cell-associated with neuronal intracytoplasmic vacuoles that often leave neurons with only a nucleus and fine wisps of cytoplasm. White matter is usually spared, except for secondary Wallerian degeneration. There is, however, considerable topographic variation. Atrophy may occur diffusely or may be predominantly confined to cortex, cerebellum, thalamus, basal ganglia, or specific tracts. Inflammatory changes are absent. CJD may be divided into three main pathologic variants: frontopyramidal (33%), occipitoparietal (18%), and diffuse cortical/basal ganglia type (49%). Any of these variants may also reveal lesions of the thalamus, midbrain, cerebellum, or spinal cord (85–87).

Various MRI abnormalities have been described in CJD (86–89). These include diffuse cortical atrophy, focal symmetrical lesions of occipital gyri, bilateral basal ganglion and thalamus lesions, and periventricular abnormality. Although white matter lesions are rarely present, diffuse white matter changes have been described. The lesions appear hyperintense on T2- and isointense on T1-weighted images. Rarely, T1 hyperintensity in the globus pallidus has been described with no abnormality on T2-weighted images in this region and is probably due to the higher concentration of prions in the region with minimal gliosis and spongiform changes on histopathology (88). This is thought to be responsible for the T1 shortening effect. Serial studies on MRI usually show progression of the disease. Shyu et al. have described a case of biopsy proven CJD where initial MRI showed hyperintensity of the cortex and basal ganglia on T2 and progressed to hypointensity of these regions on T2 with hyperintensity of the earlier normal-appearing white matter (Fig.12). They showed high iron contents on pearl staining of the cortical tissues confirming that the large amount of iron deposits as the patient progresses to the vegetative state (89). Though MRI is superior to CT in demonstrating lesions in CJD, the conventional spin-echo imaging may be normal in a large number of patients especially with early cases.

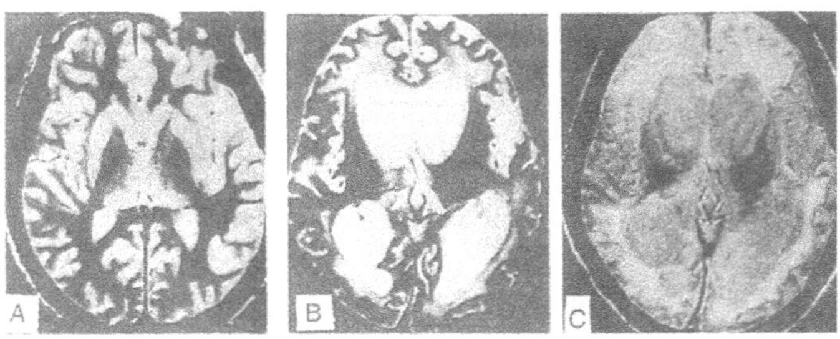

Figure 12. Creutzfeldt–Jakob disease. Axial MR images obtained 2 months after symptom onset show increased signal intensity within the head of caudate nuclei, putamen, and thalami on T2-weighted image (a). Repeat images obtained 1 year after onset, demonstrate marked brain atrophy, thinning of cortex, diffuse white matter degeneration including subcortical U fibers, and hypointense signals at basal ganglia on T2-weighted (b) and proton density (c) images. (Reprinted from Shyu WC, Lee C, Hsu Y, et al. Panencephalitis Creutzfeldt–Jakob disease: Unusual presentation of magnetic resonance imaging and proton magnetic resonance spectroscopy. J Neurol Sci 1996; 138: 157–160 with permission from Elsevier Science.)

Recently, FLAIR and diffusion-weighted imaging (DWI) have been found to be more sensitive in the detection of lesions in CJD where routine imaging is normal (Fig.13) (89–91). The lesions appear hyperintense on diffusion-weighted images and simulate hyperacute stroke. The mechanisms underlying these changes are uncertain. DWI has been shown to be sensitive to the changes in the compartmentalization and diffusion of water and measures alterations in energy metabolism of tissue indirectly. Restricted diffusion seen in CJD has been attributed to several factors associated with cytotoxic edema, including changes in the organization of the extracellular space and increased intracellular viscosity. It has been suggested that DWI appears to improve the *in vivo* diagnosis of CJD and may thus decrease the number of false-negative MRI examinations when relying on conventional spin-echo imaging (90,91).

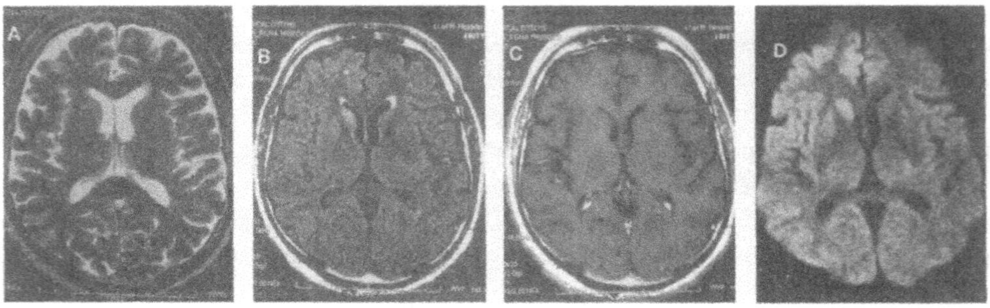

Figure 13. Creutzfeldt–Jakob disease with normal conventional MRI. T2 (a), FLAIR (b), and postcontrast T1-weighted (c) images through the lateral ventricles show no visible abnormality. DWI shows (d) hyperintensity in the right hemispheric cortex with abnormal signal in the right caudate nucleus. Histopathology confirmed the diagnosis subsequently. (Courtesy Dr. Suresh Patel, Henry Ford Hospital, Detroit, Michigan.)

Recently, *in vivo* proton MR spectroscopy has been performed in a case of panencephalitic CJD and is found to be characteristic of this condition. Proton spectrum obtained from the T2 hyperintense frontal region did not show any metabolite signal. This has been attributed to the lack of neuronal activity in the region (86).

6. VIRAL INFECTIONS OF THE SPINAL CORD

Viruses may affect the spinal cord or the nerve roots. The infections related to HIV viruses will be discussed in Chapter 7. Herpes group of viruses most commonly causes encephalitis; some of these may produce myelitis and polyradiculitis. Myelitis associated with herpes zoster is an unusual complication. The pathogenesis may involve a direct viral invasion of the spinal cord, a vasculitic process with ischemic necrosis, or an immunologic, parainfectious mechanism. The onset of myelitis can occur weeks to months after an episode of herpes zoster. Some of these viruses (EBV or CMV) have been associated with acute transverse myelitis and Guillain–Barré syndrome.

Spinal cord hyperintensity on T2-weighted images is a nonspecific feature on MRI and is seen in other parainfectious myelitides. T1-weighted images may show diffuse swelling of the cord with mild hypointensity. Postcontrast study may show focal areas of cord enhancement with enhancement of the roots suggesting polyradiculitis with myelitis.

6.1 Poliomyelitis

The term *poliomyelitis* is derived from the Greek words *polios* (gray) and *myelos* (spinal cord). The disease is characterized by aseptic meningitis and lytic infection of the anterior horn cells of the spinal cord leading to transient or permanent paresis of one or more extremities. The human polioviruses are the only epidemiologically important cause of poliomyelitis. In developed nations, the naturally occurring poliovirus has been completely eradicated due to active immunization. In contrast, poliomyelitis persists in many underdeveloped regions of the world (77–79).

Ingested polioviruses implant in the oropharynx and small bowel. The intestinal submucosal tissue appears to be the most efficient site of replication. Spread to the regional lymph nodes gives rise to minor viremia that is transient and clinically silent, and during which virus spreads to systemic reticuloendothelial tissues. Further replication of the virus in these tissues results in clinical viremia. Polioviruses are recoverable from the spinal cord for only the first several days of paralysis. The necrotic lesions and inflammatory infiltrates are distributed within the gray matter of the anterior horns of the spinal cord and motor nuclei of the pons and medulla. The severity of clinical paralysis rests more on the intensity of the lesions than on their distribution. In the later stages, cavitation and atrophy of the anterior horns appear. Demonstration of the virus on stool culture and serology helps in confirming the diagnosis of acute poliomyelitis (77).

Wasserstrom et al. described the MRI findings in the chronic phase of bulbar poliomyelitis in a child, demonstrating an exact correlation between areas of high signal intensity on T2-weighted images and areas of necrosis in the pathologic sections of the midbrain and medulla oblongata (80). During the acute phase of the disease, MRI shows an enlarged cervical spinal cord without evidence of an intrinsic mass compatible with edema. Postcontrast T1-weighted images show the enhancement of the anterior horn cell region confirming the infection of the anterior horn cells. Follow-up has shown a circular area of low signal on T1-weighted images that appeared hyperintense on T2-weighted images. The location of the lesions is typically in the anterior horns of the spinal cord, corresponding to the side and distribution of the residual weakness (81,82). These findings may also be observed in poliolike syndrome due to other enteroviruses. MRI of the spine is valuable in the investigation of sudden-onset paralysis; it can exclude spinal cord compression or an intrinsic mass (81,82).

6.2. Guillain–Barré Syndrome

Guillain–Barré syndrome, also known as acute inflammatory demyelinating neuropathy, is characterized by rapid on-set weakness, hyporeflexia or areflexia, and elevated levels of protein in the CSF without pleocytosis. Weakness may progress for up to 4 weeks, is usually symmetric, and involves the lower extremities before spreading to the upper extremities and face. Involvement of cranial nerves and autonomic dysfunction are common. Although the cause of Guillain–Barré syndrome is unclear, most evidence now supports an immune-mediated phenomenon. MRI in Guillain–Barré syndrome can show marked enhancement of thickened nerve roots of the conus medullaris and cauda equina (92). However, MRI is usually unremarkable in this entity.

6.3. Tropical Spastic Paraparesis

Tropical spastic paraparesis (TSP) is a neurologic disorder endemic in many tropical and subtropical countries, characterized by a progressive myelopathy associated with a retrovirus, human T-lymphocytotropic virus type 1 (HTLV1), which was found in 68% of patients with TSP in Martinique (93). Neuropathologic studies in these patients revealed an inflammatory myelopathy with focal spongiform demyelinating and necrotic lesions, perivascular and meningeal infiltrates, and focal gray matter destruction with a predilection for the posterior columns and corticospinal tracts. The spectrum of MRI findings includes nonspecific cord swelling and hyperintensity on T2-weighted images to cord atrophy in the late stages of the disease.

REFERENCES

1. Johnson RT. The pathogenesis of acute viral encephalitis and postinfectious encephalomyelitis. J Infect Dis 1987; 155: 359–364.
2. Cassady KA, Whitley RJ. Pathogenesis and pathophysiology of viral infections of the central nervous system. In *Infections of the central nervous system*, WM Scheld, RJ Whitley, DT Durack, eds. Philadelphia: Lippincott–Raven, 1997; 7–23.
3. Boos J, Kim JH. Biopsy histopathology in herpes simplex encephalitis and in encephalitis of undefined etiology. Yale J Biol Med 1984; 57: 751–755.
4. Gracia JH, Colon LE, Whitley RJ, et al. Diagnosis of viral encephalitis by brain biopsy. Semin Diagn Pathol 1984; 1: 1334–1342.
5. Gressens P, Langston C, Martin JR. In situ PCR localization of herpes simplex virus DNA sequences in disseminated herpes encephalitis. J Neuropathol Exp Neurol 1994; 53: 469–482.

6. Lakeman FD, Whitley RJ. Diagnosis of herpes simplex encephalitis: Application of polymerase chain reaction to cerebrospinal fluid from brain-biopsied patients and correlation with disease. J Infect Dis 1995; 171: 857–873.
7. Demaerel P, Wilms G, Robberecht, et al. MRI and herpes simplex encephalitis. Neuroradiology 1992; 32: 490–493.
8. Schmidbauer M, Podreka I, Wimberger D, et al. SPECT and MR imaging in herpes simplex encephalitis. J Comput Assist Tomogr 1991; 15: 811–815.
9. Schroth G, Gawehn J, Thron A, Vallbracht A, Voigt K. Early diagnosis of herpes simplex encephalitis by MRI. Neurology 1987; 37: 179–183.
10. Tien RD, Felsberg GJ, Osumi AK. Herpesvirus infections of the CNS: MR findings. Am J Roentgenol 1993; 161:167–176.
11. Tsuchiya K, Inaoka S, Mizutani Y, Hachiya J. Fast fluid-attenuated inversion recovery MR of intracranial infections. Am J Neuroradiol 1997; 18: 909–913.
12. Burke JW, Mathews VP, Elster AD, Ulmer JL, McLean M, Davis SE. Contrast enhanced magnetization transfer saturation imaging improves MR detection of herpes simplex encephalitis. Am J Neuroradiol 1996; 17: 773–776.
13. Domingues RB, Fink MCD, Tsanaclis AMC, et al. Diagnosis of herpes simplex encephalitis by magnetic resonance imaging and polymerase chain reaction assay of cerebrospinal fluid. J Neurol Sci 1998; 157: 148–153.
14. Lester JW, Carter MP, Reynolds TL. Herpes encephalitis: MR monitoring of response to acyclovir therapy. J Comput Assist Tomogr 1988; 12: 941–943.
15. Koelfen W, Freund M, Guckel F, Rohr H, Schultze C. MRI of encephalitis in children: Comparison of CT and MRI in the acute stage with long-term follow-up. Neuroradiology 1996; 38: 73–79.
16. Jay V, Hwang P, Hoffman HJ, Becker LE, Zielenska M. Intractable seizure disorder associated with chronic herpes infection. Child's Nerv Syst 1998; 14: 15–20.
17. Hitosugi M, Ichijo M, Matsuoka Y, Takenaka N, Fujii H. Proton MR spectroscopy findings in herpes simplex encephalitis. Clin Neurol 1996; 36: 839–843.
18. Menon DK, Sargentoni J, Peden CJ, et al. Proton MR spectroscopy in herpes simplex encephalitis: Assessment of neuronal loss. J Comput Assist Tomogr 1990; 14: 449–452.
19. Takanashi J, Sugita K, Ishii M, Aoyagi M, Niimi H. Longitudinal MR imaging and proton MR spectroscopy in herpes simplex encephalitis. J Neurol Sci 1997; 149: 99–102.
20. Sakakibara R, Hattori T, Fukutake T, Mori M, Yamanishi T, Yasuda K. Micturational disturbance in herpetic brainstem encephalitis; contribution of the pontine micturition center. J Neurol Neurosurg Psychiatry 1998; 64: 269–272.
21. Lentz D, Jordan JE, Pike GB, Enzmann DR. MRI in varicella-zoster virus leukoencephalitis in the immunocompromised host. J Comput Assist Tomogr 1993; 17: 313–316.
22. Gilden DH, Murray RS, Wellish M, Keinschmidt-Demasters BK, Vafai A. Chronic progressive varicella zoster virus encephalitis in an AIDS patient. Neurology 1988; 38: 1150–1153.
23. Lipton SA, Schaeffer PW, Ma MJ. Case records of the Massachusetts General Hospital. N Engl J Med 1996; 335: 1587–1595.
24. Weaver S, Rosenblum MK, DeAngelis LM. Herpes varicella zoster encephalitis in immunocompromised patients. Neurology 1999; 52: 193–195.

25. McCullers JA, Lakeman FD, Whitley RJ. Human herpesvirus 6 is associated with focal encephalitis. Clin Infect Dis 1995; 21: 571–576.
26. Mookerjee BP, Vogelsang G. Human herpesvirus 6 encephalitis after bone marrow transplantation: Successful treatment with ganciclovir. Bone Marrow Transplant 1997; 20: 905–906.
27. Bosi A, Zazzi M, Amantini A, et al. Fatal herpesvirus 6 encephalitis after unrelated bone marrow transplant. Bone Marrow Transplant 1998; 22: 285–288.
28. Morgello S, Cho ES, Nielsen S, Devinsky O, Petito CK. Cytomegalovirus encephalitis in patients with acquired immunodeficiency syndrome: An autopsy study of 30 cases and a review of literature. Hum Pathol 1987; 18: 289–297.
29. Huang PP, McMeeking AA, Stempien MJ, Zagzag D. Cytomegalovirus disease presenting as a focal brain mass: Report of two cases. Neurosurgery 1997; 40: 1074–1079.
30. Silversides JL, Richardson JC. Neurological complications of infectious mononucleosis. Can Med Assoc J 1950; 63: 138–143.
31. Ross JP, Cohen JI. Epstein–Barr virus. In *Infections of the central nervous system,* WM Scheld, RJ Whitley, DT Durack, eds. Philadelphia: Lippincott–Raven, 1997; 117–127.
32. Ito H, Sayama S, Irie S, et al. Antineuronal antibodies in acute cerebellar ataxia following Epstein–Barr virus infection. Neurology 1994; 44: 1506–1507.
33. Kou K, Itoh M, Kawano Y. A case report of EB virus-induced meningoencephalitis associated with brain MRI abnormalities (basal ganglia). J Jpn Pediatr Soc 1994; 98: 2052–2059.
34. Shouji H, Kusuhara T, Honda H. Relapsing acute disseminated encephalomyelitis associated with chronic Epstein–Barr virus infection: MRI findings. Neuroradiology 1992; 34: 340–342.
35. Ono J, Shimizu K, Harada K, Mano T, Okada S. Characteristic MR features of encephalitis caused by Epstein–Barr virus: A case report. Pediatr Radiol 1998; 28: 569–570.
36. Tsai TF. Arboviral infections in the United States. Infect Dis Clin North Am 1991; 5: 73–103.
37. Piliero PJ, Brody J, Zamani A, Deresiewicz RL. Eastern equine encephalitis presenting as focal neuroradiographic abnormalities: Case report and review. Clin Infect Dis 1994; 18: 985–988.
38. Morse RP, Bennish ML, Darras BT. Eastern equine encephalitis presenting with a focal brain lesion. Pediatr Neurol 1992; 8: 473–475.
39. Faber S, Hill A, Connery ML, Dingle JH. Encephalitis in infants and children caused by the virus of the eastern variety of equine encephalitis. J Am Med Assoc 1940; 114: 1725—1731.
40. Dent JH. Pathological changes in the brains of children infected with the virus of eastern equine encephalomyelitis. Bull Tulane Med Fac 1955; 14: 85–96.
41. Hauser GH. Human equine encephalomyelitis, eastern type, in Louisiana. New Orleans Med Surg J 1948; 100: 551–558.
42. Deresiewicz RL, Thaler SJ, Hsu L, Zamani AA. Clinical and neuroradiographic manifestations of eastern equine encephalitis. N Engl J Med 1997; 336: 1867–1874.
43. Misra UK, Kalita J, Jain SK, Mathur A. A study of radiological and neurophysiological changes in Japanese encephalitis. J Neurol Neurosurg Psychiatry 1994; 57: 1484–1487.
44. Shoji H, Murakami T, Murai I, Kida H, Sato Y, Kojima K, Abe T, Okudera T. Follow-up study by CT and MRI in three cases of Japanese encephalitis. Neuroradiology 1990; 32: 215–219.
45. Shoji H, Hiraki U, Kuwasaki N, Toyomasu T, Kaji M, Okudera T. Japanese encephalitis in the Kurume region of Japan: CT and MRI findings. J Neurol 1989; 236: 255–259.

46. Kumar S, Misra UK, Kalita J, Sawlani V, Gupta RK, Gujral R. MRI in Japanese encephalitis. Neuroradiology 1997; 39: 180–184.
47. Zimmerman HM. The pathology of Japanese B encephalitis. Am J Pathol 1946; 22: 965–991.
48. Deasi A, Shanker SK, Ravi V, Chandramuki A, Gauridevi M. Japanese encephalitis virus antigen in human brain and its topographic distribution. Acta Neuropathol 1995; 89: 368–373.
49. Ishii K. Virological and serological diagnosis of Japanese encephalitis. Adv Neurol Sci 1967; 11: 300–311.
50. Pradhan S, Pandey N, Yadav S, Gupta RK, Mathur A. Parkinsonism due to predominant involvement of substantia nigra in Japanese encephalitis. Neurology 1999, 53: 1781–1786.
51. Cerna F, Mehrad B, Luby JP, Burns D, Fleckenstein JL. St. Louis encephalitis and substantia nigra: MR imaging evaluation. Am J Neuroradiol 1999; 20: 1281–1283.
52. Kipps A, Dick G, Moodie JW. Measles and the central nervous system. Lancet 1983; 2: 1406–1410.
53. Pearl PL, Abu Farsakh H, Starke JR, Drever Z, Louis PT, Kirkpatrick JR. Neuropathology of two fatal cases of measles in 1988–89 Houston epidemic. Pediatr Neurol 1990; 6: 126–130.
54. Dyken PR. Subacute sclerosing panencephalitis: Status. Neurol Clin 1985; 3: 179–196.
55. Haddad FS, Risk WS, Jabbour JT. Subacute sclerosing panencephalitis in the Middle East: Report of 99 cases. Ann Neurol 1977; 1: 211–217.
56. Case records of the Massachusetts General Hospital: Weekly clinicopathological exercises. N Engl J Med 1986; 314: 1689–1700. Case 25-1986.
57. Cattaneo R, Schmid A, Rebmann G, et al. Accumulated measles virus mutations in a case of subacute sclerosing panencephalitis: Interrupted matrix protein reading frame and transcription alteration. Virology 1986; 154: 97–107.
58. Risk WS, Haddad FS. The variable natural history of subacute sclerosing panencephalitis: A study of 118 cases from the Middle East. Arch Neurol 1979; 36: 610–614.
59. Chao D. Subacute inclusion body encephalitis: Report of 3 cases. J Pediatr 1962; 61: 501–510.
60. Gascon G, Yamani S, Crowell J, et al. Combined oral isoprinosine intraventricular alpha interferon therapy for subacute sclerosing panencephalitis. Brain Dev (Tokyo) 1993; 15: 346–355.
61. Yalaz K, Anlar B, Oktem F, et al. Intraventricular interferon and oral inosiplex in the treatment of subacute sclerosing panencephalitis. Neurology 1992; 42: 488–491.
62. Ohya T, Martinez AJ, Jabbour JT, Lemmi H, Duenas DA. Subacute sclerosing panencephalitis: Correlation of clinical, neurophysiologic and neuropathologic findings. Neurology 1974; 24: 211–218.
63. Lum GB, Williams JP, Dyken PR, et al. Magnetic resonance and CT imaging correlated with clinical status in SSPE. Pediatr Neurol 1986; 2: 75–79.
64. Takemoto K, Koizumi Y, Kogame S, et al. Magnetic resonance imaging of subacute sclerosing encephalitis. Jpn J Clin Radiol 1986; 31: 999–1004.
65. Murata R, Matsuoka O, Nakajimas S, et al. Serial magnetic resonance imaging in subacute sclerosing panencephalitis. Jpn J Psychiatr Neurol 1987; 41: 277–282.
66. Geller TJ, Vern BA, Sarwar M. Focal MRI findings in early SSPE. Pediatr Neurol 1987; 3: 310–312.
67. Tsuchiya K, Yamauchi T, Furui S. MR imaging vs CT in subacute sclerosing panencephalitis. Am J Neuroradiol 1988; 9: 943–946
68. Woodward KG, Weinberg PE, Lipton HL. Basal ganglia involvement in subacute sclerosing panencephalitis: CT and MR demonstration. J Comput Assist Tomogr 1988; 12: 489–491.

69. Dietrich RB, Vining EP, Taira RK, Hall TR, Phillpart M. Myelin disorders of childhood: Correlation of MR findings and severity of neurological impairment. J Comput Assist Tomogr 1990; 14: 693–698.
70. Winer JB, Pires M, Kermode A, Ginsberg L, Rossor M. Resolving MRI abnormalities with progression of subacute sclerosing panencephalitis. Neuroradiology 1991; 33: 178–180.
71. Brismar J, Generoso G, Steyern KV, Bohlega S. Subacute sclerosing panencephalitis: Evaluation with CT and MR. Am J Neuroradiol 1996; 17: 761–772.
72. Tuncay R, Akman-Demir G, Gokyigit A, et al. MRI in subacute sclerosing panencephalitis. Neuroradiology 1996; 38: 636–640.
73. Poon TP, Tchertkoff V, Win H. Subacute measles encephalitis with AIDS diagnosed by fine needle aspiration biopsy. Acta Cytol 1998; 42: 729–733.
74. Gazzola P, Cocito L, Capello E, Roccatagliata L, Canepa M, Mancardi GL. Subacute measles encephalitis in a young man with immunosuppressed for ankylosing spondylitis. Neurology 1999; 52:1074–1077.
75. Donohue WL, Playfair FD, Whitaker L. Mumps encephalitis: Pathology and pathogenesis. J Pediatr 1955; 47: 395–412.
76. Tarr RW, Edwards KM, Kessler RM, Kulkarni MV. MRI of mumps encephalitis: Comparison with CT evaluation. Pediatr Radiol 1987; 17: 59–62.
77. Bodian D. Poliovirus in chimpanzee tissues after virus feeding. Am J Hyg 1956; 64: 187.
78. Bodian D. Histopathological basis of clinical findings in poliomyelitis. Am J Med 1949; 6: 563–578.
79. Hayward JC, Gillespie SM, Kaplan KM, et al. Outbreak of poliomyelitis like paralysis associated with enterovirus 71. Pediatr Infect Dis J 1989; 8: 611–616.
80. Wasserstrom R, Mamaorian AC, McGary CT, Miller B. Bulbar poliomyelitis: MRI findings with pathologic correlation. Am J Neuroradiol 1992; 13: 371–373.
81. Makamato M, Morimoto T, Nagao H, Matsuda H. MRI in poliomyelitis like syndrome. Pediatr Radiol 1992; 22: 533–534.
82. Kornreich L, Dagan M, Grunebaum M. MRI in acute poliomyelitis. Neuroradiology 1996; 38: 371–372.
83. Prusiner SB. Prions and neurodegenerative diseases. N Engl J Med 1987; 317:1571–1581.
84. Masters CL, Harris JO, Gajdusek DC, Gibbs CJ Jr, Bernoulli C, Asher DM. Creutzfeldt–Jakob disease: Patterns of worldwide occurrence and the significance of familial and sporadic clustering. Ann Neurol 1979; 5: 177–188.
85. Brown P, Cathala F, Castaigne P, Gajdusek DC. Creutzfeldt–Jakob disease: Clinical analysis of a consecutive series of 230 neuropathologically verified cases. Ann Neurol 1986; 20: 597–602.
86. Iwasaki Y, Ikeda K, Tagaya N, Kinoshita M. Magnetic resonance imaging and neuropathological findings in two patients with Creutzfeldt–Jakob disease. J Neurol Sci 1994; 126: 228–231.
87. Urbach H, Klisch J, Wolf HK, Gass S. MRI in sporadic Creutzfeldt –Jakob disease: Correlation with clinical and neuropathological data. Neuroradiology 1998; 40: 65–70.
88. De Priester JA, Jansen GH, de Kruijk JR. Wilmink JT. New MRI findings in Creutzfeldt –Jakob disease: High signal in the globus pallidus on T1- weighted images. Neuroradiology 1999; 41: 265–268.
89. Shyu WC, Lee C, Hsu Y, et al. Panencephalitic Creutzfeldt–Jakob disease: Unusual presentation of magnetic resonance imaging and proton magnetic resonance spectroscopy. J Neurol Sci 1996; 138: 157–160.
90. Demaerel P, Heiner L, Robberecht W, Sciot R, Wilms G. Diffusion- weighted MRI in sporadic Creutzfeldt – Jakob disease. Neurology 1999; 52: 205–208.

91. Yee AS, Simon JH, Anderson CA, Sze CI, Filley CM. Diffusion weighted MRI of right hemisphere dysfunction in Creutzfeldt-Jakob disease. Neurology 1999; 52: 1514–1515.
92. Baren GB, Sowell MK, Sharp GB, Glasier CM. MR imaging in a child with Guillain-Barré syndrome. Am J Roentgenol 1993; 161: 161–163.
93. Victor M, Adams RD. Tropical spastic paraparesis and HTLVI infection. In *Principles of neurology*, M Victor, RD Adams, eds. New York: McGraw–Hill, 1993; 1109.

Chapter 5

FUNGAL INFECTIONS

Manoj K. Kathuria and Rakesh K. Gupta

Most of the almost 50,000 species of fungi are beneficial to humankind. They reside in nature and are essential in breaking down and recycling organic matter. Some contribute to the production of food and spirits while others have served humanity by providing useful secondary metabolites such as antibiotics and immunosuppressive drugs (e.g., cyclosporine). Fungi have also been manipulated to study eukaryotic genetics and molecular biology. The organisms cause major problems to humans mainly as phytopathogens, inflicting serious damage to crops. Fortunately, only a few hundred species of fungi have been implicated in human disease, and 90% of human infection can be attributed to a few dozen fungi.

Fungi have been placed in a separate biological kingdom because they differ from bacteria and other prokaryotes in many ways. They are eukaryotic organisms and are unable to photosynthesize. They possess a rigid cell wall containing chitin, mannose, and other polysaccharides. The cytoplasmic membranes contain sterols. They possess true nuclei with a nucleolar membrane and paired chromosomes and divide asexually, sexually, or by both processes.

Manoj K. Kathuria and Rakesh K. Gupta • Department of Radiodiagnosis, Sanjay Gandhi Post-Graduate of Medical Sciences, Lucknow 226014, India.

MR Imaging and Spectroscopy of Central Nervous System Infection, edited by Gupta and Lufkin. Kluwer Academic / Plenum Publishers, New York, 2001.

Fungi grow in two basic forms, as yeasts and molds. Growth in mold form produces multicellular filamentous colonies consisting of branching tubules called hyphae. The mass of intertwined hyphae is a mycelium. Yeasts are small single cells, spherical to ellipsoid in shape and reproduce by budding. Some yeastlike fungi (*Candida*) produce buds that fail to detach and become elongated and are called pseudohyphae. Dimorphic fungi occur as filaments or yeast depending on conditions of growth.

Fungal infection of the brain is invariably secondary to primary focus elsewhere, usually in the lung or intestine. Those fungi generally affecting only the immunosuppresed are aspergillus, candida, and mucor. Other fungal diseases that may occur in both immunocompromised and normal hosts include cryptococcosis (1), coccidiomycosis, histoplasmosis, and blastomycosis (2). Aspergillus, cryptococcus, mucor, and candida are ubiquitous fungi, whereas remainders are endemic to particular geographic regions (3).

Fungal infections of the CNS used to be extremely rare. In the recent past, because of an ever-increasing pool of immunocompromised hosts, with the advent of powerful antibiotics, chemotherapeutic agents, and spread of AIDS, the spectrum of organisms causing CNS infections has changed. Although fungi can cause disease in immunocompetent hosts, they more commonly cause disease in immunocompromised patients. Today, fungal infection is not rare and can be anticipated in specific clinical settings.

We will describe the commonly encountered fungal infections of the CNS, encountered in clinical practice, under (1) cranial and (2) spinal infections.

1. CRANIAL INFECTIONS

The mycotic lesions in the brain and meninges vary in appearance with the fungal forms (4). Fungi growing in infected tissues as yeast cells (cryptococcus, histoplasmosis) spread hematogenously to the microvasculature of the meninges, penetrate the vessel walls, and result in acute or chronic leptomeningitis. Less often parenchymal disease such as granulomas or abscesses is seen (4). These pathologic changes may simulate tuberculosis on imaging (5). Fungi grown as hyphae only (aspergillus, mucor) or pseudohyphae (candida) involve the parenchyma to a greater degree than the meninges, because the larger forms have limited access to the meningeal circulation (4). Therefore, the organism tends to spread to the CNS in retrograde fashion. The source of infection is frequently the sinuses or orbits.

Hyphae form mycelial colonies capable of vascular invasion and obstruction of large, medium, and small arteries, resulting in infarction and cerebritis. Pseudohyphae are adherent yeast cells, larger than individual cells, but smaller than true hyphae. Thus, CNS infection with candida often results in scattered parenchymal microabscesses secondary to small vessel (arteriole) occlusion and tissue breakdown (6). Individual or small groups of yeast cells may also penetrate the meningeal microvasculature and cause meningitis.

We next describe the common fungal infections of the brain.

1.1. Aspergillosis

Aspergillosis is a ubiquitous mold found in soil, water, decaying vegetation, or any location with organic debris (7). *Aspergillus fumigatus* accounts for most human infection out of more than 200 species of aspergillus (8). Infection with *A. flavus* or *A. niger* is also frequently seen.

Intracranial aspergillosis occurs worldwide. There is no racial or sex predilection. Most cases have been reported in adults, although the disease is also seen in children and neonates. Cerebral aspergillosis affects primarily the immunocompromised host. Its prevalence has increased with the practice of intensive chemotherapeutic regimens, use of corticosteroids, and transplantation procedures (9). Intracranial aspergillosis can occur in patients without the usual risk factors (10).

Aspergillosis is one of the most common causes of fungal infections affecting the CNS. Humans are infected by inhalation of *Aspergillus* spores or fragmented hyphae, making the lungs the primary site of infection. Intracranial seeding occurs during dissemination. Cerebral aspergillosis can also occur by direct inoculation at surgery, or with trauma or by extension from the paranasal sinuses or orbit (11). The latter route may resemble the rhinocerebral form of mucormycosis, but it may present as an indolent disease with symptoms lasting for years.

The pathology of CNS aspergillosis can be divided into three forms: infarction, granulomas, and meningitis. Aspergillosis exists only in hyphal form at body temperature (11). It tends to involve the parenchyma rather than the meninges as these hyphal forms have limited access to the meningeal microcirculation. The hyphal elements lodge in intracerebral blood vessels, grow through the vessel walls causing occlusion resulting in infarct, commonly hemorrhagic (12). This sterile infarct is converted to a septic infarct, when the fungus erodes through the vessel wall into the ischemic brain parenchyma, causing a mixed inflammatory reaction and necrosis (12,13). This erosion of the vessel wall results in fungal vasculitis and subsequently mycotic aneurysm. Aspergillosis is the most common cause of mycotic aneurysm (10, 15).

Clinically, most commonly the patients present with stroke of the involved area (14). Headache, seizures with or without fever may be present. Patients with frontal lobe involvement may develop papilledema, visual field defects, and proptosis. Signs of meningeal irritation are rare.

The different neuroimaging patterns in cases of cranial aspergillosis vary depending on the immune status of the patients and can be divided into infarcts, ring or nodular enhancing lesions consistent with abscesses or granulomas, and localized meningitis.

In severely immunocompromised patients with an aggressive form of cerebral aspergillosis, several studies (13,16,17) have reported lesions consistent with infarcts on imaging (Figs.1,2). These patients have a rapidly progressive course with fatal outcome. On T2-weighted MR images, these lesions demonstrate inhomogeneous high signal intensity with a low signal peripheral rim in cases of hemorrhagic infarction (13,16,17). There was no apparent contrast enhancement in most cases, which could be explained by the rapidity of the evolution of the disease with early fatal outcome. Patients who were less severely immunocompromised and who recovered showed peripheral contrast enhancement as expected in an infarction. The finding of cerebral infarction on imaging in a patient with known risk for cerebral aspergillosis should suggest the diagnosis. There are only a few reported cases with ring- or nodular-enhancing lesions (13, 15–19). The presence of ring or nodular enhancement consistent with granuloma or abscess formation indicates that the host defence system is able to isolate or encapsulate the offending organisms (16,19). MRI may reveal a hyperintense lesion with a peripheral rim of low signal intensity surrounded by vasogenic edema on T2-weighted images. The lesions show enhancement in postcontrast T1-weighted images. Examination of the pathology of imaged abscesses demonstrates blood breakdown products and dense population of hyphal elements. The center of the lesion consisted of coagulative necrosis; the peripheral dark rim has been explained to be due to either blood products or paramagnetic metals causing T2 shortening (15).

In a recent study, 18 cases of hematogenously disseminated aspergillosis with brain involvement were reviewed retrospectively to assess the spectrum and frequency of involvement of specific anatomic sites and imaging characteristics (20). Lesions in perforating artery territory were more common than those at corticomedullary junction. The pathologies with hematogenous spread such as abscesses and metastasis have a predilection for corticomedullary junction. Though aspergillus disseminates to the CNS hematogenously, the pathology is often different. Aspergillus may cause an infectious vasculopathy, leading initially to acute infarction or hemorrhage and later extending to surrounding tissue as an infectious cerebritis or evolving into an abscess. The apparent affinity of CNS aspergillosis for perforating artery distribution is most likely due to the invasive character of aspergillus within the walls of larger parent arteries to which it has spread hematogenously,

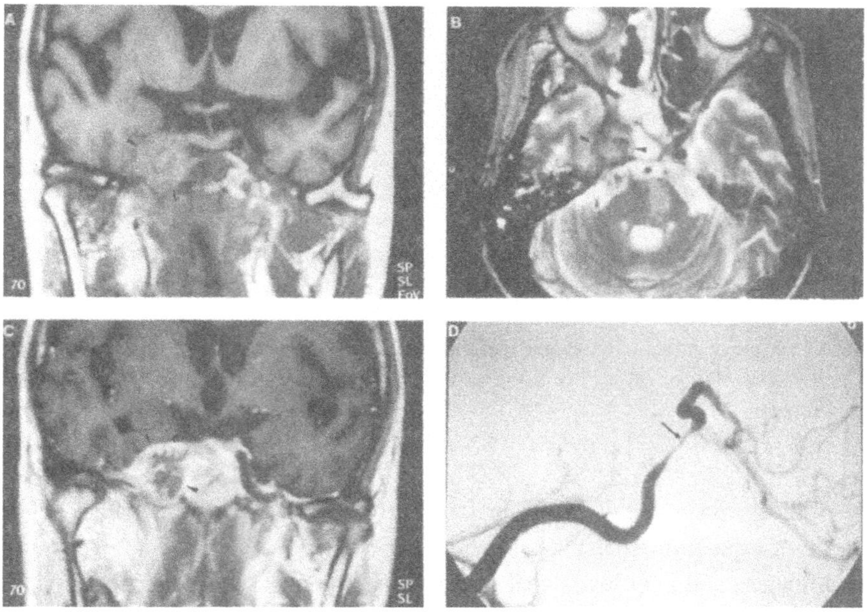

Figure 1. Aspergillosis in an adult patient with diabetes mellitus. Coronal unenhanced T1-weighted MR image (a) shows a large mass of isointensity in the area of the right cavernous sinus (arrowhead). Axial T2-weighted MR image (b) shows that the mass in the right cavernous sinus appears to be of mixed signal intensity. There is localized high intensity suggesting ischemia on the left side of the pons. Note some inflammation within the sphenoid sinus and loss of normal signal void of the right internal carotid artery. Coronal contrast-enhanced T1-weighted MR image (c) shows peripheral enhancement of the mass in the right cavernous sinus (arrowhead) and homogeneous enhancement of the inflammation in the sphenoid sinus as well. Oblique view of right internal carotid angiogram (d) shows severe narrowing of the distal petrous and proximal cavernous segments of the internal carotid artery (arrow). (Courtesy, Dr. KH Chang, National University of Seoul, Seoul, South Korea.)

subsequently compromising the origins of perforating arteries. In addition, the perforating artery may be the first to lose patency because of its narrow diameter in diffuse vasculopathy (21,22).

The survival of patients depends on early diagnosis and prompt initiation of treatment. Histopathologic diagnosis requires invasive procedures, and thus is a critical problem. Kami et al. recently reported early diagnosis of CNS aspergillosis in an immunocompromised patient by diffusion-weighted MRI where conventional CT and MRI were not of much help (23). As these lesions represent septic infarction, diffusion-weighted images may be of value in identifying early lesions in the appropriate clinical setting (23).

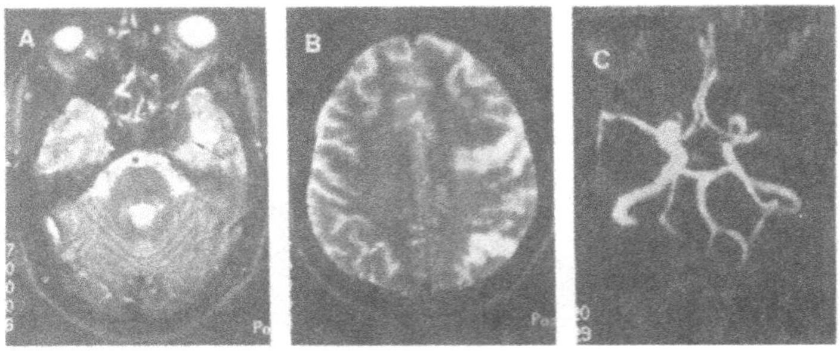

Figure 2. Aspergillosis with major vascular territory infarction. T2-weighted axial image (a) at the level of pons shows a hypointense area (arrow) in the left parasellar region with hyperintensity in the adjoining left temporal lobe. T2-weighted axial image (b) at the supraventricular level shows evidence of hyperintense areas in the frontal and parietal lobes consistent with left middle cerebral artery infarct. MR angiogram (c) shows occlusion of the left middle cerebral artery. The culture from the mass grew *Aspergillus flavus*.

Aspergillosis may spread to the brain from the paranasal sinuses (Fig.3). The MR appearance of aspergillus sinusitis has been described as decreased signal intensity on T1-weighted images, and very low signal intensity may be mistaken for air in the sinuses on T2-weighted images. Increased concentration of iron, manganese, magnesium as well as increased calcium content has been detected in the washings from the sinuses (24). It has been hypothesized that low signal on T2-weighted images is probably due to T2 shortening by these elements (24). Another report suggests that fibrous connective tissue present in the granuloma is the cause of decreased signal, as fibrous tissue has a short T2 value (25). Localized meningitis has been reported in association with initial involvement of paranasal sinuses and/or the orbits (10,13).

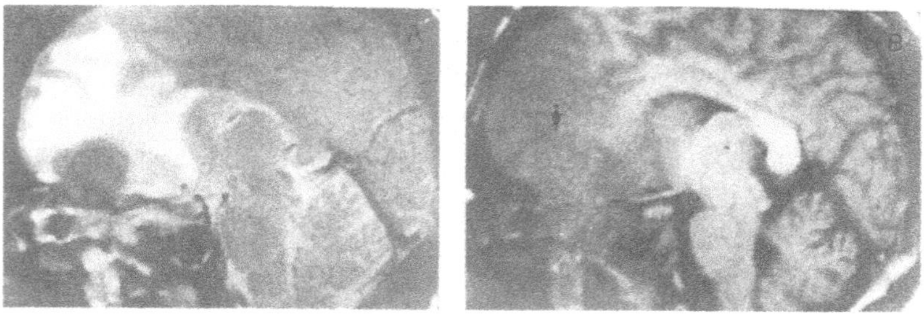

Figure 3. Rhinocerebral form of *Aspergillus* infection. T2-weighted right parasagittal (a) image shows the mass as isointense with the white matter with marked edema in the frontal lobe. The mass in the sphenoid and ethmoid sinuses shows hyperintense signal suggesting sinusitis. Right parasagittal T1-weighted image (b) shows a mass in the subfrontal region, isointense with brain parenchyma. Note the buckling of the white matter (arrow).

Proton MR spectroscopy has not been reported in aspergillus granuloma. We have performed proton MR spectroscopy in a case of aspergillus granuloma involving the left parieto-occipital region. It showed high choline, low creatine and lactate with no NAA. The spectroscopy was nonspecific and did not help in its differentiation from other neoplastic and nonneoplastic mass lesions (Fig.4).

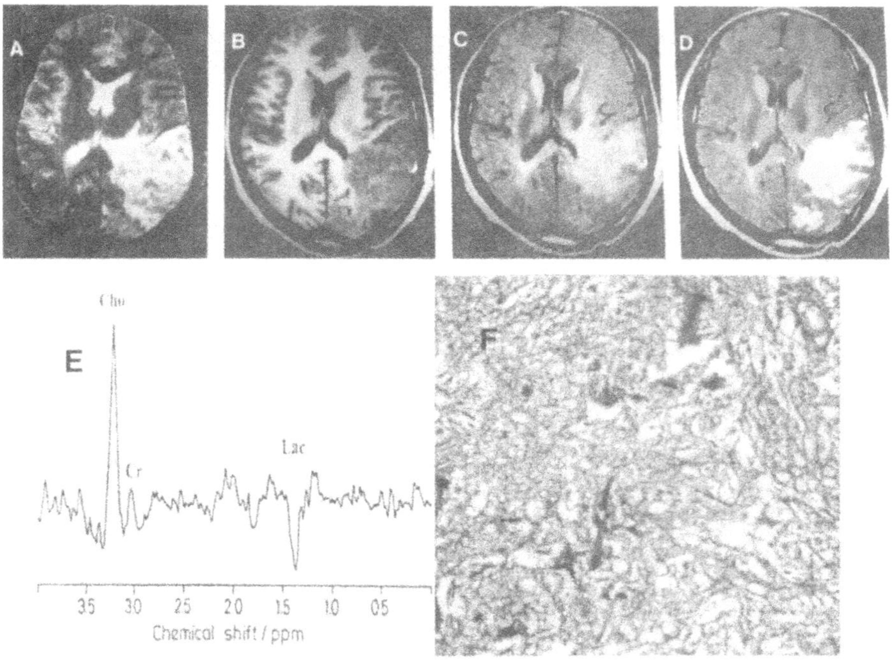

Figure 4. Aspergillus granuloma. T2-weighted axial image (a) at the level of ventricles showing a heterogeneous intensity lesion with focal hypointense areas in the left parieto-occipital region. Mass is predominantly hypointense on T1-weighted (b) and hyperintense on MTT1-weighted image (c). Postcontrast T1-weighted image (d) shows intense enhancement of the mass lesion. MR spectroscopy (e) shows raised choline, low creatine along with presence of lactate and marked decrease in NAA. Histopathology (f) reveals a fungal granuloma with hyphae clearly shown on special staining for fungus. Culture of the tissue was positive for *Aspergillus flavus*.

1.2. Cryptococcosis

Cryptococcus is the most common cause of CNS fungal infection in humans. It is a ubiquitous monomorphic fungus that primarily exists as encapsulated yeast, and reproduces by budding. A polysaccharide capsule surrounds the fungus. The physiologica stimulus to capsule enlargement is the relatively high carbon dioxide tension found in lungs and mammalian tissues. It is found in soil and occurs in high concentration in feces of pigeons and chickens. It has also been recovered from skin and mucous membranes of noninfected humans.

Men are more commonly affected than women by cryptococcal infection. In the literature, 80% of the population affected by cryptococcal infection are males (11). This trend was established long before the HIV epidemic, and holds true for both cryptococcal meningitis and cryptococcomas. The reasons are unknown, although occupational exposure to soil may be partly responsible. Cryptococcosis is nearly always an opportunistic organism and is only rarely found in immunocompetent patients. It is the most commonly identified fungal infection in the CNS in AIDS patients (26,27).

The cryptococcal infection results from inhalation or ingestion of fungus with the respiratory tract being the primary site of infection. Once established in the lungs, it may spread to the intrathoracic lymph nodes and then disseminate to seed the meninges, in a fashion similar to mycobacterium tuberculosis. Here it penetrates the meningeal vessel wall and colonizes the perivascular Virchow–Robin spaces and the CSF, producing meningitis. In the perivascular space, sheaths of fungi may lead to pseudocyst formation, typically found in basal ganglia and in the subcortical areas. This localization corresponds to the terminal part of these spaces where the pial lining meets the vessel wall. The proteolytic enzymes produced by the fungus may destroy the pial lining of the Virchow–Robin spaces so that the cryptoccoci penetrate the parenchyma causing meningoencephalitis (28). Another location is the choroid plexuses, where cryptococcus may cause formation of a voluminous mass (29).

Most patients with cryptococcal meningitis have an underlying immune disorder. In contrast, only about 5% of patients with cryptococcoma (with or without concurrent meningitis) have underlying immune disorder (21). Clinically, headache is the most common and sometimes the sole symptom. Clinical presentation may also include fever, altered mental state, stiff neck, seizures, and signs of raised intracranial pressure.

Cryptoccocal disease may be diagnosed more readily than most other systemic fungal diseases for three reasons. The infection has a unique appearance on histopathology, the yeast can be seen in the CSF, and the immunologic test for antigen in CSF and blood is highly sensitive and specific.

Imaging findings in cryptococcal meningitis are unremarkable. CT scan is mostly normal. Positive CT findings are nonspecific and include cerebral atrophy and communicating hydrocephalus (30–32). Meningeal enhancement on contrast-enhanced CT is very uncommon, as cryptococcus produces only mild inflammatory reaction in the subarachnoid space. Enlargement of the optic nerve sheath in cryptococcus meningitis either secondary to raised intracranial tension or due to direct optic nerve invasion has also been reported (33).

MRI may be negative or may uncommonly show meningeal enhancement (Fig.5). There have been reports in which a single or double dose or delayed scanning has shown meningeal enhancement in cryptococcal meningitis. Imaging findings in cryptococcal meningitis may be dependent on the immune status of the host. In immunocompromised patients with CNS cryptococcosis, hydrocephalus is uncommonly seen (31). In contrast, non-AIDS patients show hydrocephalus in approximately 24% of cases (33). The lower frequency of hydrocephalus is attributed to lack of inflammatory leptomeningeal reaction in immunocompromised hosts causing paucity of adhesions within the basal cisterns.

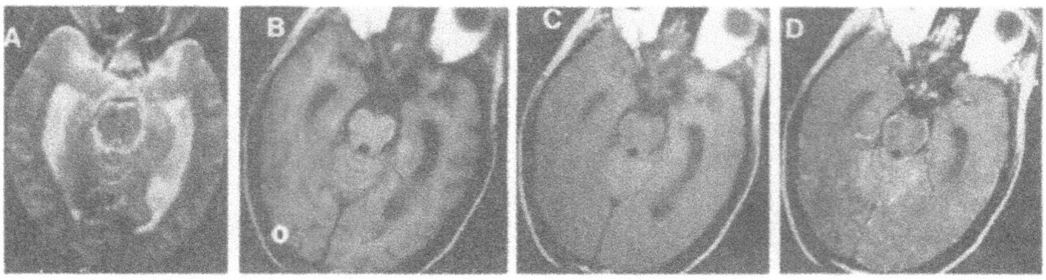

Figure 5. Cryptococcal meningitis in an immunosuppressed patient. T2-weighted (a), T1-weighted (b), and MT-T1-weighted (c) axial images at the level of upper pons do not show any obvious meningeal abnormality. Note the mildly dilated temporal horns of lateral ventricles. Postcontrast MTT1-weighted image (d) shows evidence of enhancing meninges around the pons. The MT ratio from the meninges was 34%.

A number of MR appearances have been described in CNS cryptococcosis in AIDS patients: (1) dilated Virchow–Robin spaces (Fig.6), (2) cryptococcoma, (3) miliary enhancing nodules in the parenchyma and leptomeningeal cisternal spaces (Fig.7), and (4) a mixed pattern (Figs.8,9) (34). These changes reflect the pathology of the disease. Dilated Virchow–Robin spaces filled with mucoid material, inflammatory cells, and organisms manifest on MRI as punctate hyperintense round–oval lesions on T2-weighted images and hypointense on T1-weighted images (Fig.6). These are usually less than 3 mm in size with a propensity for the basal ganglia and the brainstem. These foci do not enhance with gadolinium and are not associated with mass effect or edema.

The mass lesions or cryptococcomas have a variable appearance depending on the relative constituents of lesions. The gelatinous pseudocysts or nonreactive form of meningoencephalitis represent sequelae of cryptococcus colonization along the Virchow–Robin spaces or in the large perivascular spaces of the choroid plexuses (Fig.8). These cysts rich in fungi are filled with white gelatinous material and are located in basal ganglia and perivascular areas. They are usually hyperintense on T2- and hypointense on T1-weighted images and may show contrast enhancement on administration of Gd-DTPA.

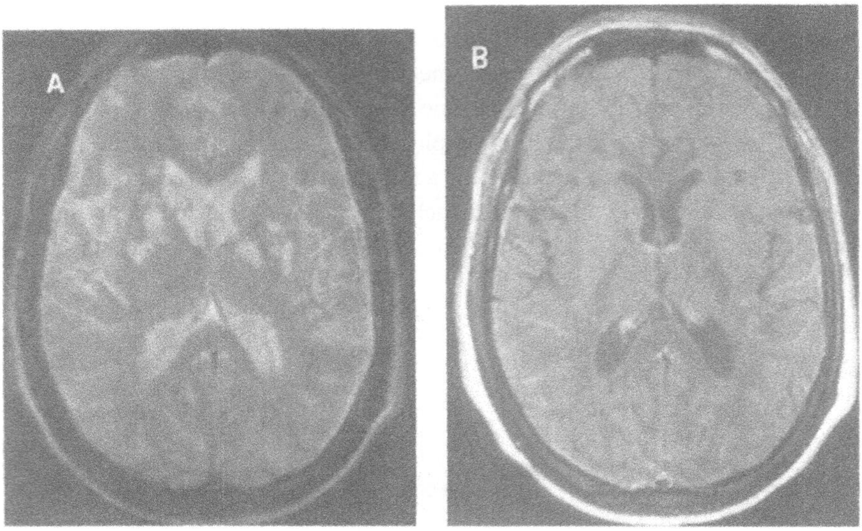

Figure 6. Cryptococcosis with dilated Virchow–Robin spaces. T2-weighted image (a) shows bilateral basal ganglia hyperintensities, which do not reveal any enhancement on postcontrast T1-weighted images (b).

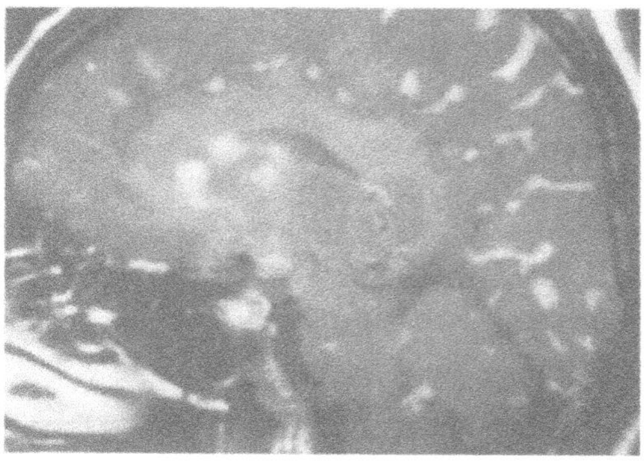

Figure 7. Cryptococcosis with cryptococcomas in an immunosuppressed patient. Parasagittal contrast-enhanced T1-weighted MR image shows diffuse corticosulcal enhancement suggesting diffuse leptomeningeal inflammation and multiple nodular enhancement in the basal ganglia suggesting cryptococcomas. (Courtesy Dr. K.H. Chang, National University of Seoul, Seoul, South Korea.)

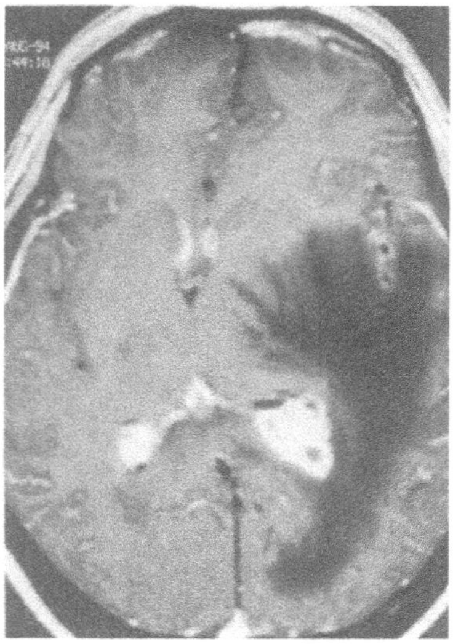

Figure 8. Cryptococcosis with choroid plexitis in an immunocompetent patient. Axial contrast-enhanced T1-weighted MR image shows massive edema of low intensity in the left periventricular area associated with marked abnormal enhancement of the enlarged choroid plexus of the left lateral ventricle. There is also diffuse corticosulcal enhancement suggesting diffuse leptomeningeal inflammation. (Courtesy Dr. K.H. Chang, National University of Seoul, Seoul, South Korea.)

The lesions that are hyperintense on T2-weighted images and are enhanced after contrast administration characterize the granulomatous type of meningoencephalitis (Figs.7,9). These granulomas are preferentially located in the ependyma of the choroids plexus and may cause localized ventricular dilatation (Fig.8).

Initial reports on MT spin-echo imaging and quantitative magnetization transfer ratio (MTR) have shown that it may be of value in differentiating fungal meningitis from other chronic meningitis (tubercular) as well from viral meningitis (35) (Fig.5).

There are isolated reports of proton MR spectroscopy in cases of intracerebral cryptococcal abscess and cryptococcomas. *In vivo* proton MR spectroscopy shows marked increase in lactate along with decrease in NAA, choline, and creatine (36). The presence of NAA, creatine, and choline appears to be due to partial volume effect from the adjacent brain parenchyma. *Ex vivo* proton MR spectroscopy has shown the presence of lipid/lactate, and alpha-alpha trehalose sugars in the regions of 3.42, 3.62, 3.83, and 5.17 ppm. This represents the polysaccharide capsule of the cryptococcal yeast (37) (fig .10).

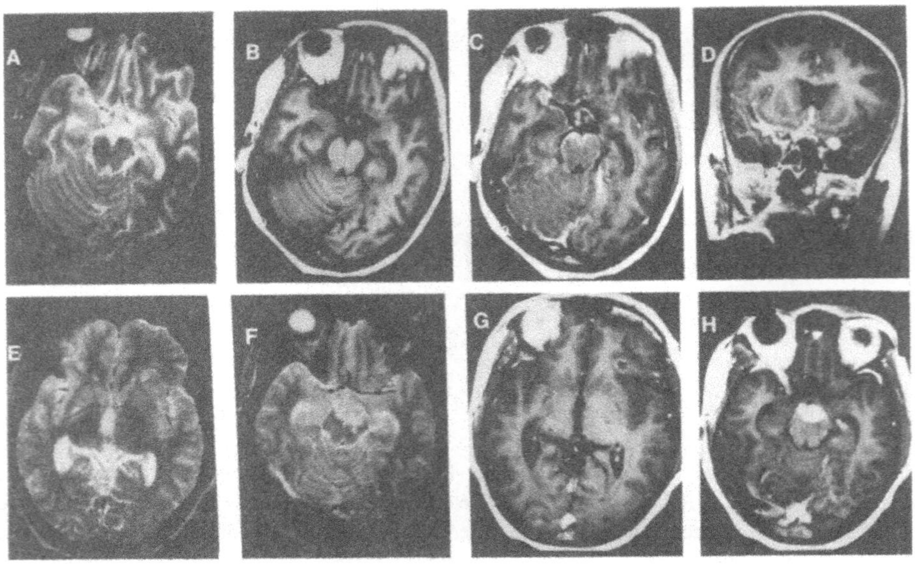

Figure 9. Cryptococcal meningitis with cryptococcomas in immunocompetent hosts. T2-weighted axial image (a) shows no obvious abnormality, and T1-weighted image (b) shows mild dilatation of the cisterns and the ventricular system. Postcontrast T1-weighted axial (c) and coronal (d) images show evidence of thick meningeal enhancement in the interpeduncular fossa, perimesencephalic cisterns and right Sylvian fissure along with an enhancing cryptococcoma in the left Sylvian fissure. In another patient, T2-weighted axial images at the level of basal ganglia (e) and upper pons (f) show evidence of a hyperintense lesion in the prepontine region and a small, mildly hyperintense area in the left frontal region. Postcontrast images (g, h) show enhancement of the cryptococcoma and thick enhancement of the prepontine lesion.

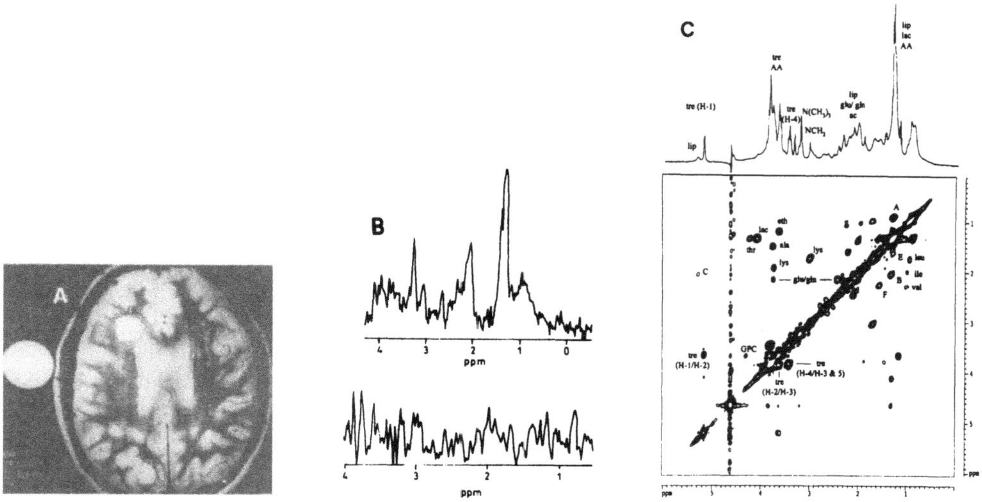

Figure 10. MR spectroscopy in a proven case of cryptococcal meningitis with cryptococcoma. T2-weighted axial image (a) showing two hyperintense areas in the right frontal (inset) and occipital regions. *In vivo* proton MR spectroscopy at TE 30 ms (b) shows broad resonances at 1.3, 2.1, and 3.4 ppm. At TE 270 ms, most of the resonances seen at TE 30 ms have disappeared suggesting metabolites with short T2 values, probably lipids. (Courtesy Drs. L. Chang and T. Ernst, Harbor University–University of California Los Angeles Medical Center, UCLA School of Medicine.) *Ex vivo* two-dimensional correlated spectroscopy (c) of a solid cryptococcoma grown in an animal model shows lipids, lactate (lac), amino acids (AA), trehalose (tre). (Courtesy Drs. U. Himmelreich, T. Sorrell, and C. Mountford, University of Sydney, Australia.)

1.3. Mucormycosis

Mucormycosis is a rare and perhaps the most aggressive fungal infection of humans.

CNS mucormycosis is a phacomycosis caused by the genus *mucor*. These common bread and fruit molds thrive in soil and vegetable matter. They can be cultured from the throat, nose, and stool of healthy individuals without existence of disease. The fungi exist as molds and form hyphae, which may become invasive, involve blood vessels, and disseminate hematogenously or extend directly into the CNS via the paranasal sinuses and orbits.

The fungus infects immunocompromised patients. Diabetics comprise at least 70% of reported cases and less than 5% of infections occur in normal hosts (11,38,39). Acidosis rather than hyperglycemia appears to be the important factor predisposing to mucormycosis (40). Diabetics uniformly develop the rhinocerebral form of the disease. Other patients who are acidemic from chromic diarrhea, chronic renal failure (CRF), or extensive burns are also prone to the rhinocerebral form. Patients may also develop CNS mucormycosis following direct trauma or with intravenous drug abuse (21).

The rhinocerebral form occurs when sinus infection extends into the CNS. The organism may spread directly through the cribriform plate, via retrograde proliferation along the vessels or via extension into the orbit and then through the optic canal or superior orbital fissure into the cavernous sinus (21,22,41–43). Mucor like pathogenic molds invade arteries causing thrombosis and infarction (22). MRI of rhinocerebral mucormycosis has shown nonspecific inflammatory changes involving the paranasal sinuses and orbits as well as the pattern described for *Aspergillus* infection (23). Low signal intensity involving paranasal sinuses on all imaging sequences has been described. This low intensity is unusual in benign sinus secretions or bacterial or allergic acute sinusitis. It is presumed to be due to a high concentration of paramagnetic metals within the infectious mass.

Hematogenous dissemination to the CNS is referred to as isolated cerebral mucormycosis and is less common than the rhinocerebral form. The usual source of infection is a focus in the lung. It is more common in patients with hematological malignancies and intravenous drug abusers. Basal ganglia are very commonly involved in patients who are intravenous drug abusers. A study reported involvement of basal ganglia in 82% of intravenous drug abusers while non-drug users showed basal ganglia involvement in only 9% of cases (21).

Involvement of cavernous sinuses, internal carotid and basilar artery in the form of abnormal vascular signal and abnormal enhancement has been shown in rhinocerebral mucormycosis secondary to thrombus formation in the vessels (39,40). Signal intensity pattern suggestive of infarcts has been reported in large vessel, small vessel as well as watershed territories.

Involvement of brain parenchyma by the fungus can be separated from bland infarction caused by the vascular involvement due to the organism on MRI (Fig.11). Parenchymal changes secondary to mucormycosis may or may not correlate with a vascular distribution (41). Contrast-enhanced MRI may help in distinguishing between these two possibilities.

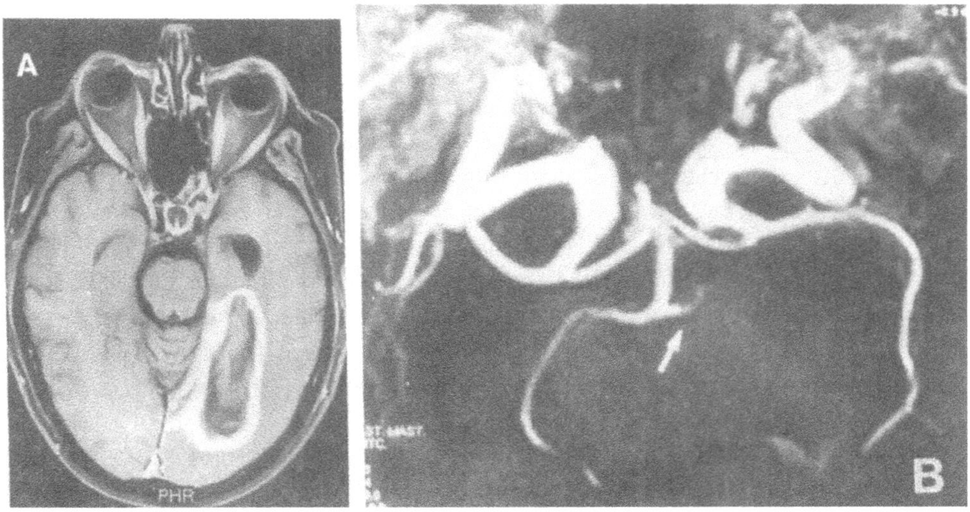

Figure 11. Cerebral mucormycosis. A case of lymphoblastic leukemia and receiving treatment with amphotericin for pulmonary mucormycosis. Postcontrast T1-weighted image (a) shows left temporal horn entrapment and a thick ring-enhancing area consistent with septic infarction. MR angiogram (b) shows occlusion of the left posterior cerebral artery. (Reprinted from J Neurol Neurosurg Psychiatry 1999, Vol. 66, p404-405, Birchall D, et al. with permission from BMJ Publishing Group.)

MR findings in isolated cerebral mucormycosis have been reported in very few cases (42,43). Basal ganglion lesions with increased signal on T1-weighted images suggesting hemorrhage have been reported. T2-weighted images revealed hyperintense signal with vasogenic edema. Survival is mostly poor and occurs with aggressive antifungal treatment and surgical debridement (22).

1.4. Blastomycosis

Blastomycosis is an uncommon fungal infection seen in endemic regions of the United States (42) and has been reported in Africa. Even in endemic regions, the incidence is low. The causative agent, *Blastomyces dermatitides,* has been isolated from soil, which appears to be the reservoir for human infection. Low soil pH promotes growth of the organism. Infection is most common in men particularly those over 50 years of age, mainly agricultural workers.

B. dermatitides is a dimorphic fungus that causes suppurative and granulomatous lesions in the lung, skin, bone, genitourinary tract, and CNS. It grows as a mycelial form at room temperature and as a yeast form at body temperature. It is the yeast form that disseminates hematogenously, causing disease (22).

Blastomycosis is a systemic disease with a variety of clinical presentations. Pulmonary disease may be active with or without extrapulmonary manifestations. CNS infection is the result of either direct extension from sinus or orbital infection or hematogenous spread. Blastomycosis involving the CNS is seen in 5% of cases and up to 25% of cases of disseminated disease. The rate of CNS blastomycosis may be higher in patients with AIDS (11). Solitary or multiple intracranial abscesses, cranial and spinal extradural abscesses with associated osteomyelitis, and meningitis, characterize it. Meningitis is the most common form of CNS blastomycosis, which usually occurs late in the course of the disease. It may also present as single or multiple abscesses or granulomas (21, 22). CNS blastomycosis is almost invariably associated with involvement of other organs.

CT and MR imaging findings are nonspecific. CT may demonstrate a hypodense or hyperdense mass, which on contrast-enhanced scans reveals variably enhancing lesion with surrounding vasogenic edema. MR imaging features are also nonspecific, and the literature is comprised of a few sporadic case reports. MR imaging of a reported midbrain blastomycosis on postcontrast T1-weighted MRI revealed homogeneous enhancing lesions with involvement of nasopharynx and paranasal sinuses (43). Dural enhancement may also occur from direct extension from paranasal sinuses. A mass lesion masquerading as a suprasellar neoplasm has also been reported. It showed a T1 isointense lesion that enhanced after administration of contrast, with a few punctate foci of low-intensity signals, and was hyperintense on T2-weighted images (44). The MR appearance of parenchymal blastomycosis abscess is indistinguishable from other causes of abscesses (Fig.12).

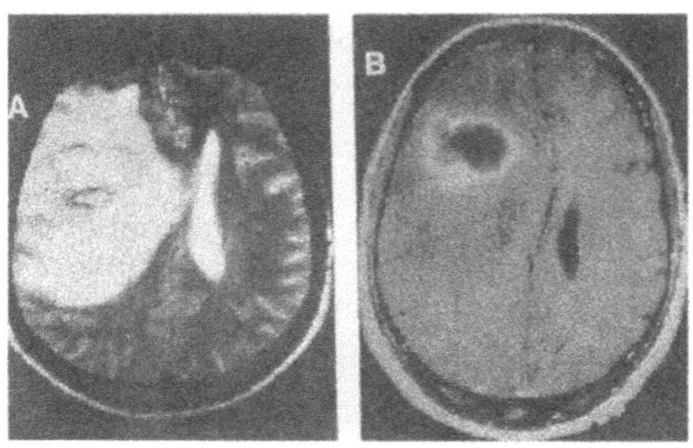

Figure 12. Blastomycosis. T2-weighted axial image (a) shows a hyperintense area with a peripheral hypointense rim with perifocal edema in the right frontal region. Postcontrast T1-weighted image (b) shows rim enhancement of the lesion consistent with an abscess. (Courtesy Drs. Suresh Mukherji and M. Castillo, North Carolina.)

Isolating the organism is required for a definitive diagnosis of blastomycosis. To date, no reliable serologic tests have been established, although several reports of newer tests appear promising (11).

1.5. Coccidiodomycosis

Coccidiodomycosis is a highly infectious dimorphic fungus that is endemic to semiarid regions. The ecologic distribution is limited to the southwestern United States and parts of Mexico and Central and South America. However, it is frequently being seen elsewhere because of frequent travel into and outside these regions.

Coccidiodomycosis results from inhalation of spores of the fungus, which cause a variable pulmonary response in affected individuals. Most cases of pulmonary coccidiodomycosis are asymptomatic and have a self-limiting respiratory illness. Rarely it may disseminate to other organs including the CNS. Dissemination occurs in 0.2 to 1.0% of patients and is much more common in immunocompromised patients (45,46).

CNS involvement is the most serious consequence of disseminated coccidiodomycosis. It may manifest as basal granulomatous meningitis or as microabscesses. Meningitis may occur with or without underlying cerebritis. Meningitis alone occurs in 13% of disseminated cases and up to 50 to 90% of patients with CNS disease (44). Meningitis with concurrent cerebritis is seen in 78% of cases. About 9% present with miliary granulomas resembling tuberculosis (47). It may be complicated by ventriculitis and communicating hydrocephalus. Focal abscesses have been reported in HIV patients although this is an uncommon finding. Vasculitis has been observed in 40% of cases of meningitis; however, this typically involves the small penetrating branches of major cerebral vessels, resulting in deep white matter infarctions (48). Vascular occlusion of major cerebral vessel is rare (49). Subarachnoid hemorrhage secondary to rupture of mycotic aneurysm has rarely been reported as a fatal complication of coccidiodomycosis meningitis (50,51).

On MRI, typically coccidiodomycosis meningitis presents as leptomeningeal enhancement, which may be intense, greatest in the basal, Sylvian, and interhemispheric cisterns (52). The leptomeningeal enhancement may also extend to the spinal canal (53). Occlusion of the anterior spinal artery due to meningeal inflammatory process has been reported (52). Abnormal T2 hyperintense signal seen most commonly in the cerebral deep white matter secondary to ischemia possibly from vasculitis is also seen (22,52). Entrapment of any portion of the ventricular system may occur secondary to ependymitis, resulting in obstructive hydrocephalus. Obstructive hydrocephalus is also seen secondary to meningeal inflammatory reaction. Less commonly focal white matter or deep gray matter, enhancing lesions are identified (Fig.13), and these represent granulomas. Focal abscesses have been reported in HIV patients, although this is an uncommon finding (22).

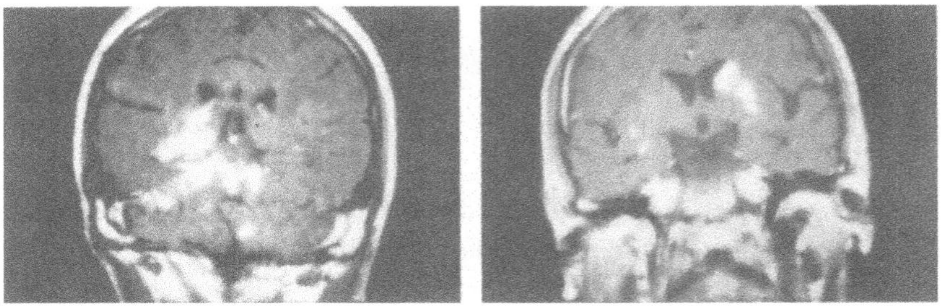

Figure 13. Coccidiodomycosis granulomas. Postcontrast T1-weighted coronal images at the level of ventricles show enhancing lesions in the right periventricular region. This was subsequently confirmed as coccidiodomycosis. (Courtesy Drs. Suresh Mukherji and M. Castillo, North Carolina.)

1.6. Histoplasmosis

Histoplasmosis is a systemic fungal infection caused by *Histoplasma capsulatum*. The disease is endemic in river valleys of the Ohio, the Mississippi, and the St. Lawrence. It is also endemic in North Carolina and Virginia (54). Most adults living in endemic areas have been shown to be infected, by skin test surveys. There is a bimodal incidence of the disseminated disease; about half of the cases occur in infants and young children, and the remainder in people past the third decade of life (11).

Entry occurs by inhalation of airborne spores and most cases of histoplasmosis are limited to the lungs. Pulmonary infection presents clinically as either an acute infiltrative form or a chronic inflammatory process. Acute infection is usually asymptomatic or manifests as a flulike syndrome. The chronic cavitating form is clinically and radiologically similar to tuberculosis. Hematogenous spread is rare in immunocompetent hosts. In immunocompromised patients, dissemination can occur to any organ in the body, most commonly the liver, spleen, lymph nodes, and bone marrow (21).

CNS involvement in histoplasmosis is seen in 7.6% of all cases and in 24% of patients with disseminated disease (55). Nearly half of the cases are subclinical meningitis (56) and many of the others are limited to meningitis without detectable focal lesions. Three types of CNS involvement based on pathologic observations have been described (57): (1) miliary granulomas in brain, meninges, and choroid plexus; (2) meningitis and associated vasculitis usually involving the basilar portion of the brain; and (3) scattered focal lesions of cerebritis. Goodwin et al. have added a fourth category of spinal cord involvement (55). A granuloma large enough to present as intracerebral mass is known as histoplasmoma (58).

Imaging features of CNS histoplasmosis are nonspecific. Contrast-enhanced CT scan in patients with intracranial mass lesions caused by histoplasmosis demonstrates lesions of variable density and nonspecific ring enhancement with surrounding edema (58,59). With MR imaging the lesions are usually of low intensity on T1-weighted images and show high signal on T2-weighted images with a variable degree of surrounding edema and ring enhancement. A low signal rim may be present on T2-weighted images that may be due to presence of paramagnetic free radicals in macrophages in the walls of abscess capsules (60). The lesions may occur anywhere in the brain but mainly at the gray–white matter junction. Lesions in thalamus (60) and spinal cord have been described in the literature. Diffuse leptomeningeal enhancement may be present (60,61). A report on a solitary case of biopsy-proven choroid plexus histoplasmosis showed an enlarged hyperdense choroid plexus on CT (62). MRI in this case showed an enlarged choroid plexus that was isointense on T1-weighted and hypointense on T2-weighted images. Gadolinium-enhanced T1-weighted images revealed peripheral enhancement and mottled central enhancement. These imaging findings are different from choroid plexitis caused by bacteria and other fungi, which show marked increase in size and contrast enhancement with extensive edema of the surrounding parenchymal tissue.

1.7. Candidiasis

Candida species are small round yeasts that reproduce by budding. *Candida albicans* accounts for most of the cases of candidiasis. Candida species are part of the normal intestinal flora. In most of the clinical cases, candida acts as an opportunist and produces a disseminated illness. Invasive candidiasis occurs only in patients with some form of immunodeficiency or immunocompromise. Antibiotic-induced alteration in the balance of gut flora allows seeding of candida from the intestinal lumen into the bloodstream. Disruption of normal skin barriers by indwelling catheters, burns, or knife or gunshot wounds may also predispose to infection by candida (21,45,63). There is an increased incidence of both disseminated candidiasis and CNS involvement in newborns, especially premature infants (64).

Candida may involve the CNS in many ways. CNS infection is reported in 18–52% of patients with disseminated spread (22). Most commonly multiple microabscesses are seen, typically in the distribution of middle cerebral artery (65). Noncaseating granulomas are also found but are less common. Full sized abscesses are seen in about 15% of cases (66). Vascular involvement by direct involvement of vessel wall and mycotic aneurysms as seen with molds such as aspergillus and mucor, is also seen in candida infection. We have observed candida involving the paranasal sinuses and extending to brain parenchyma with multiple abscess formation (Fig.14).

On CT, the microabscesses appear iso- or hypodense on nonenhanced studies and show multiple punctate enhancing nodules on contrast-enhanced scans (67). Granulomas may appear as hyperdense nodules on CT (67) with surrounding nodular or ring enhancement (68). CT often underestimates the extent of pathology in patients with widespread fungal disease (69). A few cases of small lesions representing microabscesses or noncaseating granulomas at gray–white junctions have been reported on MRI. As with other pathologies that have hematogenous dissemination the location may be secondary to more prominent blood flow at gray–white interfaces. A "target" appearance may be seen on MRI (67). These lesions show a central core of hypointense signal surrounded by hyperintensity representing edema on T2-weighted images (69). MR may also demonstrate evidence of meningitis, abscess, vasculitis, and infarction but is not able to differentiate the type of fungal disease (Fig.14).

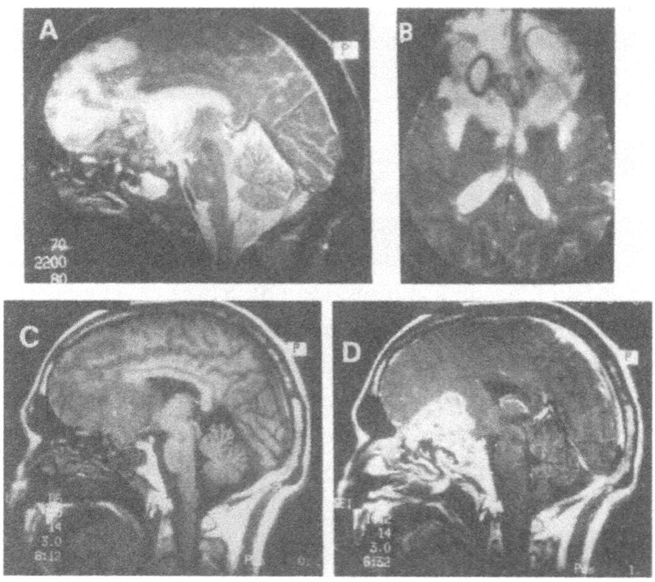

Figure 14. Rhinocerebral candidiasis. Sagittal T2-weighted image (a) shows mixed intensity area in the basal frontal region with edema. Axial T2-weighted image (b) shows multiple hyperintense areas with hypointense rims in the bilateral frontal lobes. Sagittal T1-weighted image (c) shows the lesion to be isointense to the gray matter. Postcontrast T1-weighted image (d) shows heterogeneous enhancement of the lesion along with enhancing sphenoid and ethmoid sinuses. Pus culture from the brain grew *Candida albicans*.

2. SPINAL INFECTIONS

Spinal infections caused by fungi and funguslike organisms are rare. They have been reported with candida, aspergillosis, coccidiomycosis, and maduromycosis (70–74). Routes of contamination vary from direct implantation by external or internal injuries to hematogenous spread from primary infection of lungs or skin (75). Immunosuppresed patients are at higher risk than immunocompetent individuals.

Some fungal species are distributed worldwide, whereas others are endemic in certain geographic locations. A few organisms (blastomycosis, aspergillosis) may produce radiologic findings that are indistinguishable from those of tuberculosis, such as bone destruction, abscess formation, and gibbus deformity; others (cryptococcosis, actinomycosis, coccidiodomycosis, mycetoma) affect the vertebral bodies by producing patchy destruction or sclerosis but will spare the disks (76).

As these infections are rare, little literature is available on MR appearance of fungal spine infections. Candida and aspergillus spondylitis are characterized by low signal intensity on T1-weighted and high signal intensity on T2-weighted images with intervening disk (70,71). The bone marrow in the affected vertebral bodies was noted to be of low signal intensity on both T1- and T2-weighted images suggesting an underlying fibrotic process compared to the more aggressive pyogenic infections, reflecting the indolent nature of infection. In some cases, hyperintensity of the vertebral disk may be absent and in cases where disk height is maintained, the intranuclear cleft is preserved (71). The absence of T2 hyperintensity may reflect an absence of fungal invasion, an altered inflammatory reaction within the disk or an intrinsic characteristic either of the disk (degenerative changes) before infection or of the invading fungi (paramagnetic substances) (71). These variations in imaging findings are probably due to host immune response (70,71). In some cases paraspinal abscesses may also be found in association with spondylitis. Epidural abscess may accompany spinal aspergillus infection in 20% of the cases (77). Coccidiodomycosis spondylitis may be unifocal or multifocal and usually involves intervertebral discs, bone marrow, and adjacent epidural space and soft tissues. MR imaging appearance is variable and nonspecific. The imaging pattern most commonly observed is disc involvement with extensive heterogeneous marrow and soft-tissue involvement. These involved structures appear as hyperintense on T2- and hypointense on T1-weighted images. Postcontrast study may show heterogeneous enhancement. Extraosseous involvement is typical and usually extensive. The abnormal extensive soft tissue involvement is commonly observed along with disk and extensive heterogeneous marrow involvement with lack of bony deformities (Fig.15) are the features that help to differentiate coccidiodomycosis from other causes of infective spondylitis (73). MR imaging screening of the entire vertebral column often reveals occult areas of involvement. The findings resembled tuberculosis with regard to the large paraspinal mass, posterior, lateral element disease, abnormal disk enhancement, and multifocal involvement.

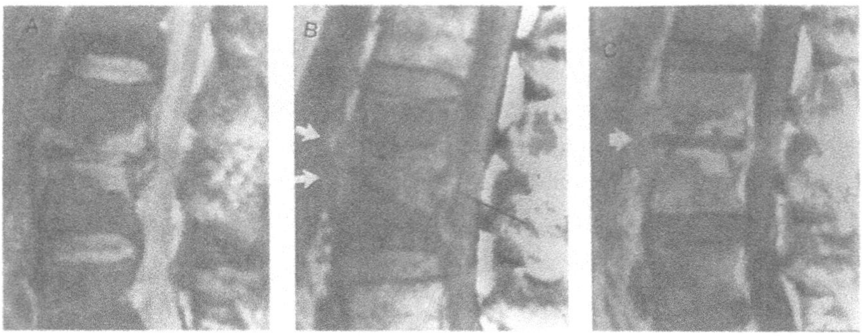

Figure 15. Coccidiodomycosis spondylitis. Sagittal T1-weighted image (a) shows a hyperintensity within disc and adjacent vertebral body endplates with bone marrow edema in third and fourth lumbar vertebral bodies. Also seen are the anterior (white arrows) and posterior subligamentous spread and anterior epidural disease (black arrow). Sagittal T2-weighted image (b) shows narrowing of intervertebral disc space with high signal intensity within the disc and adjacent vertebral bony endplates. Postcontrast T1-weighted sagittal image (c) shows enhancement of the involved vertebrae, including region adjacent to the disc. Note the mild enhancement of the subligamentous phlegmon (arrow). (Reprinted with permission from Olson EM, Duberg AC, Herron LJ, Kissel P, Smilovitz D. Coccidiodal spondylitis: MR findings in 15 patients. Am J Roentgenol 1998; 171: 785–789.)

Blastomycosis spondylitis is endemic in the river valleys of United States and resembles tuberculosis. It shows thoracolumbar predilection, anterior vertebral body destruction and collapse, posterior element and disk space involvement, and paravertebral masses (78,79). This and many other fungal infection also involve the ribs (78).

Spinal cord involvement with histoplasmosis may appear as subtle decreased intensity on T1-weighted images and increased intensity on T2-weighted images (59). Postcontrast study may show focal enhancement (Fig.16).

Fungal meningitis of the spine is exceedingly rare and indistinguishable from those of other meningitides (80). Intramedullary lesions caused by fungi are rare. A case of spinal intramedullary abscess caused by *C. albicans* has been reported (81). The imaging appearance was similar to intramedullary abscesses caused by other infective agents.

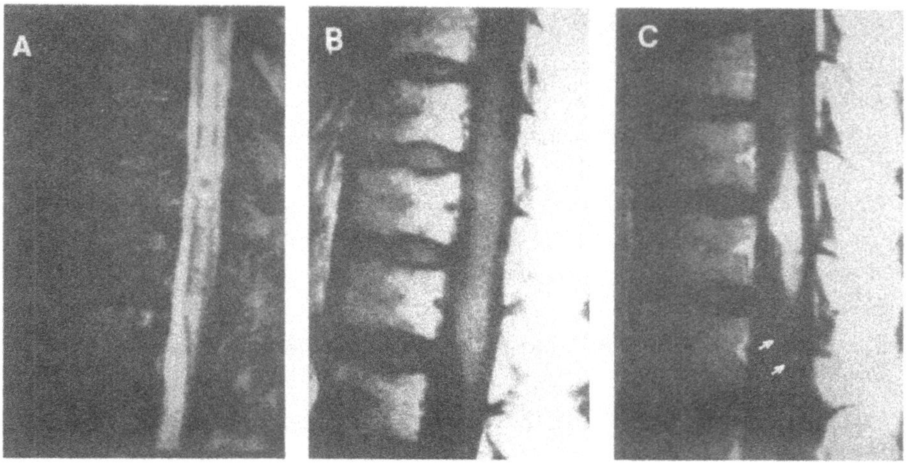

Figure 16. Intramedullary histoplasmosis. Sagittal T2-weighted image (a) shows diffuse increased signal within lower thoracic cord and conus. Sagittal T1-weighted image (b) shows minimal cord irregularity. Postcontrast T1-weighted image (c) shows large enhancing lesion at T11 level with two satellite enhancing lesion in conus (arrows) (Reprinted with permission from Desai SP et al. Disseminated histoplasmosis. Am J Neuroradiol 1991; 12:290–292.)

REFERENCES

1. Carey RM, Kimball AC, Armstrong D. Toxoplasmosis. Clinical experiences in a cancer hospital. Am J Med 1973; 54:30–38.
2. Lyons RW, Andriole VT. Fungal infections of the CNS. Neurol Clin 1986; 4:159–170.
3. Kobayashi GS. Fungi. In *Microbiology*, BD Davis, R Dullbecco, HN Eisen, HS Ginsbert, eds. New York: Harper & Row, 1980; 817–850.
4. Parker JC Jr, Dyer MC. Neurological infection due to bacteria, fungi and parasites. In *Textbook of neuropathology*, RL Doris, DM Robertson, eds. Baltimore: Williams & Wilkins, 1985; 632–703.
5. Sze G, Zimmerman RD. The magnetic resonance imaging of infections and inflammatory diseases. Radiol Clin North Am 1988; 26: 839–859.
6. Zimmerman RA, Bilaniuk LT, Sze G. Intracranial infection. In *Magnetic resonance imaging of the central nervous system*, M Brant-Zawadzki, D Norman, eds. New York: Raven Press, 1987; 235–257.
7. Rinaldi MG. Invasive aspergillosis. Rev Infect Dis 1983; 5:1061–1077.
8. Young RC, Jennings A, Bennett JE. Species identification of invasive aspergillus in man. Am J Clin Pathol 1972; 58: 554–557.
9. Torre-Cineros J, Lopez OL, Kusne S, et al. CNS aspergillosis in organ transplantation: A clinicopathological study. J Neurol Neurosurg Psychiatry 1993; 56:188–193.

10. Walsh TJ, Hier DB, Caplan LR. Aspergillosis of central nervous system: Clinicopathological analysis of 17 patients. Ann Neurol 1985:18:574–582.
11. Spekowitz K, Armstrong D. Space-occupying fungal lesions. In *Infections of the central nervous system*, WM Scheld, RJ Whitley, DT Durack, eds. Philadelphia: Lippincott–Raven, 1997; 7 41–762.
12. Beal MF, O'Carroll CP, Kleinman GM, Grossman RI. Aspergillosis of the nervous system. Neurology 1982; 32: 473–479.
13. Ashdown BC, Tein RD, Feisberg GJ. Aspergillosis of the brain and paranasal sinuses in immunocompromised patients: CT and MR findings. Am J Roentgenol 1994; 162: 155–159.
14. Matsumara S, Sato S, Takamatsu HFH, et al. Cerebral aspergillosis as a cerebral vascular accident. No To Shinkei 1988; 40:225–232.
15. Mikhael MA, Rushovish AM, Ciric I. Magnetic resonance imaging of cerebral aspergillosis. Comput Radiol 1985; 9: 85–89.
16. Miaux Y, Ribaud P, Williams M, et al. MR of cerebral aspergillosis in patients who have had bone marrow transplantation. Am J Neuroradiol 1995; 16: 555–562.
17. Shuper A, Levitsky HI, Cornblath DR. Early invasive CNS aspergillosis. An early missed diagnosis. Neuroradiology 1991; 33: 183 –185.
18. Jinkins JR, Siqueira E, Zuheir Al-Kawi M. Cranial manifestations of aspergillosis. Neuroradiology 1987; 29: 181–185.
19. Cox J, Murtagh FR, Wilfong A, Brenner J. Cerebral aspergillosis: MR imaging and histopathologic correlation. Am J Neuroradiol 1992; 13: 1489–1492.
20. DeLone DR, Goldstein RA, Peterman G, et al. Disseminated aspergillosis involving the brain: Distribution and imaging characteristics, Am J Neuroradiol 1999; 20: 1597–1604.
21. Ostrow TD, Hudgins PA. Magnetic resonance imaging of intracranial fungal infections. Top Magn Reson Imaging 1994; 6: 22–31.
22. Harris DE, Enterline D. Fungal infections of the central nervous system. Neuroimag Clin North Am. 1997; 7: 187–198.
23. Kami M, Shirozu I, Mitani K, et al. Early diagnosis of central nervous system aspergillosis with combined use of cerebral diffusion-weighted echo-planar magnetic resonance imaging and polymerase chain reaction of cerebrospinal fluid. Intern Med 1999; 38: 45–48.
24. Zinreich SJ, Kennedy DW, Malat J, et al. Fungal sinusitis: Diagnosis with CT and MR imaging. Radiology 1988; 169:439–444.
25. Gupta RK, Singh AK, Bishnu P, et al. Intracranial aspergillus granuloma simulating meningioma on MR imaging. J Comput Assist Tomogr 1990; 14: 467–469.
26. Mathews VP, Alo PL, Glass JD, Kumar AJ, et al. AIDS-related CNS cryptococcosis: Radiologic–pathologic correlation. Am J Neuroradiol 1992; 13: 1477–1486.
27. Riccio TJ, Hesselink JR. Gd-DTPA-enhanced MR of multiple cryptococcal brain abscesses. Am J Neuroradiol 1989; 10: S65–66.
28. Andreula CF, Burdi N, Carella A. CNS cryptococcosis in AIDS: Spectrum of MR findings. J Comput Assist Tomogr 1993; 17: 438–441.
29. Bowen BC, Post MJD. Intracranial infections. In *Magnetic resonance imaging of the brain and spine*, SW Atlas, ed. New York: Raven Press, 1991; 523–526.

30. Popovivich MJ, Arthur RH, Helmer E. CT of intracranial cryptococcosis. Am J Neuroradiol 1990; 11: 139–142.
31. Tan CT, Kaun BB. Cryptococcal meningitis, clinical-CT scan considerations. Neuroradiology.1987; 29: 43–46.
32. Cornell SH, Jacoby CJ. The varied tomographic appearance of intracranial cryptococcosis. Radiology 1982; 143: 703–707.
33. Ofner S, Baker RS, Visual loss in cryptococcal meningitis. J Clin Neuro Ophthalmol 1987; 7: 45–48.
34. Tien Rd, Chu PK, Hesselink JR, Duberg A, Wiley C. Intracranial cryptococcosis in immunocompromised patients: CT and MR findings in 29 cases. Am J Neuroradiol 1991; 12: 283–289.
35. Gupta RK, Kathuria MK, Pradhan S. Magnetization transfers MR imaging in central nervous system tuberculosis. Am J Neuroradiol 1999; 20: 867–875.
36. Chang L, Miller BL, McBride D, et al. Brain lesions in patients with AIDS: H-1 MR spectroscopy. Radiology 1995; 197: 525–531.
37. Dzendrowski T, Himmelreich U, Malik R, Dowd S, Mountford C, Sorrel T. Distinction between cerebral cryptococcomas, staphylococcus aureus infections and tumors in an animal model. Proc Intl Soc Mag Reson Med 2000; 8: 173.
38. Blitzer A, Lawson W, Meyers BR, et al. Patient survival in paranasal sinus mucormycosis. Laryngoscope 1980; 90: 635–648.
39. Straatsma BR, Zimmerman LE, Gass JDM. Phycomycosis: A clinico-pathologic study of fifty-one cases. Lab Invest 1962; 11:963–985.
40. Abramson E, Wilson D, Arky RA. Rhinocerebral phycomycosis in association with diabetic ketoacidosis. Ann Intern Med 1966; 66: 735–742.
41. Press GA, Weindling SM, Hesselink JR, et al. Rhinocerebral mucormycosis: MR manifestations. J Comput Assist Tomogr 1988; 12: 744–749.
42. Terk MR, Underwood DJ, Zee CS, et al. MR imaging in rhinocerebral and intracranial mucormycosis with CT and pathologic correlation. Magn Reson Imaging 1992; 87: 81–87.
43. Stave GM, Heimberger T, Kerkering TM. Zygomycosis of the basal ganglia in intravenous drug users. Am J Med 1989; 86:115–117.
44. Mitchell TG. General characteristics of fungi. In *Zinsser microbiology*, WK Joklik, HP Willet, DB Amos, CM Wilfret, eds. Norwalk: Appleton & Lange, 1992.
45. Bazan C, Rinaldi MG, Rauch RR, Jinkins JR. Fungal infections of the brain. In *Neuroimaging clinics of North America*, JR Hesselink, ed. Philadelphia: Saunders, 1991; 1: 57–88.
46. Hadley MN, Martin NA, Spetzler RF, Johnson PC. Multiple intracranial aneurysms due to Coccidiodes immitis infection. J Neurosurg 1987; 66: 453–456.
47. Sobel RA, Ellis WG, Nelson SC, et al. Central nervous system coccidiodomycosis. Human Pathol 1984; 15: 980—985.
48. Galgiani JN. coccidiodes immitis meningitis. In: PK Peterson, JS Remington, eds. In defence of the brain. Molden: Blackwell Science Inc; 1997: 227–238.
49. Kobayashi RM, Coel M, Niwayama G, et al. cerebral vasculitis in coccidiodal meningitis. Ann Neurol 1977; 281–284.
50. Erly WK, LabadieE, Williams PL, et al. Disseminated coccidiodomycosis complicated by vasculitis: a cause of fatal subarachnoid hemorrhage in two cases. AJNR 1999; 20: 1605–1608.

51. Hadley MN, Martin NA, Spetzler RF, et al. Multiple intracranial aneurysms due to coccidiodes immitis infection. J Neurosurg 1987; 66: 453–456.
52. Wrobel CJ, RothrockJ. Coccidiomycosis meningitis presenting as anterior spinal artery syndrome. Neurology 1992; 42: 1840.
53. Fish DG, Ampel NM, Galgiani JN, et al. Coccidiomycosis during human immunodeficiency virus syndrome. Medicine 1990; 69: 384–391.
54. Salaki JS, Lauria DB, Chmel H. Fungal and yeast infections of the central nervous system. Medicine 1894; 63: 108–132
55. Goodwin RA, Shapiro JL, Thurman GH, et al. Disseminated histoplasmosis: clinical and pathological correlations. Medicine 1980; 59: 1–33.
56. Cooper RA, Goldsein E. Histoplasmosis of the central nervous system. Report of two cases and review of the literature. Am J Med 1963; 35: 45–57.
57. Shapiro JL, Lux JJ, Sprofkin BE. Histoplasmosis of the central nervous system. Am J Pathol 1955; 31: 319–335.
58. Vakili ST, Eble JN, Richmond BD, et al. Cerebral histoplasmoma (case report). J Neurosurg 1983; 59:332–336.
59. Desai SP, Bazan C, Hummell W, et al. Disseminated histoplasmosis. Am J Neuroradiol 1991; 12:290–292.
60. Haimes AB, Zimmerman RD, Morgello S, et al. MR imaging of brain abscesses. Am J Neuroradiol 1989; 10:279–291.
61. Dion FM, Venger BH, Landon G, et al. Thalamic histoplasmoma: CT and MR imaging. J Comput Assist Tomogr 1987; 11:193–195.
62. Jaster H, Dohan FC, Bertorini TE, et al. Choroid plexus histoplasmoma: Clinical, pathologic, and imaging findings. South Med J 1996; 89: 902–905.
63. Burgert SJ, Classen DC, Burke JP, Blatter DD. Candidial brain abscess associated with vascular invasion: A devastating complication of vascular catheter related candidemia. Clin Infect Dis 1995; 21: 202–205.
64. Smego RA, Perfect JR, Durack DT. Combined therapy with amphotericin B and 5-fluorocytocine for candida meningitis. Rev Infect Dis 1984; 6: 791–801.
65. Parker JC, McClosky JJ, Lee RS. The emergence of candidiasis: The dominant post-mortem cerebral mycosis. Am J Clin Pathol 1978; 70: 31–36.
66. Lipton SA, Hickey WF, Morris JH, et al. Candidial infection in the central nervous system. Am J Med 1984; 76: 101–110.
67. DeLaPaz RL, Enzmann D. Neuroradiology of the acquired immunodeficiency syndrome. In *AIDS and the nervous system,* ML Rosenblum, ed. New York: Raven Press, 1988; 121–153.
68. Kelly WM, Brandt-Zawadzki M. Acquired immunodeficiency syndrome: Neuroradiological findings. Radiology 1983; 149: 485–491.
69. Enzmann DR, Brandt-Zawadzki M, Britt RH. CT of central nervous system infections in immunocompromised patients. Am J Roentgenol 1980; 135: 263–267.
70. Munk PL, Lee MJ, Poon PY, et al. Candida osteomyelitis and disc space infection of the lumbar spine. Skeletal Radiol 1997; 26: 42–46.
71. Williams RL, Fukui MB, Meltzer CC, Swarnkar A, Johnson DW, Welch W. Fungal spinal osteomyelitis in the immunocompromised patient: MR findings in three cases. Am J Neuroradiol 1999; 20: 381–385.

72. Mawk JR, Erickson DL, Chou SN, et al. Aspergillus infections of the lumbar disc space. J Neurosurg 1983; 58: 270–278.

73. Olson EM, Duberg AC, Herron LJ, Kissel P, Smilovitz D. Coccidiodal spondylitis: MR findings in 15 patients. Am J Roentgenol 1998; 171: 785–789.

74. Mariat F. The mycetomas: Clinical features, pathology, etiology and epidemiology. Contrib Microbiol Immunol 1977; 4: 1–39.

75. Sharif HS. Role of MR imaging in the management of spinal infections. Am J Roentgenol 1992; 158: 1333–1345.

76. Resnick D, Niwayama G. Osteomyelitis, septic arthritis and soft tissue infection; the axial skeleton. In *Diagnosis of bone and joint disorders,* D Resnick, G Niwayama, eds. Philadelphia: Saunders, 1988; 2619–2754.

77. Cortet B, Deprez X, Triki R, et al. Aspergillus spondylodiscitis. Apropos of 5 cases. Rev Rhum Ed Fr 1993; 60: 37–44.

78. Gehweiler JA, Capp MP, Chick EW. Observations on the roentgen patterns in blastomycosis of bone. Am J Roentgenol 1970; 108: 497–510.

79. Goldman AB, Freiberger RH. Localized infectious and neuropathic diseases. Semin Roentgenol 1979; 14: 19–32.

80. Garo B, Sze G, Sharif H. MR imaging of intradural inflammatory disease of the spine. Am J Neuroradiol 1991; 12: 1009–1019.

81. Lindner A, Becker G, Warmuth-Mertz M, Shalke BCG, Bogdahn U, Toyka KV. Magnetic resonance image findings of spinal intramedullary abscess caused by candida albicans: Case report. Neurosurgery 1995; 36: 411–412.

Chapter 6

PARASITIC INFECTIONS

Rakesh K. Gupta and Kee-Hyun Chang

Infection of the central nervous system (CNS) by different parasites is in endemic proportion in different parts of the world. The distribution of the type of infection depends on the food habits, prevalence of the type of infection in the region, and local hygienic conditions in many developing countries in Asia, Africa, Central and South America, and Mexico. These infections may also be seen in developed countries due to population migration from the developing countries where these infections may be prevalent. Although a definite diagnosis of CNS parasitic infection is usually made on histopathology, the clinical diagnosis is generally based on a combination of the origin of the patient, clinical features, serology, and neuroimaging characteristic features. Neuroimaging plays a critical role in diagnosis and management of such patients as it provides the accurate location and extent of parasitic load and degree of host's immune response to the parasites.

MRI is superior to CT in the evaluation of most CNS parasitic diseases, especially in detecting the lesions close to the skull bones, in characterizing the lesions, and in

---

Rakesh K. Gupta • Department of Radiodiagnosis, Sanjay Gandhi Post-Graduate of Medical Sciences, Lucknow 226014, India. Kee-Hyun Chang • Department of Diagnostic Radiology, Seoul National University College of Medicine, Seoul 110-744, South korea.

MR Imaging and Spectroscopy of Central Nervous System Infection, edited by Gupta and Lufkin. Kluwer Academic / Plenum Publishers, New York, 2001.

delineating the extent of the surrounding parenchymal changes. Sometimes, neuroimaging studies show nonspecific or ambiguous findings. The specific antibody tests for parasites by enzyme-linked immunosorbent assay (ELISA) or indirect hemagglutination of serum or cerebrospinal fluid (CSF) may be useful in the evaluation of patients with nonspecific and inconclusive neuroimaging findings.

Many kinds of parasitic infections affecting the CNS have been described worldwide. We review clinical features, laboratory and MR findings of some of these commonly encountered parasitic infections in detail.

1. CYSTICERCOSIS

Cysticercosis is an infection caused by *Taenia solium* larvae following the ingestion of eggs excreted in human feces by the adult worm. Up until the second half of the nineteenth century, the larval form of this parasite was believed to be a distinct species, and Rudolphi gave the name *Cysticercus cellulosae* to it in 1809. Sanctioned by usage, such designation persisted, even after the life cycle of *T. solium* had been demonstrated. The name *Cysticercus*, proposed by Linnaeus, is derived from the Greek *kustis* (cyst, vesicle) and *kercos* (tail), and is justified by the fact that the parasite contains a structure that is similar to a small tail. As for the term *cellulosae*, Rudolphi proposed it in 1809 based on the frequent location of the cysticercus in the connective tissue (1).

The designation *neurocysticercosis* (NCC) refers to the presence of cysticerci in the CNS. It is endemic in most developed and developing countries of Latin America, Central and South Africa, and Asia (notably China, India, and Indonesia). It is still relatively frequent in Portugal, Spain, and Eastern European countries, particularly Poland and Romania (2–9). NCC is also endemic in developed countries, among social groups including emigrants, and has been in expansion over the last decades. The disease particularly affects communities in which hygienic conditions are poor. The usual routes of contamination are (1) the use of contaminated water for the irrigation of vegetables resulting in oro-fecal route; (2) consumption of infected pork meat, and (3) endogenous autoinfection.

Life cycle of T. solium. The adult tapeworm lives in the small intestine of humans, its only definitive host. It attaches itself to the intestinal mucosa by means of a scolex equipped with four lateral suckers especially adapted for adhering, and a rostellum, located in the terminal portion of the scolex, which bears 25–50 hooklets (10). The gravid proglottids containing the eggs are released in the intestinal lumen. From there, the eggs are usually ingested by the pig (the most common intermediate host), by humans, and by other mammals. Once in the digestive tract, the eggs lose their coat by the action of gastric and pancreatic enzymes and oncospheres are liberated. Aided by their hooklets, the oncospheres cross the intestinal wall and local venules, enter the venous systemic circulation, and are carried to the heart, where they enter the arterial systemic circulation to reach the definitive organs of the host (e.g., skeletal muscles, CNS, subcutaneous tissue, eye). Here the oncospheres lose their hooklets, acquire a vesicular shape, and evolve into cysticerci by gradual evagination of the protoscolex; their evaluation takes place over a period of 2 months (11). It is accepted that, when lodged in the eye and cerebral ventricles, the parasite can repeatedly evaginate and invaginate, while searching for an appropriate location for attaching itself (12).

C. cellulosae is the most frequent form of cysticercus observed in NCC cases (13). It is characterized by a rounded or oval cyst measuring from 5 to 15 mm in diameter, with a thin, translucent, membranous wall. The cyst is filled with a clear fluid and contains a pearly white invaginated scolex (10). Larger cysts, 2 to 4 cm in diameter, are rarely found (10). The larval variety known as *C. racemosus*, a designation proposed by Zenker in 1882, is less frequently found. The coexistence of *C. cellulosae* and *C. racemosus* is observed in about 10% of the cases of NCC (13). The racemosus cysts are 4–12 cm and are devoid of scolex. It has been admitted that the racemose form arises from *C. cellulosae* segmentation and /or sprouting of new cysts, with gradual expansion of each cyst and degeneration of the scolex (13).

The number of parasites varies widely in NCC. In 20–53% of cases, solitary cysticerci are found (14). When multiple, the cysticerci are usually few in number; the finding of hundreds of parasites, characterizing the disseminated form, is rare (15). Cysticerci may be lodged in any part of the leptomeninges, brain, and spinal cord. Cysticercosis is a chronic slowly progressive and frequently asymptomatic disease and is located on imaging or autopsy. It may remain viable in the CNS for several years (usually 1–3 years), depending on the greater or lesser host immune tolerance. Morphologically, three stages of development and regression of the cysticercus in the CNS are recognized (2,13): (1) The cystic or vesicular stage is viable and composed of well-defined, fluid-filled membrane which unlike hydatid cysts contains only one scolex. When the parasite is located in the nerve tissue, discrete fibrillary astrocytosis can be seen more externally. (2) The necrotic, colloid or granular stage corresponds to parasite necrosis and associated inflammatory process. It appears as an eosinophilic structure in which components of the bladder and scolex are in various stages of disintegration. In direct apposition to the necrotic parasite are multinuclear giant cells, foamy macrophages, and a small number of neutrophils. Deposition of cholesterol crystals is also observed. Neoformation of dense, fibrous connective tissues usually with a circinate contour, permeated or surrounded by lymphoplasmocytic inflammatory infiltrate and a few eosinophils can be seen surrounding the necrotic cysticercus. The adjacent neural tissue shows moderate to intense fibrillary astrocytosis. Edema and/or necrosis of the surrounding neural tissue may be present in some cases. (3) In the fibrocalcified nodule stage, fibrosis develops with time, progressively occupying the entire lesion. This stage can be macroscopically recognized as a nodule of smaller size than the bladder in the preceding stage, with a whitish, white-grayish, or grayish central area surrounded by a thin capsule of grayish or somewhat whitish color, corresponding to the necrotic cysticercus and fibrosis, respectively. Fibrosis occupies the entire lesion, associated or not with residual cellular infiltrate. The fibrous nodule thus formed frequently calcifies, as seen in 57–64% of NCC cases subjected to CT (14,16). Calcification may result from partial dystrophic calcification of the necrotic larva or from the presence of cysticercal calcareous corpuscle. The latter may remain in the lesion for an indefinite period, thus permitting identification of the parasite in the fibrocalcified lesions (10,13). Dystrophic calcification is a longlasting process and may take 2 to 10 years to be detected on X-ray films.

Various CNS lesions associated with NCC may be found, depending on the type of cysticercus, location of the parasite, and host immune inflammatory response. When located in the basal cisterns, the intense inflammatory and fibrosing reaction induced by *C. racemosus* often results in chronic leptomeningitis, with marked whitish thickening of the pia–arachnoid around the cysticerci, obstruction of the CSF drainage orifices in the fourth ventricle, internal hydrocephalus, and increased intracranial pressure (13,17). Ventricular location of the cysticercus frequently causes granular ependymitis. As the parasite dies, the inflammatory reaction intensifies and the parasite adheres to the ventricular surface due to reactive fibrillary gliosis and granular ependymitis (13,17,18). Thickening of the leptomeninges by inflammation and fibrosis causes entrapment and occlusion of the blood vessels that arise from the circle of Willis (19). The most frequent findings are fibrous thickening of the adventitia, thinning and focal destruction of the middle layer, areas of interruption and duplication of the internal elastica limitans, and intimal fibrosis of the local vessels of small, medium, and rarely larger diameter (17). Lacunar infarcts in the territory of the lenticulostriate arteries and, less commonly, large infarcts (Fig.1) involving the superficial or deep territory of the middle cerebral, internal carotid, and anterior cerebral arteries, in this order of frequency may develop (13,19–21). Other arterial lesions, such as presence of inflammatory infiltrate in almost all of the layers, fibrinoid necrosis, atheromalike deposits, and occlusive thrombosis, may occasionally be found.

NCC may be asymptomatic or cause a variety of manifestations. The symptomatology depends on the number of cysts, location of the parasite, stage of development of the parasite, and intensity of host immune-inflammatory response. We describe the MRI features based on location and are classified as parenchymal, cisternal, intraventricular, and spinal.

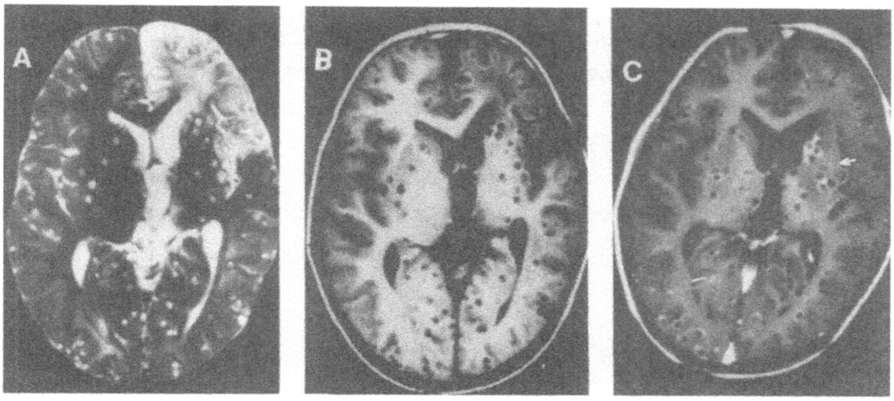

Figure 1. Neurocysticercosis with anterior cerebral artery territory infarction. T2-weighted axial image (a) shows evidence of an infarct in the left frontal region with multiple hyperintense cysts bilaterally. There is evidence of mild dilatation of left frontal horn. On T1-weighted image (b) the cysts appear hypointense. Some of the cysts show peripheral enhancement on postcontrast T1-weighted image (c). Note the focal enhancement along the left Sylvian fissure (white arrow).

1.1. Parenchymal Cysticercosis

The brain parenchyma is most frequently involved in CNS cysticercosis. CT is inferior to MR in its detection and MR may show multiple cysts while CT may appear normal (22). MRI findings of the parenchymal cysticercosis are protean, due to various evolutionary stages of the infection (23–29). In the initial stage, the embryo that has invaded the brain is noncystic; MRI usually shows no abnormalities. In the early development to the larva, focal nonenhancing areas of edema may be seen, which may progress to small homogeneously enhancing lesions in a few months (29,30). After 3 to 12 months of infection, a cysticercus is fullygrown, and its bladder contains clear fluid. In this viable vesicular stage, the larva appears as a round cyst with a mural nodule representing scolex. The cyst appears as hyperintense on T2- and hypointense on PD- and T1-weighted images and scolex is seen as eccentrically placed nodule, hypointense on T2- and hyperintense on PD- and T2-weighted images (Fig.2). This is considered as pathognomonic of the disease and represents a live innocuous larva when it is not associated with perifocal edema (23–29). Postcontrast study does not reveal any enhancement (30). The lesion may be single or multiple (23–29, 31). In the next stage, the larva begins to degenerate. The cystic fluid becomes turbid, the surrounding capsule thickens, and host inflammatory response occurs. On MRI, increased signal intensity of the cystic fluid, thickening of the cyst wall, surrounding edema, and contrast enhancement are seen at this stage (23–31) (Fig.3). The signal intensity of the cystic fluid is higher than that of CSF on T1- and PD-weighted images due to high protein contents and it may sometimes appear T1 bright (32,33) (Fig.2). On T2-weighted images the cystic fluid and surrounding edema appear as high intensity, whereas the cyst wall and scolex appear as isointense or hypointense relative to normal parenchyma. Contrast enhancement is usually ring-shaped. When a solitary cystic lesion with ring-shaped enhancement is present, differentiation from an abscess or neoplasm is difficult with MRI. Severity of inflammatory response is highly variable. A diffuse encephalitic form causing severe brain swelling may endure, especially in children (29).

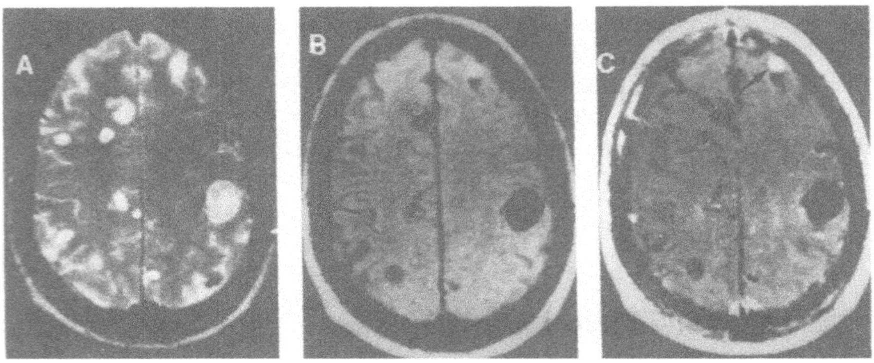

Figure 2. Cysticercosis. T2-weighted axial image (a) at supraventricular level shows multiple variously sized cystic areas bilaterally with no perifocal edema. T1-weighted image (b) shows hypointensity of the cysts and postcontrast image (c) shows no enhancement. Note the hyperintense cyst (arrow) in the left frontal region.

During or early after chemotherapy with praziquantel or albendazole, degenerative changes of the worm are accelerated, causing more marked inflammatory reaction, which is reflected by a large area of surrounding edema, increased intensity of cystic fluid on T1- and PD-weighted images, and increased contrast enhancement (29,34). This accentuation of the parenchymal inflammatory reaction correlates well with transient exacerbation of neurologic symptoms and CSF profile.

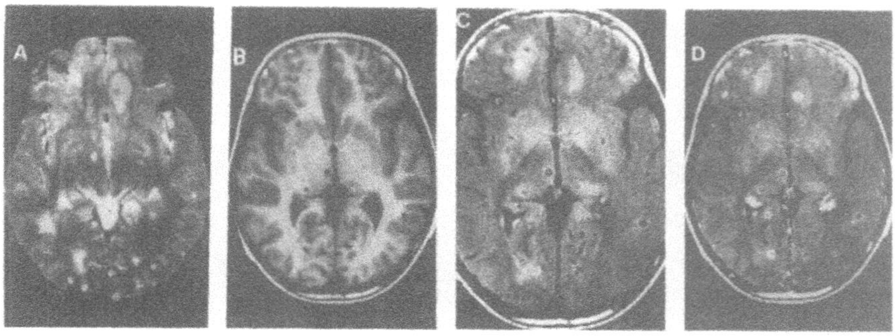

Figure 3. Cysticercosis. T2-weighted axial image (a) at the level of midbrain shows multiple hyperintense areas bilaterally with perifocal edema in some. T1-weighted image (b) shows hypointense nature of these lesions. On MT T1-weighted image (c), peripheral hyperintensity is seen in some of the lesions. Postcontrast T1-weighted image (d) shows ring enhancement of the lesions in the left frontal and right occipital region.

In the third stage, the larva is retracted with its fluid content being absorbed, and its inflammatory capsule becomes quite thick and collagenous. In this stage, signal intensity of the lesion usually appears isointense to the normal brain parenchyma on T1-weighted image and isointense to hypointense with or without central high signal intensity on T2-weighted image (Fig.4). On contrast-enhanced T1-weighted image, it appears as homogeneously enhancing or ring-shaped enhancing nodule with or without surrounding edema, mimicking a tuberculoma, other granuloma, small abscess, or metastasis (23–29). Magnetization transfer MRI has been used recently in the differentiation of a cysticercus granuloma from tuberculous granuloma (35,36). It has been noted that the magnetization transfer ratio is significantly higher in the T2 hypointense lesions of cysticercus granulomas compared to normal gray and white matter as well as tuberculoma (36). This is due to the fact that cysticercus larvae have higher protein and amino acid contents than tuberculomas resulting in higher magnetization transfer (35) (Fig.4). In the final nodular calcified stage, the lesion is further shrunken and completely mineralized. It is seen as iso/hypointense on T1- or hypointense on T2-weighted images. The hypointensity is due to either fibrosis or calcification. When the lesion is seen as hypointense nodule on T2-weighted image, it may be difficult to differentiate from occult vascular malformation or cavernous hemangioma. Demonstration of susceptibility on T2* helps in differentiating the nonmineralized lesion from calcified lesion. Use of phase imaging helps in separating vascular malformation from calcification due to the difference in phase of the two conditions (37). The calcification shows positive phase while occult vascular malformation shows negative phase. The time interval required for a lesion to evolve through all stages varies, ranging from 2 to 10 years with an average of 5 years. A patient with multiple lesions in different stages of evolution is frequently seen.

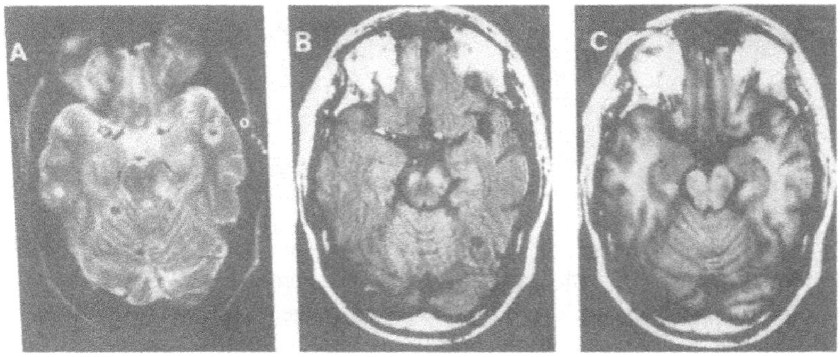

Figure 4. T2 hypointense cysticercosis lesions. T2-weighted axial image (a) through the level of midbrain/pons shows evidence of T2 hypointense areas in the cerebellar hemispheres, bilateral temporal lobes, and in the left crus region. A hyperintense lesion is also seen in the right temporal lobe. On MTT1-weighted image (b), visualization of these lesions is difficult and the MT ratio from the T2 hypointense areas was more than 35%. Only T2 hyperintense lesion is seen on T1-weighted image (c) and the rest of the lesions are isointense and not visible.

In patients with a cysticercus granuloma presenting with seizures, it is debatable whether to treat the lesion with albendazole or to leave it for its natural evolution (38,39). Most of the lesions either disappear or become calcified with time. Some scientists believe that these patients should be treated only with antiepileptics, and treating the cysticercus with albendazole does not alter the seizure control but only improves the imaging. The other group believes that treating the offending agent helps in better seizure control. Recently, a relationship has been shown between the perilesional gliosis and epileptogenesis of these healing/healed granulomas as depicted by MT MRI. Demonstration of perilesional gliosis on MT MRI may help in segregating the seizure-prone from seizure-free patients and may help in overall management of such patients (40).

1.2. Cisternal Cysticercosis

This infection is usually located in the basal cisterns, Sylvian fissures, or ventricles. The cisternal cysticercus is usually readily identified on MRI as multiple cystic masses in the basal cisterns (29). Racemose cysticercus may manifest as a large lobulated cyst compressing the adjacent structures. Signal intensity of the cystic content usually parallels that of CSF on all MR pulse sequences (Fig.5). The cyst wall of the larva may be seen as septumlike curved lines on T1-weighted image but is usually masked by high intensity of the CSF on T2-weighted image (29). Neither scolex nor ring enhancement around the cysticercus is seen. When small cysticerci with simple bladder are infected, they may only be suggested by apparent focal cisternal widening and may be difficult to diagnose with MRI. Serology for specific antibody in serum, CSF, or both by ELISA is helpful in diagnosing those cases. Racemose cysticercus tends to agglomerate in the basal cisterns. It frequently incites an extensive leptomeningeal inflammation, causing fibrotic thickening of the surrounding tissue (29). Chronic granulomatous meningitis around the basal cistern may result in communicating hydrocephalus (Fig.5). Proliferative endarteritis either due to basal exudates or due to the presence of cyst close to the wall may cause lacunar infarction or rarely territorial infarction due to major vessel occlusion (19). Rarely, territorial vascular infarction without evidence of a cyst near the vessel wall or basal meningitis has been described in NCC and is postulated to be due to the release of toxins from the dying NCC resulting in focal vasculitis (21). Contrast-enhanced MRI may reveal leptomeningeal enhancement of various degrees in the basal cisterns. A combination of large cyst(s), multilobulation of a cyst, lack of mural nodule within the cyst, cisternal location, and leptomeningeal enhancement on MRI strongly suggests that the lesion is caused by racemose cysticercus, especially in endemic areas.

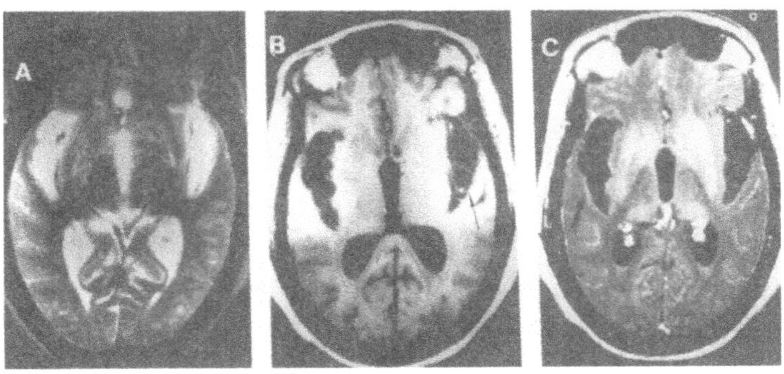

Figure 5. Cisternal cysticercosis. T2-weighted axial image (a) at the level of third ventricle shows dilated Sylvian fissures bilaterally with evidence of hydrocephalus. T1-weighted axial image (b) shows evidence of septae in the right Sylvian fissure and a well-defined cyst with scolex (arrow) in the left Sylvian fissure. Postcontrast T1-weighted image (c) shows enhancement along right Sylvian fissure (arrow) suggestive of meningitis

1.3. Intraventricular Cysticercosis

Intraventricular cysticercosis accounts for 7–20% of intracranial cysticercosis (1,29). The fourth ventricle is the most common location. On MRI, the cysticercus within the ventricle is more readily detected by visualization of the scolex, the rim of the cyst wall, the subependymal tissue reaction, and/or different intensity between the cystic content and the

CSF (29,41) (Fig.6). The cyst wall of the cysticercus may be adherent to the ventricular wall, causing ependymal and subependymal inflammation that is reflected by subependymal rim of high intensity on PD- and T2-weighted images (29). The signal intensity of the cystic content of the larva may be higher than that of CSF on T1- and PD-weighted images due to turbid cystic fluid containing cellular debris and high protein content in the degenerating stage or may become hypointense on T2-weighted images when in the healing stage and may be associated with ependymitis (42) (Fig.7). In general, T1- and PD-weighted images are better than T2-weighted images in the evaluation of intraventricular cysticercus, because on T2-weighted images, the high signal intensity of the cystic fluid is indistinguishable from the CSF and subependymal edema (Fig.6). Aqueductal stenosis may occur because of fibrotic adhesions secondary to ependymal inflammation. Sagittal T1-weighted imaging is particularly useful for the evaluation of the aqueductal stenosis and for discriminating the fourth ventricular cysticercosis from dilated fourth ventricle. However, even MRI occasionally may not distinguish an intraventricular cysticercus from the dilated ventricle (29). Therefore, in cases of hydrocephalus of unknown cause, specific antibody testing by ELISA is advisable in the endemic regions.

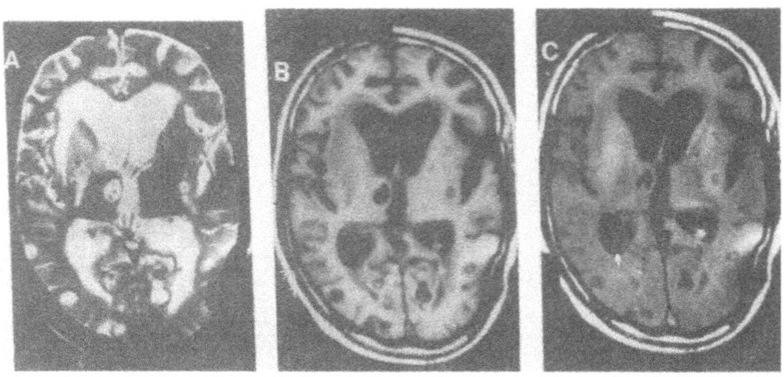

Figure 6. Intraventricular cysticercosis. T2-weighted axial image (a) at the level of third ventricle shows evidence of multiple hyperintense cysts with hypointense eccentrically placed scolices bilaterally with hydrocephalus. An infarct is seen in the right basal ganglia region. On T1-weighted image (b), the cysts appear hypointense with hyperintense scolices. On MTT1-weighted image (c), the wall of the cyst is well seen within the trigone of right lateral ventricle (white arrow).

Proton MR spectroscopy. In vivo proton MR spectroscopy has been described in a case of a large cysticercus cyst (43). The metabolites observed are lactate, succinate, acetate, Alanine, and an unassigned resonance at 3.3 ppm (Fig.8). Some of the metabolites seen in cysticercus cyst are also observed in hydatid cyst and brain abscess.

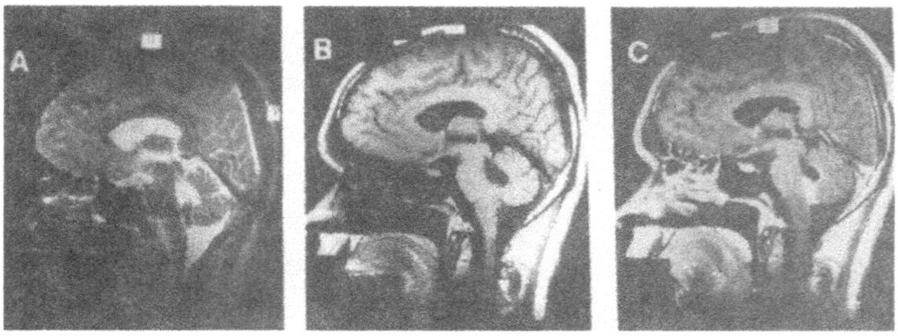

Figure 7. Intraventricular healing cysticercosis with ependymitis. T2-weighted sagittal section (a) shows evidence of a hypointense area in the inferior part of fourth ventricle with mildly dilated fourth ventricle. On T1-weighted image (b) the lesion appears isointense. Postcontrast image (c) shows enhancement of the lesion along with roof of fourth ventricle suggestive of ependymitis.

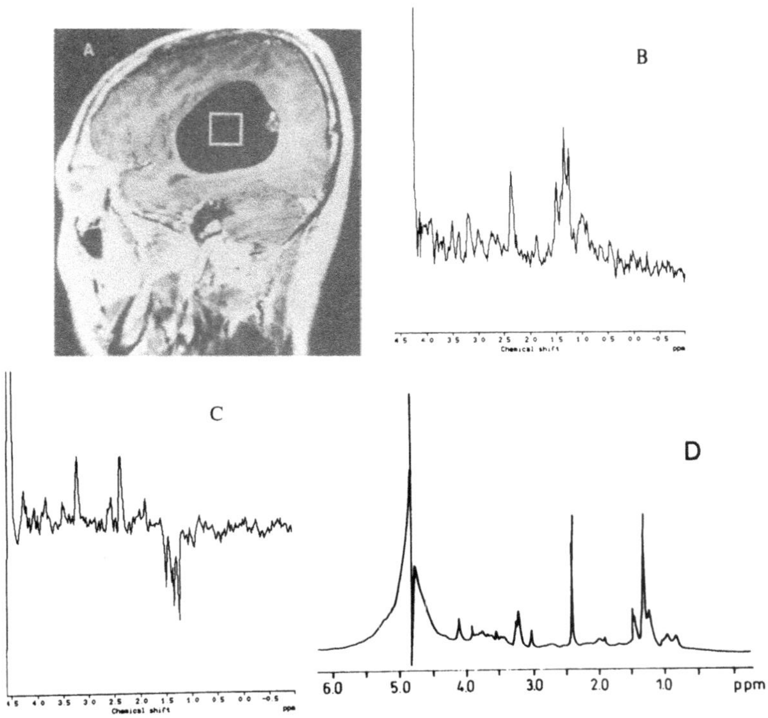

Figure 8. Proton MR spectroscopy of cysticercosis. Sagittal T1-weighted MR image (a) shows a large cyst with a mural nodule. An ROI of 8 ml is seen within the cyst. Proton MR spectra obtained at TR/TE of 2000/270 (b) and 2000/135 (c) show succinate at 2.4 ppm, lactate at 1.3 ppm, and unassigned metabolites. *Ex vivo* proton MR spectrum (d) of the cysticercal cysts obtained from the subcutaneous cyst of another patient confirms the assignments seen *in vivo*.

1.4. Spinal Cysticercosis

Cysticercosis within the spinal canal is extremely rare. Intraspinal cysticercosis involves the subarachnoid space and less often the cord or epidural space. Subarachnoid cysticercus presents as intradural extramedullary cyst and/or as arachnoiditis (29) (Fig.9). The cysticercus appears as a cystic mass compressing the spinal cord or the nerve roots in association with leptomeningeal enhancement of various degrees. There is no or minimal contrast enhancement in the wall of the cysticercus cyst. At lumbosacral level, it may be seen as a multiseptate cyst or multiple cysts with leptomeningeal enhancement. Intramedullary cysticercosis manifests as small round cyst(s) with or without eccentric nodule (scolex) and ring-shaped enhancement within the cord (44,45) (Fig.10). The spinal cord may be swollen and edematous, due to either edema in the intramedullary cyst or ischemia/infarct caused by compromise of the anterior spinal artery secondary to cysticercal arachnoiditis (29).

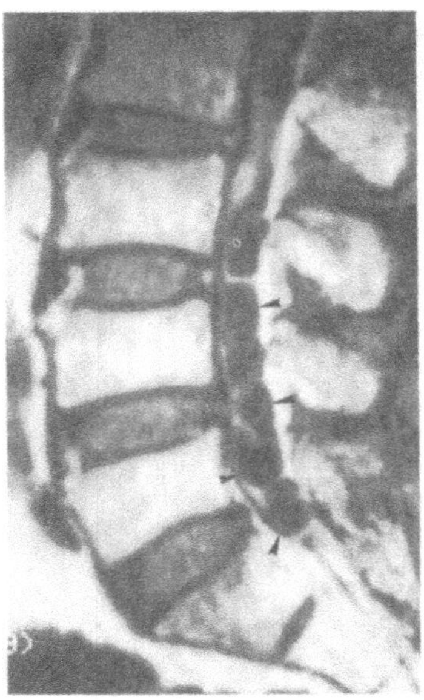

Figure 9. Spinal cysticercosis within the subarachnoid space. Sagittal contrast-enhanced T1-weighted MR image of the lumbar spine demonstrates multiple cysts of low intensity occupying the thecal sac at the lumbar level (arrowheads). There are some linear and nodular foci between the cysts, suggesting adhesive inflammation of the leptomeninges and nerve roots. Multiple cysticerci were surgically excised.

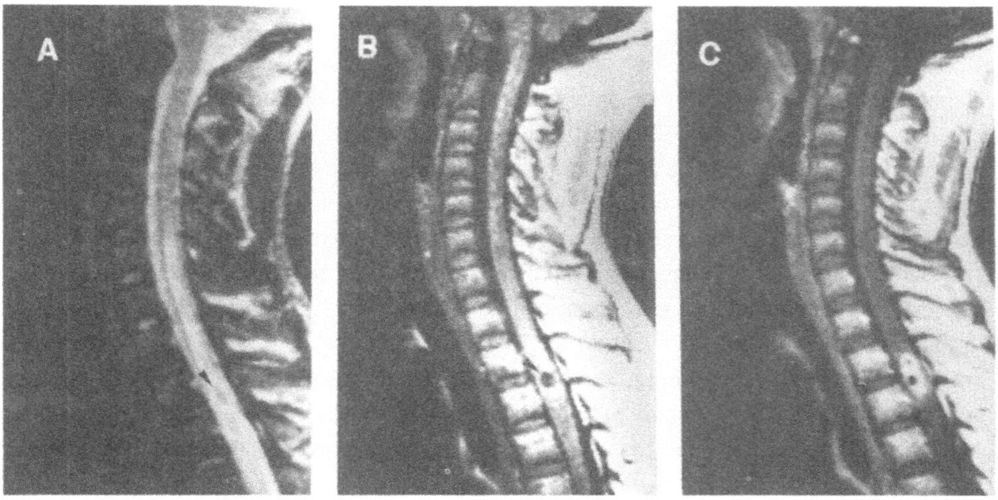

Figure 10. Intramedullary cysticercosis. Sagittal T2-weighted image (a) of the cervicothoracic region shows a hypointense area (arrowhead) with surrounding hyperintensity in the cord. On T1-weighted image (b) the lesion appears hypointense with a hyperintense nodule suggestive of scolex (arrowhead). Postcontrast T1-weighted image (c) shows thick peripheral enhancement.

2. PARAGONIMIASIS

This is an infestation caused by a lung fluke of the genus *Paragonimus*. The most important species is *P. westermani*. It is endemic in eastern and Southeast Asia, West Africa, and Latin America. Human infection occurs by ingestion of uncooked freshwater crabs or crayfish infected with the encysted larvae (metacercariae) of the fluke. When humans are infected, the metacercariae of the paragonimus excyst in the small intestine and the larvae penetrate the intestinal wall, the peritoneum, and the diaphragm, finally invading the pleural cavity and the lung parenchyma in 2 to 8 weeks. The main habitat of adult paragonimus is the lung. Some larvae may bypass the lungs and migrate to the intracranial cavity, probably through the perivascular spaces in the foramina of the skull base, where the worms penetrate meninges and invade brain parenchyma directly. The cerebral paragonimiasis comprises approximately 1% of the pulmonary paragonimiasis cases (46,47). The disease usually involves one cerebral hemisphere, most commonly in the posterior part of the brain. The frontal lobe is involved more frequently than reported. The cerebellum, brainstem, and spinal canal are uncommon sites of involvement. Clinically, patients with cerebral paragonimiasis have epilepsy, headache, hemiparesis, hyperesthesia, blurred vision, diplopia, homonymous hemianopsia, and meningitic symptoms.

As the patients with early active paragonimiasis can be treated successfully by chemotherapy with bithionol or praziquantel, diagnosis at an early stage is regarded as crucially important (47–49). Cerebral paragonimiasis is definitely diagnosed when eggs of paragonimus are seen in the brain lesion on the pathology examination. In addition to neuroimaging findings, important clues leading to presumptive diagnosis of cerebral paragonimiasis include the origin of the patient, history of eating undercooked freshwater crustaceans, associated pulmonary symptoms and abnormal findings on chest radiographs suggesting leukocytosis and eosinophilia pulmonary paragonimiasis, increased immunoglobulin in serum, positive reaction to intradermal test, and CSF findings of eosinophilia and elevated protein levels. Antibody test by ELISA for paragonimus-specific antibody (IgG) in CSF is highly sensitive and specific in the diagnosis of an early active cerebral paragonimiasis, but the reaction usually is negative in patients with chronic calcification (50).

The initial intracranial lesion of cerebral paragonimiasis manifests as exudative aseptic inflammation. Local meningitis may occur when the lesion is formed in the subarachnoid area. The migrating worm may also cause congestion, vasculitis, and capillary rupture, which result in infarction, hemorrhage, and necrosis, usually in the subcortical region (48). In time, granulomatous lesions are formed around the adult paragonimus. The granulomatous lesions can be classified into two forms: abscesslike cystic lesions and solid granulomas. Most commonly, the lesions are multiple conglomerated and interconnected granulomas, which are located around a focus. The contents are caseous necrotic debris with scattered paragonimus ova. In this early active stage, the most common and characteristic MRI finding is conglomerated, multiple ring-shaped enhancement resembling a "grape cluster" with surrounding edema of variable degrees (Fig.11), which strongly suggests early active cerebral paragonimiasis, particularly in endemic areas (49,50). This finding represents multiple crowded abscesslike granulomas. Each "ring" on MRI usually is smooth and round or oval. The size of each "ring" is variable, ranging from a few millimeters to more than 3 cm in diameter, usually 1 to 3 cm. The center of each "ring" shows signal intensity similar to or slightly higher than that of CSF on MR images of all pulse sequences. The wall of each "ring" is usually isointense relative to the brain parenchyma on T1-weighted images and isointense or hypointense on T2-weighted images. If there are no intervening normal brain tissues among the conglomerate granulomas, the lesion may be mistaken for a single lobulated neoplasm with multifocal necrotic centers. Nodular and/or irregular enhancing lesions representing solid granulomas may be seen with or without the conglomerates of the ring-shaped lesions. Only a solitary ring-shaped lesion may also be seen. Edema of various degrees usually surrounds the enhancing lesions. If there is a small ring-enhancing lesion, differentiation from other granulomas such as tuberculoma or cysticercosis in degenerating stage or abscess is impossible. Rarely, ependymal and periventricular invasion can cause choroiditis, which appears as a thickened, enhancing choroid plexus with the periventricular-enhancing lesion on MRI (49). There may be a poorly defined area of hypointensity on T1-weighted images and hyperintensity on T2-weighted images with fingerlike projections along the white matter but no contrast enhancement, suggesting vasogenic edema or cerebritis; this finding results from an initial exudative aseptic inflammation around the worm. Localized foci of hemorrhagic focus can be readily seen as an area of high attenuation on CT, which may, however, be overlooked on MR images. On MR images, the hemorrhage in subacute or early chronic stage appears as foci of high signal intensity on T1-weighted images (49,50). In endemic areas, paragonimiasis should be included as a possible cause of cerebral hemorrhage and infarction.

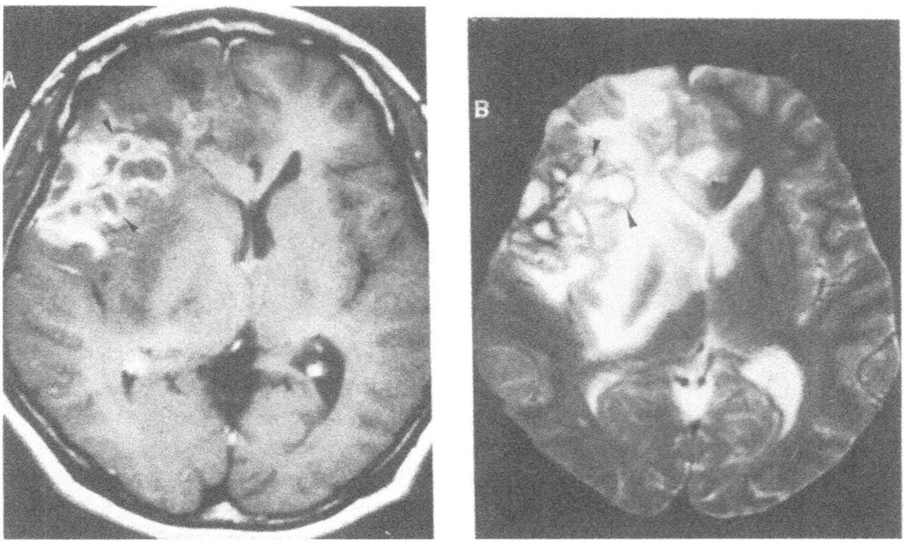

Figure 11. Cerebral paragonimiasis in early active stage. Axial T2-weighted image (a) shows multiple ring-shaped lesions (arrowheads) with massive edema of high intensity extending to the right basal ganglia and thalamus. Axial contrast-enhanced T1-weighted image (b) shows multiple conglomerated, abscesslike ring-enhancing lesions (arrowheads) in the right frontal lobe with extensive surrounding edema.

In the chronic stage, liquefaction necrosis and fibrinous gliosis occur around the granulomas, leading to cortical atrophy, subcortical cerebromalacia, secondary ventricular dilatation, and porencephaly. Nearly all of the granulomas shrink and invariably become calcified. On MRI, localized cerebromalacia is usually seen around the calcification (50,51). The calcification may appear to be punctate and amorphous, spotty round nodular, solitary round cystic, congregated multiple round or oval cystic, (so-called soap-bubble appearance) on plain skull radiograph (52) and CT (53). Soap-bubble calcification is considered to be most specific of the disease. On MRI, the calcifications are seen as signal void or hypointense nodules and not infrequently as "egg-shell appearance" with central content of low or high intensity (Fig.12). CT scan is more sensitive than MRI in detection of calcifications. MRI, however, is better than CT in the evaluation of exact extent of cerebromalacia and provides more detailed information on the soft tissue contents within the calcified granuloma.

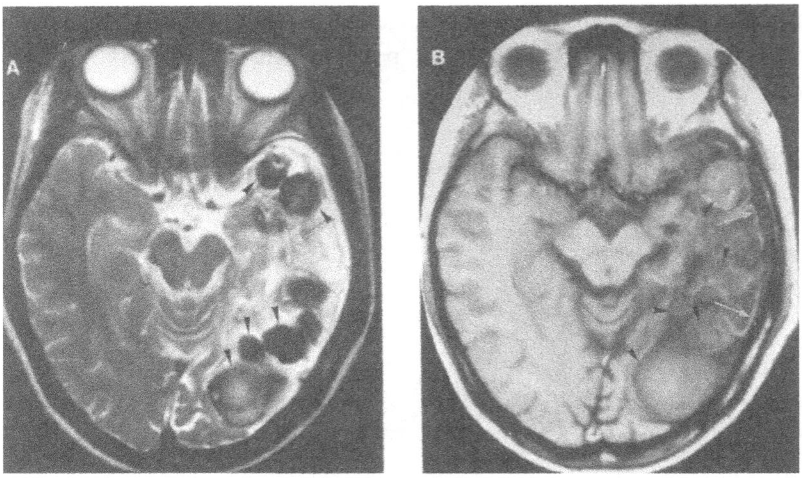

Figure 12. Cerebral paragonimiasis in chronic stage. Axial T2-weighted image (a) shows multiple round nodules of hypointensity (arrowheads) associated with adjacent cerebromalacia in the left temporal and occipital lobes. The round nodules of hypointensity represent inactive calcified abscesslike lesions of "soap bubble appearance." T1-weighted image (b) shows that the areas of cerebromalacia in the left temporal and occipital lobes appear hypointense. The signal intensity of the round nodules suggesting calcification is slightly hypointense or isointense relative to normal brain parenchyma. There is focal linear high intensity in some cortices (arrows), suggesting either cortical laminar necrosis or microhemorrhages.

3. SPARGANOSIS

Human sparganosis is a rare parasitic infection caused by a migrating tapeworm larva (sparganum) of the genus *Spirometra*. The disease has been reported worldwide but is more common in eastern Asia. It usually involves the subcutaneous tissue or muscle of the chest, abdominal wall, or limbs. Intracranial involvement is relatively rare (54).

Humans can be infected in three ways. By drinking untreated water with infected copepods (first intermediate host), by ingesting raw or inadequately cooked flesh of snakes or frogs (second intermediate host) infected with the sparganum, and by applying the flesh of an infected intermediate host as a poultice to a wound. Most patients are rural inhabitants and have a history of ingestion of frogs, snakes, or both (54–57). After a sparganum is introduced into a human, it penetrates the gastrointestinal wall and begins to migrate systemically. The route of intracranial invasion is not known yet, but foramina of the skull base might be used as an entry. Cerebral sparganosis most frequently involves the cerebral hemispheres, particularly the frontoparietal lobes, and rarely occurs in the cerebellum, brainstem, and spinal canal. The major presenting symptoms are seizure, hemiparesis, and headache, with a chronic course (57).

On pathologic examination, a thick granulomatous wall surrounds the worm. The live worm is 1 cm to 1 m long and characteristically is white and ribbon-shaped, moving by itself with transverse wrinkling and an actively protruding head. The worm can survive for 5–20 years in humans. The live or degenerated worm contains a large number of calcospherules. White matter around the worm or worm tract shows organizing edema and fibrillary gliosis associated with a considerable loss of myelin (55).

Neuroimaging findings reflect a longstanding inflammation with both active granulomatous lesions and irreversible degeneration of the brain tissue, probably related to a combined effect of long survival and migration of the worm and histotoxic effects of proteases secreted by the worm. The following imaging findings are characteristic of cerebral sparganosis: (1) widespread degeneration of the white matter with adjacent ventricular dilatation and cerebral atrophy; (2) multiple punctate or small nodular calcifications; (3) coexisting enhancing lesion of irregular nodular, amorphous, or beaded shape (Fig.13). A combination of the above imaging triad seems to be specific for cerebral sparganosis and is seen in about two-thirds of cases. Occasionally, the enhancing lesion is associated with surrounding edema, compressing the lateral ventricle (55–57). Follow-up MRI may show progression of the lesion and a change in location of the enhancing nodule. Sometimes, MRI reveals linear or curvilinear enhancing lesions conforming to the shape of the worm on coronal or sagittal plane. Occasionally, the areas of hyperintensity on T1- and hypointensity on T2-weighted images in subcortical areas, which probably represent subacute or early chronic petechial hemorrhages, observed on MR images. With MRI alone, however, it may be difficult to determine the presence of small calcific foci, which is one of the diagnostic clues of cerebral sparganosis. Therefore, CT may be better than MRI as an initial imaging modality in the diagnosis of cerebral sparganosis. It is important to determine whether the worm is alive or dead before surgery, because there is no effective chemotherapeutic drug for sparganosis and surgical removal of a live worm is the treatment of choice. Surgical removal of the worm is indicated only in cases with a live worm because the granuloma around a dead worm spontaneously regresses over time. With a single imaging study, it is difficult to determine whether the worm is alive, although a relatively large enhancing lesion associated with edema compressing adjacent structures is highly suggestive of a live worm within the lesion. A change in location of the enhancing nodule or worsening of other imaging findings on sequential studies would indicate that the worm is alive and that the patient is a surgical candidate (56,57). In patients with nonspecific imaging findings, antisparganum antibody testing by ELISA is helpful in the differential diagnosis in endemic areas.

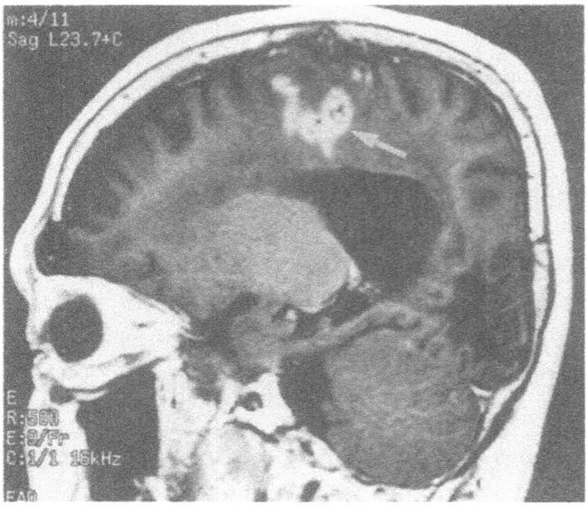

Figure 13. Cerebral sparganosis. Sagittal contrast-enhanced T1-weighted image shows a tortuous tubular enhancing lesion (arrow) in the left posterior frontal area. At surgery, a live sparganum was found in the same area. There is a focal cerebromalacia in the left occipital lobe associated with adjacent ventricular dilatation.

4. ECHINOCOCCOSIS

Echinococcosis or hydatidosis presents in endemic foci and poses serious health hazard, causing high morbidity and considerable mortality. Humans become accidentally infected as intermediate hosts, harboring the larval form of the organism or hydatid cyst, acquired by food-borne exposure to eggs in the feces of definitive hosts, mainly dogs or other animals. Definite hosts, in turn, acquire the disease by eating viscera infected with hydatid cysts, mainly from sheep and cattle.

The two most widespread species in human disease are *E. granulosus* and *E. multilocularis*, responsible for cystic echinococcosis and alveolar echinococcosis respectively, with dissimilar prevalence rates throughout the world. The Mediterranean and the Middle East are endemic regions with diverse focal prevalence rates. In Latin America, high focal prevalence rates of *E. granulosus* are found in Chile, Argentina, Uruguay, and southern Brazil as well as in Peru with wide dissimilar distribution within each country. Alveolar echinococcosis is relatively rare, but still represents a serious problem in cold regions, such as Alaska, several northern central states of the United States, central Europe,

China, Russia, and Turkey due partly to *E. multilocularis* being transmitted between foxes or dogs and rodents, but reliable epidemiologic data are scanty (58,59).

The *E. granulosus* taenia is a platyhelminth belonging to the order Cestoda, family Taeniidae, genus *Echinococcus*, measuring 3–5 mm in length, and parasitizing the gut of the dog, other canids, and other less common definitive hosts. The taenia consists of a small scolex, with four suckers and a double row of hooklets, followed by the neck and three rings. When gravid, the third ring that is roughly 2 mm in length carrying 500–800 embryophores or eggs within the uterus, becomes detached and sheds eggs into the feces of the dog. Each egg consists of a hexacanth embryo (or scolex) that is resistant to environmental changes and remains fertile for almost 20 days. The intermediate host, mostly humans and sheep, becomes infected by the ingestion of eggs shed by the definitive host, the dog for the domestic cycle, which in turn ingests infected mutton harboring hydatid cysts. In the definitive host, the scolex attaches to the gut wall, forming the adult worm (57,58).

In the intermediate host, the hexacanth embryo is released in the small intestine, penetrating mucosa and migrating by the portal system to the liver where hydatid cysts are formed and may metastasize mostly to the lungs as well as the brain, periorbital tissue, vertebrae, kidney, and pericardium. A central cyst develops from the embryo with thick nonnucleated laminated external layer or cuticle. This external layer behaves as a semipermeable membrane and is lined on the inside by a thin germinal nucleated membrane (proligerous) rich in glycogen. The brood capsules or daughter cysts arise remaining attached to this germinal layer, while scolices develop from the germinal layer within the brood capsules. Hydatid cysts are filled with fluid rich in scolices that together with the brood vesicles form the hydatid sand (58).

For the life cycle to be completed the hydatid cyst from the intermediate host liver, lung, or brain must be ingested by the definitive host. The definitive host, the dog for the domestic cycle, in turn, ingests infected mutton harboring hydatid cysts. In addition to the domestic dog–cattle cycle, there are also sylvatic cycles involving wild animals such as foxes and wolves as definitive hosts. *E. granulosus* infection of the CNS has been estimated at below 2%, and may be primary, often presenting as solitary parenchymal cysts, while intraventricular and meningeal location is most exceptional. Secondary involvement is due to cyst rupture or hydatid fluid dissemination from a distant organ such as the heart. Damage to the CNS is uncommon and due to cyst formation within the brain, cerebellum, skull, or vertebrae and more rarely in the brainstem or spinal cord through vertebral or rib infection (58,59).

Signs and symptoms include increase in intracranial pressure, headache, vomiting, papilledema, and hemiparesis as well as occasional seizures, while meningeal or cranial nerve involvement is most unusual. Most cases with spinal cord and/or radicular compression develop sensory and motor deficits, together with pain and impaired bladder function.

4.1. Intracerebral Hydatid Disease

MRI usually shows a large well-defined cystic mass with no perifocal edema. If the cyst is large, the mass effect on the adjoining structures is evident. The cyst shows CSF intensity on T1-, PD-, and T2-weighted images. Postcontrast study usually does not show any peripheral rim enhancement (Fig.14) (60–62). Occasionally, there is perifocal edema (Fig. 15) around the cyst and rim enhancement on injection of Gd-DTPA and it becomes difficult to differentiate it from the abscess. It is probably due to cyst degeneration resulting in perifocal edema and reactive gliosis. The cyst may be seen as unilocular, multilocular, multiple unilocular, or cyst with septation (63). *In vivo* proton MR spectroscopy has shown biochemical signatures that are specific to the parasitic cyst and are different from brain abscess. The dominant resonances usually observed in hydatid cyst are acetate, succinate, lactate, and glycine at 1.92, 2.4, 1.33, and 3.56 ppm, respectively (64). The presence of these resonances suggests the anaerobic glycolysis seen in parasitic cysts (Fig.14). In addition, small contributions from alanine and isoleucine, leucine and valine are also seen at 1.5 and 0.9 ppm (64,65).

It is known that the presence of a parasite-derived laminated cuticle, consisting predominantly of carbohydrates and proteins, characterizes hydatid cysts and forms a barrier between host and parasite. However, the passage of host macromolecules into hydatid cysts has been experimentally demonstrated (58).

MRI features of *E. multilocularis* have been rarely reported. It shows low signal intensity on T1- and multilocular high-intensity mass on T2-weighted images with perifocal edema. Postcontrast study shows heterogeneous enhancement. The imaging features are nonspecific and may simulate oligodendroglioma and brain abscess. The diagnosis should be supported by the serology in suspected cases coming from the endemic region (66).

Primary treatment of cerebral hydatid cyst is surgical. In patients with recurrent hydatid and poor risk patients, albendazole has been tried but its role has not been established in CNS involvement (59).

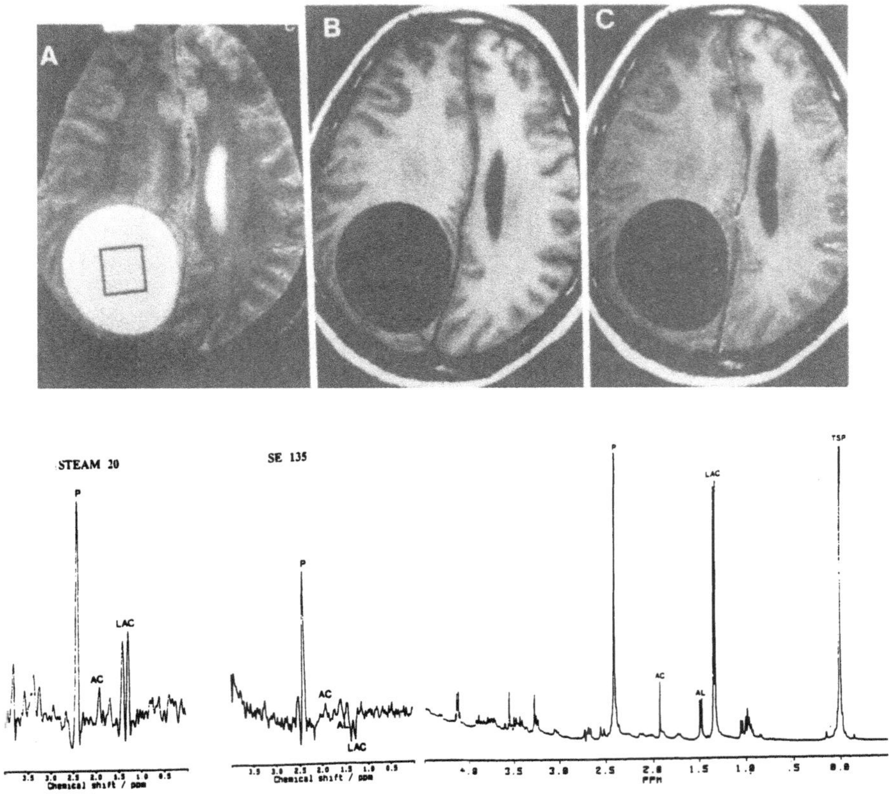

Figure 14. Cerebral hydatid cyst. T2-weighted axial image (a) shows evidence of a well-defined, rounded hyperintense area in the right parieto-occipital region with mass effect. On T1-weighted image (b) the cyst appears hypointense similar to CSF and no enhancement is seen on postcontrast T1-weighted image (c). Proton MR spectroscopy with STEAM 20 (d) from the inset in (a) shows lactate (LAC) at 1.33 ppm, alanine at 1.5 ppm, acetate at 1.92 ppm (AC) and succinate (P) at 2.4 ppm; at SE 135 (e) LAC, AL show phase reversal while AC and P show normal phase. *Ex vivo* study (f) of the cyst fluid confirms the assignments *in vivo*.

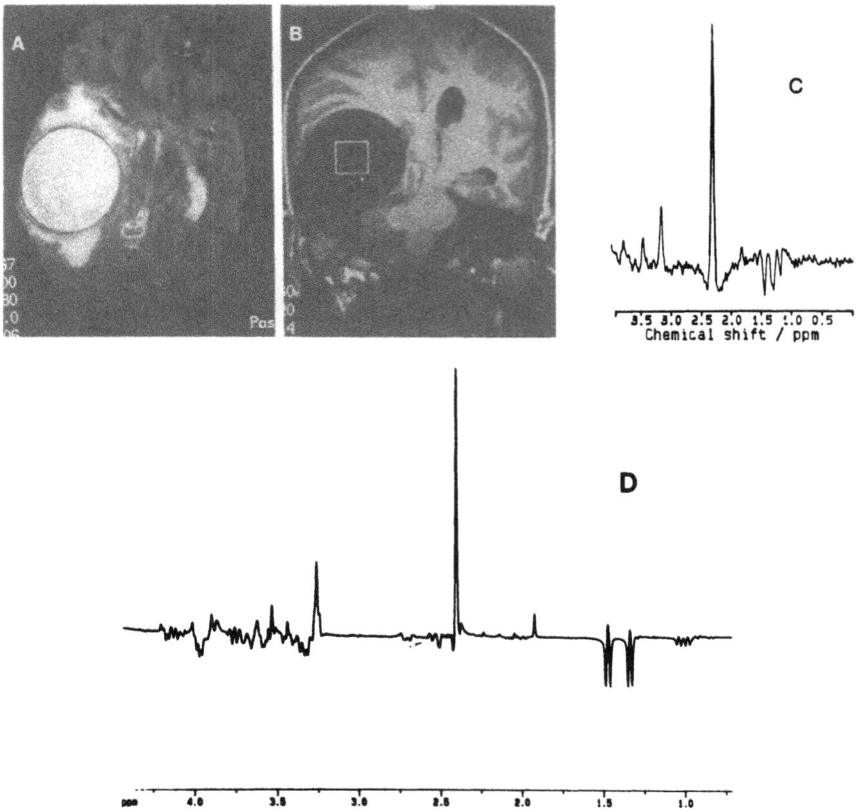

Figure 15. Cerebral hydatid cyst with perifocal edema. Axial T2-weighted image (a) shows a large hyperintense area in the right temporal region with a well-defined hypointense rim and surrounding edema. Coronal T1-weighted image (b) shows the hypointense cyst with mass effect on ipsilateral lateral ventricle and midline shift. MR spectroscopy from the inset shown in (b) done with spin echoes 135 ms (c). It demonstrates lactate (1.33 ppm), alanine (1.5 ppm), acetate (1.92 ppm), succinate (2.4 ppm), choline (3.2 ppm), and a resonance at 3.56 ppm assigned to glycine. *Ex vivo* spin-echo spectrum (d) of the fluid confirms the assignments seen *in vivo*.

4.2. Spinal Hydatid Disease

Hydatid disease of the spine occurs in 1% of all cases with hydatidosis and is most commonly located in the dorsal spine. MRI demonstrates bony destruction. The signal from the cyst is inhomogeneous; hypointense on T1- and hyperintense on T2-weighted images (Fig.16). CT is better than MR in delineation of calcification in the wall of the cyst. The cyst may rarely be completely extradural without involvement of the vertebrae; the embryo possibly being carried through the porto-vertebral venous shunts. The primary site of the infestation and precise extent of the spinal hydatid disease are very difficult to verify. Braithwaite and Lees classified the spinal disease into five types: (1) primary intramedullary hydatid cyst, (2) intradural extramedullary hydatid cyst, (3) extradural intraspinal hydatid cyst, (4) hydatid disease of the vertebra, (5) paravertebral hydatid disease (67). The first three are indeed rare. The viable cyst is of low intensity on T1- with iso- to mildly hyperintense cyst wall (68–71). A decrease in hyperintensity on follow-up T2-weighted images and increase in hypointensity from collapsed cyst wall are indicative of succumbed cyst (68,71).

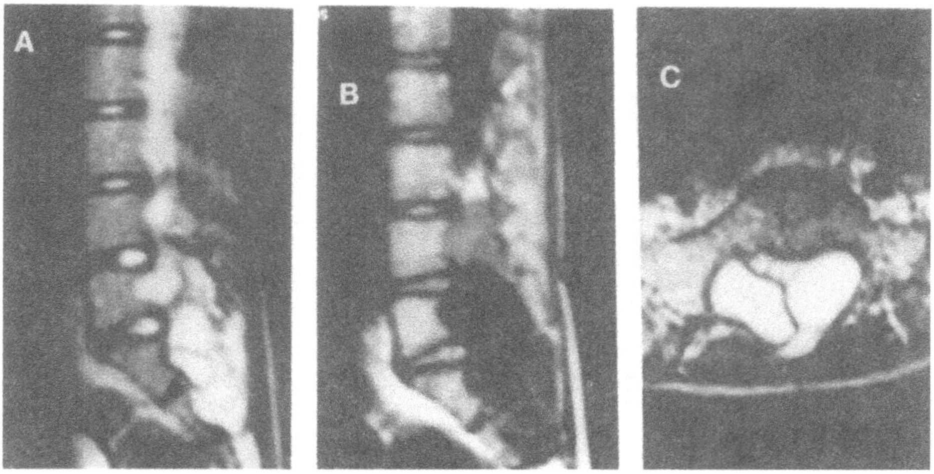

Figure 16. Hydatid cyst of spine. T2-weighted sagittal image (a) of the lumbosacral spine shows evidence of lobulated hyperintense area with hypointense septae in the spinal canal with scalloping of fifth lumbar to second sacral vertebrae. On T1-weighted image (b) the lesion appears hypointense with hyperintense septa. The hypointense septa are well seen in axial image (c).

5. AMEBIC BRAIN ABSCESS

Entamoeba histolytica is a major human pathogen, with possibly 10% of the world population infected and annually causing up to 100,000 deaths worldwide (72). Primary infection involves the colon directly following ingestion of the organism while spread to the brain is thought to be hematogenous from the colon via hepatic pulmonary or vertebral veins. The reported incidence of amebic brain abscesses in patients with confirmed amebic liver abscess varies from 0.6 to 8.1%, while infection of the CNS without hepatic involvement is rare (72).

Although the intestinal form is the most common form of the disease, its involvement is seen in 50% of patients with cerebral amebiasis (72,73). Amebic brain abscesses may be single or multiple, and commonly involve cortical and deep gray matter. The commonest sites of involvement of the brain are the frontal lobe and basal ganglia regions with the left side being more common than the right with a ratio of 2.3:1 (73).

MR appearance does not differ significantly from that described in pyogenic abscess (Fig.17). T1-weighted imaging shows central hypointensity with a surrounding thin rim of isointensity, presumably representing a capsule of granulation tissue. The T2-weighted images show central hyperintensity, presumably representing central necrosis/liquefied brain with hypointense peripheral rim (74–76).

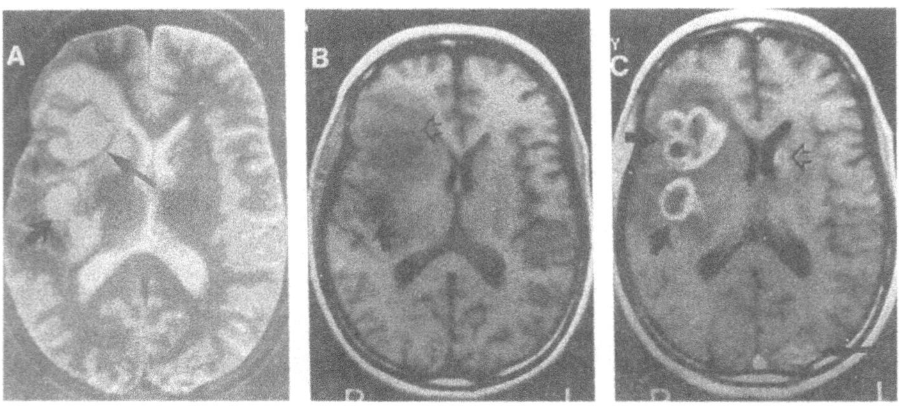

Figure 17. Cerebral amebic abscess. T2-weighted image (a) with hypointense area (arrow) and low-intensity capsule (long arrow) and surrounding edema. T1-weighted image (b) shows mixed intensity changes in right frontoparietal region (open arrow) and possible cystic areas (arrow). Postcontrast T1-weighted image (c) shows multiple ring-enhancing lesions (arrows). Note also the enhancing lesion in the left occipital region (arrow). (Reprinted from Clinical Radiology, Vol. 6, De Villiers JP, et al., pp. 307–309. Amoebic abscess of the brain, 1998, by permission of the publisher W.B. Saunders.)

The difficulty in obtaining histologic confirmation of the disease is well described in other anatomic sites. Serology is helpful in nonendemic areas only. Cerebral amebiasis is an important and potentially lethal condition in the differential diagnosis of brain abscess especially when no recognized pathogen is isolated from the pus. This is particularly true in tropical and subtropical regions where the disease is prevalent.

6. CEREBRAL MALARIA

Cerebral malaria is a life-threatening complication of *P. falciparum* infestation that occurs in approximately 2% of cases (77). In endemic areas, it affects mainly children. Occurrence in adults is far less frequent, yet it is seen among persons who have lived away from endemic areas for a sufficient time to lose their immunity. Progressive clinical changes occur along with high fever and chills. The neurologic manifestations are nonspecific because of diffuse involvement of the brain. Transient extrapyramidal and neuropsychiatric manifestations as well as isolated cerebellar ataxia may occur, but localizing signs are rare (77–79). Coma may ensue and approximately one-third of patients die.

There are few reports describing the MRI features of cerebral malaria (Fig.18). In a prospective MR study of 24 patients with cerebral malaria imaged on a 0.2-T system, only increase in brain volume was recorded. It was attributed to an increase in cerebral blood flow caused by vasodilatation and sequestration of infected erythrocytes (80). In another study, cortical and subcortical ischemic brain lesions have been reported in a case of cerebral malaria (81–84). Other imaging findings reported are central pontine and upper medulla myelinolysis and microinfarcts of the cerebellar hemispheres (82,83).

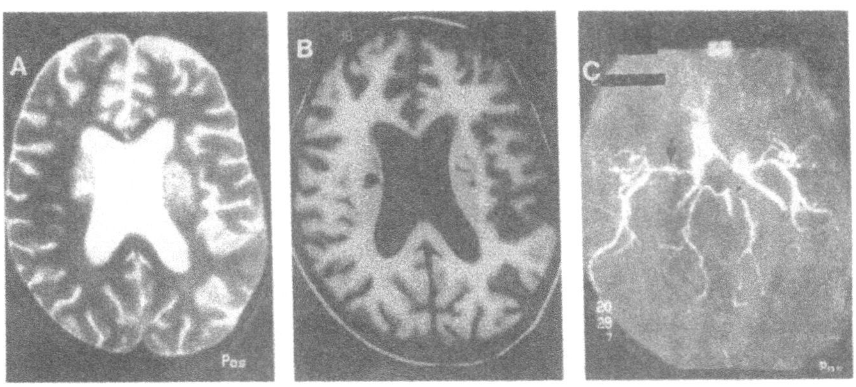

Figure 18. Cerebral malaria. T2-weighted axial image (a) at the level of ventricles shows hyperintense areas in the periventricular location that appears hypointense on T1-weighted image (b) suggestive of infarcts. MR angiogram (c) shows narrowing of left middle cerebral artery (large arrow) and posterior cerebral artery (small arrow). Patient had clinical presentation of cerebral malaria and peripheral blood smear was positive for *Plasmodium falciparum*.

The histologic findings in cerebral malaria include sequestration of infected erythrocytes in brain vessels, mainly cortical and perforating arteries, with perivascular ring hemorrhages and white matter necrosis (78,79). Presence of edema is more difficult to document pathologically because, in postmortem studies, brain edema may not be appreciated. In some patients, diffuse or focal hyperintensity has been reported better seen on FLAIR imaging and considered characteristic of cerebral edema (83,84). The imaging abnormalities reverse on successful treatment (84). In some cases, the irreversible myelin damage may finally occur. Recent pathologic and experimental studies have focused on the effect of endothelial activation, nonspecific immune inflammatory response, and subsequent release of cytokines. These factors could lead to vascular engorgement and vasodilatation with alteration of cerebral blood flow and edema, which might explain the presence of raised intracranial pressure and brain swelling (78,79). The hypothesis of brain swelling resulting from vasodilatation and edema may explain the reversal of clinical and MRI signs with successful treatment (84).

7. OTHER MISCELLANEOUS PARASITIC INFECTIONS INVOLVING THE CNS

Rarely, trichinosis, *Toxocara canis*, and schistosomiasis may infect the CNS. Isolated cases with MRI features have been described and are nonspecific for these conditions. We will describe in brief, the clinical and imaging features to complete the parasitic CNS infections.

7.1. Neurotrichinosis

Trichinosis is a parasitic disease acquired by ingestion of undercooked meat contaminated with larvae of *Trichinella spiralis*. Mild and moderate forms are usual. The symptoms of overt trichinosis consist of fever, headache, myalgia, periorbital edema, nausea, and diarrhea. Several forms, including neurologic complications, occur in 5–10% of the cases (85,86). Pathologic findings include small ischemic cavities in the white matter and small cortical infarcts (85). Neurologic disturbances consist of both diffuse encephalopathy and focal neurologic deficits and usually occur simultaneously a few days after the first general symptoms. There are few reports describing the MRI features of neurotrichinosis (87,88). MRI may show high signal intensity in corona radiata and/or periventricular white matter on T2-weighted images with cortical infarcts. Postcontrast study shows enhancement of cortical lesions with no enhancement of deep white matter lesions (88). On follow-up MRI, there is disappearance of lesions and some manifest as decrease in size and disappearance of the enhancement paralleled by partial remission of symptoms (88).

7.2. *Toxocara canis*

T. canis is a common roundworm in the dog that can be responsible in humans for a so-called "visceral larva migrans syndrome" of generalized illness with eosinophilia and symptoms arising from larval invasion of different organs. Involvement of the liver, lungs, and eyes is common, but CNS infestation is rare (85).

Only a few cases of neurotoxocarosis have been investigated with MRI (89–91). The first case involving the brain was reported in 1991 and described approximately 25 homogeneously enhancing parenchymal lesions (89), which decreased dramatically in size and number following antihelminthic treatment. A second report published in 1994 described solitary right-sided intensely and homogeneously enhancing focus at the T2/3 level (90). Sensory disturbances including paresthesiae, hyperpathia, and dysesthesiae were the patient's prominent complaints and hyperesthesia in one leg was the only residual impairment after treatment. The lesion involved predominantly the posterior column (Fig.19) although other areas are also involved. Preferential location of residual damage within the posterior columns may be a clue to a new etiologic hypothesis. This concept has to be validated by clinical and MRI monitoring of other cases of spinal neurotoxocaral infection. However, clearly symmetrical residual damage confined to both posterior columns strongly suggests a specific sensitivity of these structures to the organism (or its toxins). The usefulness of serial MRI during antihelminthic chemotherapy for CNS infections must be emphasized, since eosinophilia and serum or even CSF titers of Toxocara antibodies may be negative or borderline at the time of infestation and, on the other hand, may remain positive for months or years after clinical improvement. Assessment of resolution of the lesions by serial MRI confirms the diagnosis and the adequacy of treatment (91).

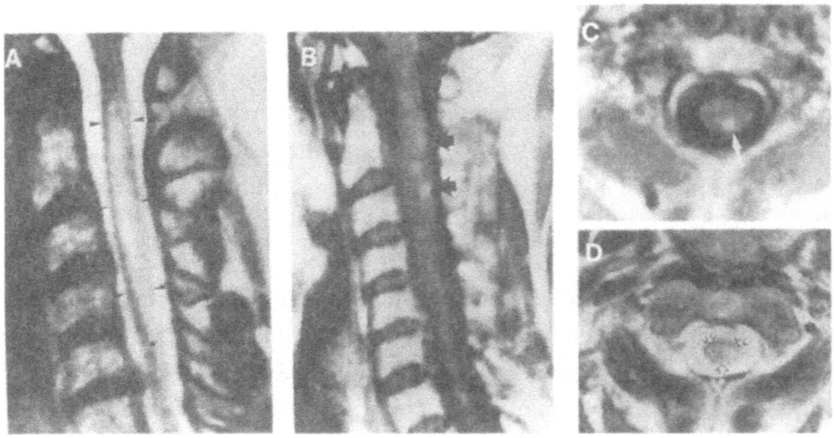

Figure 19. Toxocara canis myelopathy. Sagittal T2-weighted fast spin-echo image of the cervical spine (a) showing a large high-signal area within the swollen cord (arrowheads) from C2 to C5 (stars). Contrast-enhanced T1-weighted spin-echo image (b) shows spotty enhancing foci in the posterior segment of cord at C2–C3 (arrows). Contrast-enhanced axial T1-weighted image (c) at C2 level showing a large enhancing focus within the posterior column (arrow). Axial T2 FSE (d) at C2 level demonstrated swollen cord with a large high-signal area throughout the cord (open arrows). (Reprinted from Duprez TPJ, et al. MRI of cervical cord lesions and their resolution in Toxocara canis myelopathy. Neuroradiology 1996; 38: 792–795, by permission of Springer-Verlag.)

7.3. Schistosomiasis

Schistosomiasis is one of the most important parasitic diseases, occurring in more than 200 million people worldwide. Three different species of the trematode *Schistosoma* cause human disease: *S. japonicum* in Japan, China, the Philippines, and Southeast Asia; *S. mansoni* in Africa, parts of Southwest Asia, the Caribbean, and South America; and *S. haematobium* in Africa and Southwest Asia. The disease occurs predominantly in countries between the latitudes of 36° north and 34° south (92).

Humans are the definitive host for these three main species. Adult worms live in blood vessels in specific anatomic sites according to the species. Adult worms mate in the blood vessels of the vertebrate host, after which the female worm migrates against normal blood flow into venules around the intestine or bladder. She lays hundreds to thousands of incompletely embryonated eggs and about half of them migrate to the intestine or the urinary bladder. Some of the eggs are discharged through feces or urine into the environment. The eggs on reaching freshwater hatch into miracidia that enter the freshwater snail, an intermediate host (92–94).

CNS schistosomiasis usually follows egg migration into the spinal cord or brain vasculature leading to microinfarction or granuloma formation. Because of their small size, *S. japonicum* eggs are more likely to reach the brain; they cause 60% of all intracranial lesions. The larger eggs of *S. mansoni* cause most spinal cord lesions, being rarely found in the brain. Schistosome eggs are presumed to reach the spinal cord venules by means of retrograde flow from the iliac veins and inferior vena cava via the venous plexus of Batson. Localization of eggs is predominantly in the lumbosacral venous plexus. Deposition of eggs in the brain appears to occur from embolization through the vertebral venous plexus, particularly in patients with cor pulmonale (92–94).

The tissue response to schistosome eggs may be absent or it may be a moderate inflammation characterized by an infiltrate with lymphocytes, eosinophils, and macrophages with perivascular infiltration or a granulomatous reaction. Focal and diffuse vasculitis may also be found. The regions where the eggs may localize are leptomeninges, parietal, occipital, and temporal lobes, basal ganglia, hippocampus, and cerebellum. Significant granulomas rarely develop in the brain. In the spinal cord granulomas, hypersensitivity reaction, and anterior spinal arterial occlusions may occur. Neurologic symptoms depend on the site of involvement in the brain or spinal cord. Four manifestations of schistosomiasis occur in the spinal cord: (1) asymptomatic deposition of eggs in the spinal cord, (2) confluent granulomatous intramedullary mass in the distal cord, (3) radicular involvement with granulomas around the conus and cauda equina, and (4) acute transverse myelitis subsequent to diffuse granulomatous changes. Granulomatous intramedullary mass in the distal cord is responsible for 78% of the reported cases (93,94).

Cranial MRI may show cerebral edema and multifocal infarcts (95). In late stages, cerebral atrophy is seen. MR of schistosomal myelitis secondary to *S. mansoni* reveals nonspecific enlargement of lower thoracic cord and conus and increased signal intensity on T2-weighted images (Fig.20). Postgadolinium injection, there is enhancement of the intramedullary foci and follow-up study after treatment with praziquantel reveals resolution of the lesion (96,97). T2 hyperintensity is out of proportion to the focus of contrast-enhancing area. The MRI changes are nonspecific and can be seen in myelitis due to other bacterial and viral agents (97).

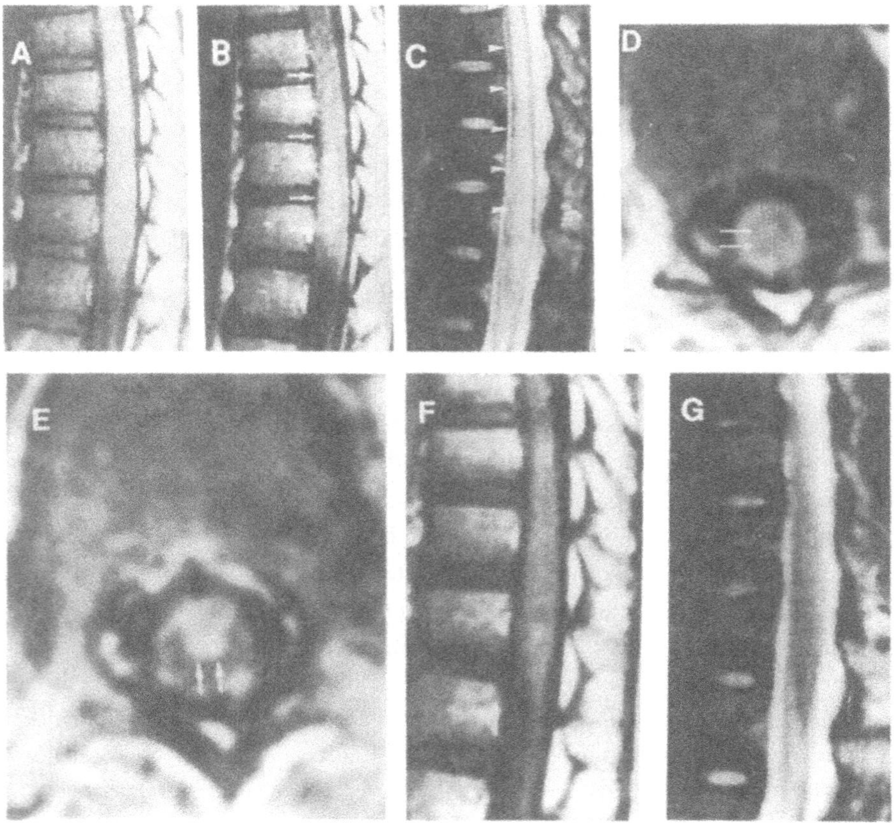

Figure 20. Schistosoma mansoni intramedullary abscess. T1-weighted sagittal and axial images before (a and d) and after (b and e) contrast administration show widened anteroposterior and transverse dimensions of the spinal cord with low central signal intensity (arrows, d) and with intramedullary contrast enhancement (arrows, b and e), most prominent anteriorly. The initial T2-weighted image (c) shows abnormal high signal in the distal cord and conus (arrowheads). Five months later, there is no persisting enhancement after contrast enhancement (f), and signal intensity on T2-weighted image is normal (g). (Reprinted with permission from Murphy KJ, et al. Am J Neuroradiol 1998; 19: 341–348.)

REFERENCES

1. Pitella JEH. Neurocysticercosis. Brain Pathol 1997; 7: 681–693.
2. Ahuja GK, Roy S, Kamla G, Virmani V. Cerebral cysticercosis. J Neurol Sci 1978; 35: 365–374.
3. Del Brutto OH, Sotelo J. Neurocysticercosis: An update. Rev Infect Dis 1988; 10: 1075–1087.

4. Lobato RD, Lamas E, Portillo JM, et al. Hydrocephalus in cerebral cysticercosis. Pathogenetic and therapeutic considerations. J Neurosurg 1981; 55: 786–793.
5. Monteiro L, Coelho T, Stocker A. Neurocysticercosis—A review of 231 cases. Infection 1992; 20: 61–65.
6. Nash TE, Neva FA. Recent advances in the diagnosis and treatment of cerebral cysticercosis. N Engl J Med 1984; 311: 1492–1496.
7. Schantz PM. Surveillance and control programs for cestode diseases. In *Parasitic diseases:Treatment and control*, MJ Miller, EJ Love, eds. Boca Raton, Florida: CRC Press; 275–290.
8. Scharf D. Neurocysticercosis. Two hundred thirty eight cases from a California hospital. Arch Neurol 1988; 45: 777–780.
9. Sparks AK, Neafie RC, Connor DH. Cysticercosis. In *Pathology of tropical and extraordinary diseases*, CH Binford, DH Connor, eds. Washington, DC: Armed Forces Institute of Pathology, 1976; 539–543.
10. Orihel TC, Ash LR. *Parasites in human tissues.* Chicago: American Society of Clinical Pathologists, 1995.
11. Escobar A, Nieto D. Parasitic diseases. In *Pathology of the nervous system*, J Minckler, ed. New York: McGraw–Hill, 1972; 2503–2521.
12. Flisser A, Madrazo I. Evagination of Taenia solium in the fourth ventricle. N Engl J Med 1996; 335: 753–754.
13. Rabiela Cervantes MT, Rivas-Hernandez A, Rodrigues-Ibarra J, Castillo-Medina S, Cancino FM. Anatomopathological aspects of human brain cysticercosis. In *Cysticercosis: Present state of knowledge and perspectives,* A Flisser, K Sillms, JP Laclette, C Larralde, eds. New York: Academic Press, 1982; 179–200.
14. McCormick GF, Zee C-S, Heiden J. Cysticercosis cerebri. Review of 127 cases. Arch Neurol 1982; 39: 534–539.
15. Wadia N, Desai S, Bhatt M. Disseminated cysticercosis. New observations, including CT scan findings and experience with treatment by praziquantel. Brain 1988; 111: 597–614.
16. Sotelo Jguerrero V, Rubio F. Neurocysticercosis: A new classification based on active and inactive forms. A study of 753 cases. Arch Intern Med 1985; 145: 442–445.
17. Trelles JO, Trelles L. Cysticercosis of the nervous system. In *Handbook of clinical neurology*, PJ Vinken, GW Bruyn, eds. Amsterdam: North–Holland, 1978; 291–320.
18. Zee CS, Segali HD, Destian S, Ahmadi J, Apuzzo ML. MRI of intraventricular cysticercosis: Surgical implications. J Comput Assist Tomogr 1993; 17: 932—939.
19. Del Brutto OH. Cysticercosis and cerebrovascular disease: A review. J Neurol Neurosurg Psychiatry 1992; 55: 252–254.
20. Rodriguez-Cabajal J, Del Brutto OH, Panagos P, Huebe J, Escobar A. Occlusion of the middle cerebral artery due to cysticercotic angiitis. Stroke 1989; 20: 1095–1099.
21. Kohli A, Gupta RK, Kishore J. Anterior cerebral artery territory infarction in neurocysticercosis: Evaluation by MR angiography and in vivo proton MR spectroscopy. Pediatr Neurosurg 1997; 26: 93–96.
22. Sharma K, Gupta RK. Negative scan in Neurocysticercosis. Pediatr Neurosurg 1993; 19:206–207.
23. Teitelbaum GP, Otto RJ, Lin M, et al. MR imaging of neurocysticercosis. Am J Neuroradiol 1989; 153: 857–866.
24. Zee CS, Segall HD, Boswell W, Ahmadi J, Nelson M, Colleti P. MR imaging of neurocysticercosis. J Comput Assist Tomogr 1988; 12: 927–934.
25. Del Brutto OH, Zenteno MA, Salgado P, Sotelo J. MR imaging in cysticercotic encephalitis. Am J Neuroradiol 1989; 10: S18–S20.

26. Suss RA, Maravilla KR, Thompson J. MR imaging of intracranial cysticercosis: Comparison with CT and anatomopathologic features. Am J Neuroradiol 1986; 7: 235–242.
27. Jena A, Sanchetee PC, Gupta RK, Khushu S, Chandra R, Lakshmipathi N. Cysticercosis of brain shown by magnetic resonance imaging. Clin Radiol 1988; 39: 542–546.
28. Creasy J, Alacron J. Magnetic resonance imaging of neurocysticercosis. Top Magn Reson Imaging 1994; 6: 59–68.
29. Chang KH, Cho SY, Hesselink JR, Han MH, Han MC. Parasitic diseases of the central nervous system. Neuroimaging Clin North Am 1991; 1: 159–178.
30. Chang KH, Lee JH, Han MH, et al. The role of contrast enhanced MR imaging in the diagnosis of neurocysticercosis. Am J Neuroradiol 1991; 12: 501–513.
31. Rajshekhar V, Chandy MJ. Comparative study of CT and MRI in patients with seizures and a solitary cerebral cysticercus granuloma. Neuroradiology 1996; 38: 542–546.
32. Spickler EM, Lufkin RB, Teresi L, Lanman T, Levesque M, Bentson JR. High-signal intraventricular cysticercosis on T1-weighted MR imaging [Case Report]. Am J Neuroradiol 1989; 10: S64.
33. Suh DC, Chang KH, Han MH, Lee SR, Han MC, Kim CW. Unusual MR manifestations of neurocysticercosis. Neuroradiology 1989; 31: 396–402.
34. Jena A, Sanchetee P, Tripathi R, Jain R, Gupta A, Sapra M. MR observations on the effects of praziquantel in neurocysticercosis. Magn Reson Imaging 1992; 10: 77–80.
35. Kathuria MK, Gupta RK, Roy R, Gaur V, Hussain N, Pradhan S. Measurement of magnetization transfer in different stages of neurocysticercosis. J Magn Resom Imaging 1998; 8: 473–479.
36. Gupta RK, Kathuria MK, Pradhan S. Magnetization transfer MR imaging in central nervous system tuberculosis. Am J Neuroradiol 1999; 20: 867–875.
37. Yamada N, Imakita S, Sakuma RT, Takamiya M. Intracranial calcification on gradient-echo phase images depiction of diamagnetic susceptibility. Radiology 1996; 198:171–178.
38. Kramer LD. Medical treatment of cysticercosis: Ineffective. Arch Neurol 1995; 52: 101–102.
39. Del Brutto OH, Santibanez R, Noboa CA, Aguirre R, Diaz E, Alarcon TA. Epilepsy due to neurocysticercosis: Analysis of 203 patients. Neurology 1992; 42: 389–392.
40. Gupta RK, Kathuria MK, Pradhan S. Magnetization transfer MR imaging demonstration of perilesional gliosis: Its relationship with epilepsy in treated or healed neurocysticercosis. Lancet 1999; 354: 41–42.
41. Rhee RS, Kumasaki DY, Sarwar M, Rodriguez J, Naseem M. MR imaging of intraventricular cysticercosis. J Comput Assist Tomogr 1988; 11: 598–601.
42. Gupta RK, Jain VK, Kumar S, Gupta S, Haque I, Gujral RB. Unusual MRI appearance of cysticercus within the fourth ventricle. Neuroradiology 1993; 35: 457–458.
43. Chang KH, Song IC, Kim SH, et al. In vivo single voxel proton MR spectroscopy in intracranial cystic masses. Am J Neuroradiol 1998; 19: 401–405.
44. Castillo M, Quencer RM, Post MJD. MR of intramedullary spinal cysticercosis. Am J Neuroradiol 1988; 9: 393–395.
45. Gaur V, Dev R, Kathuria M, Husain M, Gupta RK. MRI of isolated intramedullary cysticercosis. Clin Radiol 2000; 55: 311–314.
46. Bia FJ, Barry M. Parasitic infections of the central nervous system. Neurol Clin 1986; 4: 171–206.
47. Chang KH, Cho SY, Hesselink JR, Han MH, Han MC. Parasitic diseases of the central nervous system. Neuroimaging Clin North Am 1991; 1: 159–178.

48. Chang KH, Han MH. MRI of CNS parasitic diseases. J Magn Reson Imaging 1998; 8: 297–307.
49. Cha SH, Chang KH, Cho SY, et al. Cerebral paragonimiasis in early active stage: CT and MR features. Am J Roentgenol 1994; 162: 141–145.
50. Chang KH, Cha SH, Han MH, et al. An imaging diagnosis of cerebral paragonimiasis: CT and MR findings and correlation with ELISA antibody test. J Korean Rad Soc 1993; 29: 345–354.
51. Kadota T, Ishikura R, Tabuchi Y, et al. MR imaging of chronic cerebral paragonimiasis. Am J Neuroradiol 1989; 10: S21–S22.
52. Oh SJ. Roentgen findings in cerebral paragonimiasis. Radiology 1968; 90: 291–299.
53. Udaka F, Okuda B, Okada M, Tsufi T, Kameyama M. CT findings of cerebral paragonimiasis in the chronic state. Neuroradiology 1988; 30: 31–34.
54. Cho SY, Bae JH, Seo BS, et al. Some aspects of human sparganosis in Korea. Korean J Parasitol 1975; 13: 60–77.
55. Chang KH, Cho SY, Chi JG, et al. Cerebral sparganosis: CT characteristics. Radiology 1987; 165: 505–510.
56. Moon WK, Chang KH, Cho SY, et al. Cerebral sparganosis: MR imaging versus CT features. Radiology 1993; 188: 751–757.
57. Kim DG, Paek SH, Chang KH, et al. Cerebral sparganosis: Clinical manifestations, treatment and outcome. J Neurosurg 1996; 85: 1066–1071.
58. Taratuto AL, Venturiello SM. Echinococcosis. Brain Pathol 1997; 7: 673–679.
59. Kammerer WS. Echinococcosis affecting the central nervous system. Semin Neurol 1993; 13:144–147.
60. Coats R, von Sinner W, Rahm B. MR imaging of an intracerebral hydatid cyst. Am J Neuroradiol 1990; 11: 1249–1250.
61. Turgut M, Benli K, Eryilmaz M. Secondary multiple intracranial hydatid cysts caused by intracerebral embolism of cardiac echinococcosis: An exceptional case of hydatidosis. J Neurosurg 1997; 86: 714–718.
62. Nurchi G, Floris F, Montaldo C, Mastio F, Peltz T, Coraddu M. Multiple cerebral hydatid disease: Case report with magnetic resonance imaging study. Neurosurgery 1992; 30: 436–438.
63. Aksungur EH, Oguz M, Bicakci K, Altey M. Cerebral hydatid lesion with daughter cysts: MR and CT findings. Am J Roentgenol 1994; 163: 1269–1270.
64. Kohli A, Gupta RK, Poptani H, Roy R. In vivo proton magnetic resonance spectroscopy in a case of intracranial hydatid cyst. Neurology 1995; 45: 562–564.
65. Novak M, Hameed N, Buist R, Blackburn BJ. Metabolites of alveolar echinococcus as determined by 31-P and 1H nuclear magnetic resonance spectroscopy. Parasitol Res 1992; 78: 665–670
66. Bensaid AH, Dietemann JL, Filippi de la Palavesa MM, et al. Intracranial alveolar echinococcosis: CT and MRI. Neuroradiology 1994; 36: 289–291.
67. Braithwaite PA, Lees RF. Vertebral hydatid disease. Radiological assessment. Radiology 1981; 140: 763–766.
68. Jena A, Tripathi RP, Jain AK. Primary spinal echinococcosis causing paraplegia. Case report with MR and pathological correlation. Am J Neuroradiol 1991; 12: 560.
69. Ogut AG, Kanberoglu K, Altug A, Cokyuksel O. CT and MRI in hydatid disease of cervical vertebrae. Neuroradiology 1992; 34: 430–432.
70. Marani SAD, Canossi GC, Nicoli FA, Alberti GP, Monni SG, Casolo PM. Hydatid disease: MR imaging study. Radiology 1990; 175: 701–706.

71. Tekkok IH, Benli K. Primary spinal extradural hydatid disease: Report of a case with magnetic resonance characteristics and pathological correlation. Neurosurgery 1993; 33: 320–323.
72. Campbell S. Amoebic brain abscess and meningoencephalitis. Semin Neurol 1993; 13: 153–160.
73. Lombardo L, Alonso P, Saenz-Arroyo L, et al. Cerebral amoebiasis—report of 17 cases. J Neurosurg 1964; 21: 704–708.
74. Dietz R, Schanen S, Kramann B, Erpelding J. Intracranial amoebic abscess: CT and MRI findings. J Comput Assist Tomogr 1991; 15: 168–170.
75. De Villiers JP, Durra G. Amoebic abscess of the brain. Clin Radiol 1998; 6: 307–309.
76. Zagardo MT, Castellani RJ, Zoarski GH, Bauserman SC. Granulomatous amebic encephalitis caused by leptomyxoid amoebae in an HIV-infected patient. Am J Neuroradiol 1997; 18: 903–908.
77. Marsden PD, Bruce-Chwatt LJ. Cerebral malaria. In *Topics on tropical neurology*, RW Hornabrook, ed. Philadelphia: Davis, 1975; 29–44.
78. Janota I, Doshi B. Cerebral malaria in United Kingdom. J Clin Pathol 1979; 32: 769–772.
79. Turner G. Cerebral malaria. Brain Pathol 1997; 7: 569–582.
80. Millan JM, San Millan JM, Munoz M, Navas E, Lopez-Veles R. CNS complications in acute malaria: MR findings. Am J Neuroradiol 1993; 14: 493–494.
81. Kampfl AW, Birbamer GG, Pfausler BE, Haring HP, Schmutzhard E. Isolated pontine lesion in algid malaria: Clinical features, management and MRI findings. Am J Trop Med Hyg 1993; 48: 818–822.
82. Saissy JM, Pats B, Renard JL, Dubayle P, Herve R. Isolated bulb lesion following mild Plasmodium falciparum malaria diagnosed by MRI. Intens Care Med 1996; 22:610.
83. Looareesuwan S, Wilairatana P, Krishna S, et al. Magnetic resonance imaging of the brain in patients with cerebral malaria. Clin Infect Dis 1995; 21: 300–309.
84. Cordoliani YS, Sarrazin JL, Felten D, Caumes E, Leveque C, Fisch A. MR of cerebral malaria. Am J Neuroradiol 1998; 19: 871–874.
85. Pawlowski ZS. Clinical aspects in man. In *Trichinella and trichinosis*, WC Campbell, ed. New York: Plenum Press, 1983; 367–401.
86. Fourestie V, Douceron H, Brugieres P, Ancelle T, Lejonc JL, Gherardi RK. Neurotrichinosis. A cerebrovascular disease associated with myocardial injury and hypereosinophilia. Brain 1993; 116: 603–616.
87. Ellrodt A, Halfon P, LeBras P, et al. Multifocal central nervous system lesions in three patients with trichinosis. Arch Neurol 1987; 44: 432–434.
88. Feydy A, Touze E, Miaux Y, et al. MRI in a case of neurotrichinosis. Neuroradiology 1996; 38: S80–S82.
89. Ruttinger P, Hadidi H. MRI in cerebral toxocaral disease. J Neurol Neurosurg Psychiatry 1991; 54: 361–362.
90. Kumar J, Kimm J. MR in Toxocara canis myelopathy. Am J Neuroradiol 1994; 15: 1918–1920.
91. Duprez TPJ, Bigaignon G, Delgrange E, et al. MRI of cervical cord lesions and their resolution in Toxocara canis myelopathy. Neuroradiology 1996; 38: 792–795.
92. Liu L. Spinal and cerebral schistosomiasis. Semin Neurol 1993; 13:189–200.
93. Pittella JE, Lana-Peixoto MA. Brain involvement in hepatosplenic schistosomiasis mansoni. Brain 1981; 104: 621–632.
94. Gelfand M. Neurological complications of parasitic disease. In *Tropical neurology*, JD Spillane, ed. London: Oxford University Press, 1977; 247–258.

95. Scrimgeor EM, Gajdusek DC. Involvement of the CNS in schistosomiasis mansoni: A review. Brain 1985; 108:1023–1038.
96. Silbergleit R, Silbergleit R. Schistosomal granuloma of the spinal cord: Evaluation with MR imaging and intraoperative sonography. Am J Roentgenol 1992; 158:1351–1353.
97. Murphy KJ, Brunberg JA, Quint DJ, Kazanjian PH. Spinal cord infection: Myelitis and abscess formation. Am J Neuroradiol 1998; 19: 341–348.

Chapter 7

AIDS AND ITS RELATED INFECTIONS

K. E. Sakaie and Robert G. Gonzalez

Human immunodeficiency virus (HIV)-infected individuals often suffer from central nervous system (CNS) damage. Ten percent have a CNS lesion as the first clinical manifestation of AIDS while 40% suffer a neurologic complication at some point. Approximately 80% of those examined on autopsy show CNS abnormalities (1–5). In the CNS, HIV infection predominantly results in opportunistic infections, lymphoma, and neuropsychiatric disorders. Other sources of morbidity and mortality are vascular, toxic, metabolic, and autoimmune phenomena. Neuroimaging plays a key role in diagnosis and guiding therapy.

Even in the absence of opportunistic infections or lymphoma, HIV-induced damage of the CNS leads to a range of symptoms from mild cognitive-motor deficits to profound dementia. The AIDS dementia complex (ADC) scale describes the degree of neurologic dysfunction (Table 1) (6) and ranges from ADC 0 (normal mental and motor function) to ADC 4 (nearly vegetative). Another scheme, proposed by the World Health Organization and the American Academy of Neurology, splits the disorder into two major groups: HIV-associated minor motor cognitive disorder, equivalent to ADC 1, and HIV-associated cognitive-motor complex, equivalent to ADC 2–4 (7).

K.E. Sakaie and Robert G. Gonzalez • MGH-NMR Center, Charlestown, Massachusetts 02129.

MR Imaging and Spectroscopy of Central Nervous System Infection, edited by Gupta and Lufkin. Kluwer Academic / Plenum Publishers, New York, 2001.

Table 1. Staging of AIDS Dementia Complex

Stage	Description
0 (normal)	Normal mental and motor function.
1 (mild)	Able to perform all but the more demanding work or activities of daily living but with unequivocal evidence (symptoms or signs, including performance on neuropsychological testing) of intellectual or motor impairment. Can walk without assistance.
2 (moderate)	Able to perform basic activities of self-care but cannot work or maintain more demanding aspects of daily life. Ambulatory but may require a single prop.
3 (severe)	Major intellectual incapacity (cannot follow news or personal events, cannot sustain complex conversation, considerable slowing of all output) or motor disability (cannot walk unassisted, requiring walker or personal support, usually with slowing and clumsiness of arms as well).
4 (end stage)	Nearly vegetative. Intellectual and social comprehnesion and output are at a rudimentary level. Nearly or absolutely mute. Paraparetic or paraplegic with double incontinence.

Of the neuroimaging probes, the most common in a clinical setting, MRI and CT, reveal structural abnormalities and play a critical role in initial diagnosis of new-onset neurologic symptoms in general. In the absence of opportunistic infections or lymphoma, structural imaging typically finds no abnormalities until later stages of HIV disease. Advanced imaging probes reveal metabolic or functional changes in the CNS and can add diagnostic sensitivity and specificity. Among these probes are positron emission tomography (PET), single-photon-emission computed tomography (SPECT), dynamic contrast functional MRI (fMRI), and magnetic resonance spectroscopy (MRS). Such techniques show potential for early detection of CNS injury, even before the onset of neurologic symptoms, and for rapid evaluation of drug therapy. This paper focuses on MR-based methods—MRI, fMRI, and MRS.

1. MRI IN HIV

MRI is the most sensitive noninvasive imaging procedure for the detection of CNS parenchymal lesions. T2-weighted images are particularly sensitive to the changes in the concentration of water that occur with edema present in the majority of the lesions that affect the CNS in neurologically symptomatic HIV-positive individuals. T1-weighted images are useful for delineation of CNS anatomy and the presence of subacute hemorrhage. Use of intravenous, gadolinium (Gd)-based contrast agents aids evaluation of CNS lesions in symptomatic HIV-positive patients. Contrast agents shorten relaxation times of lesions that have an altered blood–brain barrier, making their appearance more conspicuous. Gd-based contrast is also useful for evaluating meningeal disease (8).

CNS infection by HIV may occur concurrently with or soon after systemic infection. Clinical signs during this acute period include acute meningoencephalitis, Guillain–Barré syndrome, and peripheral mono- or polyneuropathy. Occasionally, patients may present with generalized adenopathy and/or asymptomatic parotid swelling due to the presence of lymphoepithelial cystic or solid lesions. Cerebrospinal fluid (CSF) may show abnormal quantities of immune response factors.

Neurologic complications, the severity of which is often described with the ADC scale, may take months to years to develop. Accompanying damage to neural tissues occurs mainly in the subcortical gray matter, especially the basal ganglia, leading to progressive subcortical dementia that may be detected in early stages by neuropsychological testing. Histopathology shows mononuclear cell infiltrates, multinucleated giant cells, reactive astrocytosis, abnormalities of the white matter, and vacuolar changes in the infected tissues.

Despite clinical and neurologic evidence of CNS injury, results of structural neuroimaging studies are usually normal. Rarely, patients who present a severe encephalopathy demonstrate multifocal abnormalities as seen inT2-weighted MRI (Fig.1). Antiretroviral therapy may reverse these abnormalities (9). More often, HIV-induced brain injury appears on MRI only at later stages of the disease as whitematter disease and cortical atrophy. White matter disease, resulting from oligodendroglial damage, leads to abnormally increased signal inT2-weighted images (Fig.2). Cortical atrophy manifests itself in MRI as enlarged sulci and ventricles (Fig.2).

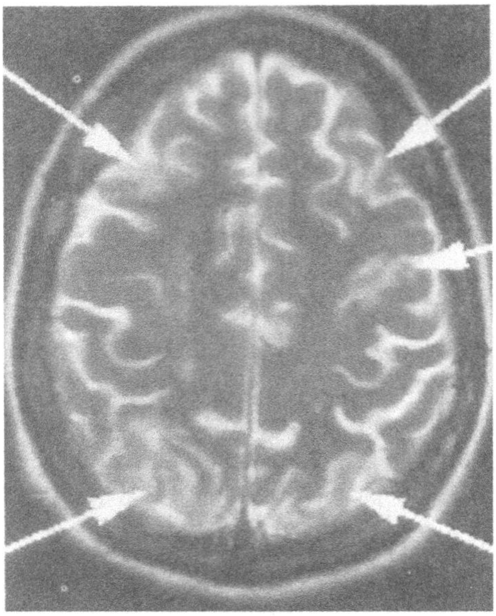

Figure 1. Multifocal abnormalities.

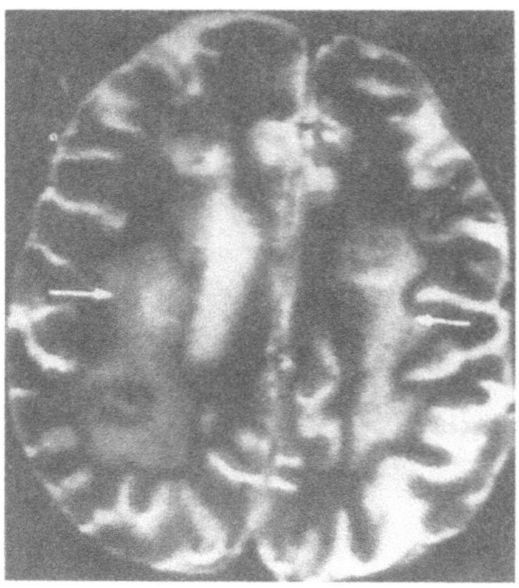

Figure 2. HIV encephalitis. AxialT2-weighted MR image at the level of the centrum semiovale displays extensive, abnormal increased signal throughout the white matter. Atrophy is manifested by enlarged sulci.

1.1. Dynamic Contrast Functional MRI

The term functional MRI (fMRI) encompasses two general types of experiments. The first exploits endogenous contrast incurred by regional hemodynamic changes accompanying brain activation. This method has been used to study basic brain functions such as vision, audition, speech, and memory, but no such studies of HIV-related neurologic disease have been published. The second type, dynamic contrast fMRI, requires an injectable contrast agent and has been used to study the effects of HIV on the CNS.

Dynamic imaging after the first pass of a bolusinjection of intravascular paramagnetic contrast agents is possible on time scales comparable to vascular mean transit times. The passage of contrast agent through the cerebral capillary network is monitored by taking a series of T2*-weighted (gradient echo) images. The resultant signal intensity–time curves obtained from analysis of these images are converted to concentration–time curves. These curves are used to calculate relative cerebral blood volume (rCBV) and flow (rCBF).

Using dynamic contrast fMRI, Tracey et al. (10) found a generalized increase in rCBV in gray matter, especially subcortical gray matter, but not in white matter in HIV-infected patients when compared with controls (Fig.3). An increase in rCBV correlates with worsening Centers for Disease Control (CDC) criteria and ADC stage. The prominent differences in rCBV among the basal ganglia may correspond to involvement found in this area by histopathology (11, 12). White matter may exhibit pathologic abnormalities, but rCBV in white matter appeared normal in this study. The relatively small absolute CBV in white matter may render changes undetectable. Alternatively, the underlying physiologic response to HIV infection resulting in increased rCBV may not occur in white matter. In one interesting case, rCBV images from an AIDS patient were obtained before and after 6 months of antiretroviral treatment (Fig.3). Along with clinical improvement from ADC 2 to ADC 0, rCBV returned to normal.

Chang et al. have recently reported an rCBF analysis using dynamic contrast fMRI (13). They found that HIV-positive subjects had areas of decreased rCBF in the inferolateral frontal cortices and increased rCBF in the posterior inferior white matter. The authors concluded that these findings are similar to those described by investigators employing SPECT.

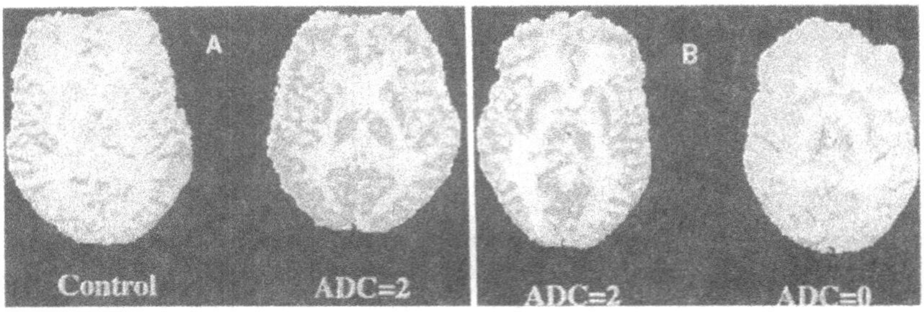

Figure 3. Representative dynamic contrast functional MRI images of normalized cerebral blood volumes at the level of the basal ganglia. The scale goes from white to yellow to red, which represents low to moderate to high relative blood volume. (a) Dynamic cerebral blood volume (CBV) image from a healthy control subject (left) and an HIV-positive patient (right) with moderate dementia (ADC Stage 2). Increase in the cortical and deep gray dynamic CBV in the patient with dementia is apparent. (b) Functional MRI images from the same patient obtained before (left) and after (right) 6 months of treatment with zidovudine. Improvement in dementia (ADC Stage 2 to 0) with treatment was accompanied by a reduction in the cortical and deep gray matter dynamic CBV.

2. *IN VIVO* MR SPECTROSCOPY IN HIV

MRS has long served as an essential tool in analytic chemistry for determining the structure of molecules and is now emerging as a powerful tool for noninvasive study of *in vivo* pathology. MRS can be performed on a standard clinical MRI scanner, yielding quantitative measurements of metabolite concentrations in 5 to 10 minutes. Such measurements provide biochemical information not available from analyses of blood, urine, or CSF, augmenting structural measurements by MRI and neuropsychological tests.

A number of comparisons have found abnormalities on MRS of HIV patients with normal MRI, indicating the benefit provided by spectroscopy (14–22).

In vivo brain MRS can augment the information provided by other methods. MRS may be helpful in sorting out the differential diagnosis of HIV dementia and other neuropathologies associated with AIDS, including certain metabolic encephalopathies. MRS also offers extra sensitivity, as neurochemical abnormalities can occur early in the course of CNS HIV infection, before structural abnormalities appear on MRI or CT and even before the onset of neurologic symptoms.

Another emerging role of *in vivo* MRS in medicine is the assessment of CNS metabolism in therapeutic outcome studies. This role of MRS has important clinical implications for drug efficacy trials, with several studies reporting reversals in brain metabolic alterations after treatment (23–25).

2.1. MRS-Detectable Metabolites of Interest in HIV

In vivo ^{1}H MRS studies have focused on the following metabolites: creatine/phosphocreatine (Cr), *N*-acetyl aspartate (NAA), choline-containing compounds (Cho), glutamate (Glu), glutamine (Gln), lactate (Lac), and myoinositol (Myo). These compounds are relatively abundant in the CNS and can usually be identified in a spectrum measured *in vivo*. Furthermore, each metabolite represents different cellular properties (26), so MRS measurements provide insight into the physiologic processes underlying CNS injury.

The Cr peak (3.0 ppm) is comprised of signal from creatine and phosphocreatine. These compounds play a role in the transfer of a phosphate group from adenosine diphosphate (ADP) to adenosine triphosphate (ATP). The importance of Cr for *in vivo* spectroscopy is its relative stability over time in various diseases. Its peak intensity is therefore used as an internal intensity reference to determine alterations in concentrations of other metabolites. A recent study, however, indicates that Cr levels may change with the progression of HIVdisease (27).

The NAA peak (2.0 ppm) is prominent because NAA occurs in large concentrations (8–15 mM) in the brain. Though the function of NAA is unknown, *in vitro* studies indicate that, in adults, NAA resides exclusively within neurons (28, 29). NAA thus serves as a neuronal marker, with a drop in NAA levels indicating neuronal loss or injury. Decreases in NAA have been found among a variety of clinical conditions that result in injury, death, or displacement of neurons, such as tumors, infarcts, abscesses, and neurodegenerative diseases. Reduction of NAA is a near-universal feature of ^{1}H MRS in HIV-related dementia, particularly during later stages of the disease (15–21, 24, 27, 30–45).

The Cho peak (3.2 ppm), as measured by *in vivo* ^{1}H MRS, encompasses signal primarily from phosphocholine and glycerophosphocholine. To a lesser extent, free choline, which is supplied exclusively from the diet, is also included. Phosphocholine is a precursor and glycerophosphocholine is a by-product of membrane phosphatidylcholine metabolism. Phosphatidylcholine is bound in cell membranes and consequently has a short T2, rendering the compound undetectable by standard *in vivo* MRS. Pathologic disturbances in cell membrane turnover may be reflected as alterations in the Cho signal (46).

An elevation in Cho may also reflect reactive gliosis—a general response of the brain to injury—because astrocytes, which proliferate during gliosis, contain higher concentrations of choline-containing compounds than neurons (29). Elevation in Cho levels found subsequent to HIV infection may therefore reflect increased membrane turnover and/or gliosis (14–22, 24, 27, 30, 33–37, 39–42, 45, 47–49). Elevated Myo (3.5 ppm), a putative glial marker (50), may also reflect gliosis. Myo is typically detected in studies performed at short TE (19, 27, 35, 36).

A number of other compounds, particularly Lac (15), lipids (19), Glu, and Gln have been observed in HIV-infected individuals by MRS, but reproducible trends have not arisen. Lac (1.3 ppm) is an end-product of anaerobic glycolysis. Potential pathologic sources for LAC in the brain include hypoxic–ischemic encephalopathy, abscess, infarction, tumor, and perhaps even infiltrating leukocytes and macrophages (51–54).

Glu, the primary excitatory neurotransmitter in the brain, also plays important roles in the metabolism of neurons and in the detoxification of ammonia in astrocytes. Gln is produced during the reuptake from the synapse of Glu by astrocytes and is the end product of the reaction of Glu and ammonia. The assessment of brain Gln concentration by ^{1}H MRS may be useful clinically in HIV-positive patients in differentiating those who are demented versus those who are encephalopathic (37).

Despite high concentrations of Gln (3 mM) and Glu (8–10mM), the aforementioned internuclear coupling effects render their spectra broad and therefore hard to quantify. Observed elevations in a broad "marker" peak [14] likely reflect changes among a number of compounds in addition to Glu and Gln.

2.1.1. Proton MRS and HIV-Induced CNS Injury

Proton MRS provides valuable information about CNS injury caused by HIV. The first results indicating the utility of MRS for the study of HIV-related CNS injury (38, 55) found depressed levels of NAA. These findings have subsequently been confirmed. A multicenter trial found reproducibility among results from different sites (48). In general, abnormalities in NAA, Cho, and Myo levels seem to correlate most strongly with the degree of neurologic impairment but less with systemic measures of disease severity. Also important is the enhanced sensitivity of MRS to CNS damage as compared with structural imaging by MRI and CT. Early examination by MRS may therefore speed the evaluation of drug therapy. Of potential importance are findings of differences among regions in the brain and the effects of different acquisition parameters. MRS has not found differences between high-risk and low-risk HIV-negative controls (40, 45), nor has MRS found differences as a function of mode of transmission (14, 48).

The most consistent findings are depressed levels of NAA and elevated levels of Cho in HIV-infected individuals with respect to controls, especially at later stages of the disease and with the onset of severe dementia (17, 21, 30, 35, 36, 38, 40, 41). This in *vivo* evidence of neuronal loss and gliosis is consistent with quantitative neuropathologic studies of postmortem samples. Abnormalities found by MRS also concur, after the onset of moderate to severe dementia, with evidence of neuronal loss and injury manifested in MRI as atrophy and diffuse white matter disease (17, 30, 35).

MRS offers enhanced sensitivity to CNS damage. NAA and Cho levels are often abnormal in neurologically impaired individuals showing normal MRI (14–22, 34). Mildly impaired and even neurologically healthy HIV-positive individuals can show elevated Cho (21, 35, 41) and Myo levels (19, 35, 36). Such elevations are consistent with astrogliosis found by histology (56–58).

Cross-sectional (27) and longitudinal (Fig.4) (31, 37, 59) studies of HIV-positive patients with cognitive abnormalities find progressive decreases in NAA and increases in Cho and Myo as dementia worsens. Correlations have also been found between metabolite levels and CD4 counts, CSF viral load, neuropsychological tests, and the degree of encephalopathy (17, 41, 60). Other systemic measures of disease severity, particularly CDC classification, tend to have little correlation with metabolite abnormalities until the most severe stage of disease is reached. Similar trends are found in HIV-positive children (16, 19). These progressive changes of metabolite levels with other measures of disease severity suggest that MRS may provide a host of surrogate markers for different types of CNS damage.

The potential for determining drug efficacy for the CNS by MRS is shown by studies in which zidovudine was found to return NAA to normal levels (23) with subsequent NAA loss accompanying progression of neurologic dysfunction (24). Such reversals in NAA abnormalities also indicate that NAA levels may reflect reversible neuronal injury as well as death. Another study indicates that individuals displaying mainly a rise in Cho levels do not respond to zidovudine (18). A different pathology may thus exist in such individuals. Because MRS can detect CNS damage early in the progression of disease, it may prove useful for the rapid evaluation of drugs, especially in light of the fact that many antiretrovirals fail to cross the blood–brain barrier.

Most MRS studies have focused on a single region of the brain, typically white matter in the parieto-occipital lobe (18, 20, 22–24, 30, 31, 33, 42, 44, 45, 48, 61), because the relatively high homogeneity of magnetic field in this region facilitates the acquisition of reliable spectra. Other studies have focused on the centrum semiovale (14, 19, 34, 39), parietotemporal regions (38), and mesial cortex (39).

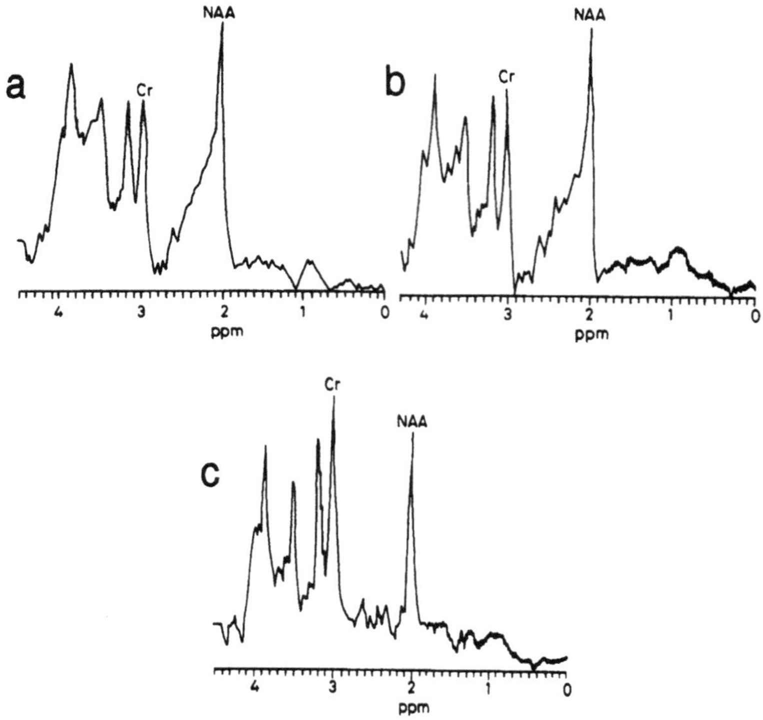

Figure 4. Serial brain MR spectra (a–c) of an HIV seropositive patient through an interval of 10 months. The NAA/Cr ratios steadily decreased through the study period. This finding suggests progressive neuronal injury. It correlates with the progressive cognitive decline in this patient who expired 3 weeks after the last MRS when he manifested ADC Stage 4. Serial brain MRI showed no morphologic change through the study period.

Some studies have attempted to determine if metabolite levels reflect the regional differences found using histopathology and neurologic studies. Multiple single voxels measured in the same patient (27, 35, 36) and chemical shift imaging (CSI) schemes (15, 17, 39–41) have been applied to the study of HIV damage of the CNS. Differences among regions have been hard to detect, though one recent study has found elevations in Cho and Myo only in frontal white matter arising among mildly neurologically impaired patients. Severely impaired patients showed elevations of Cho and Myo and a decrease in NAA in frontal gray matter and basal ganglia as well as white matter (27).

Several differences can be found between spectra measured at different TE. Spectra measured at short TE values of 20–30 ms contain more peaks than those acquired at longer TE values in the range from 100 to 200 ms. Myo, Glu, Gln amino acids, among others, appear at short TE but not long TE (14, 19, 27, 34–37, 42). Myo seems to rise in white matter even with mild neurologic dysfunction (27, 36).

The additional information provided at short TE comes at the cost of increased complexity in the spectrum. The breadth of the Glu, Gln, and amino acid peaks *in vivo* makes measurement difficult (14, 33, 34, 37). In the CNS of AIDS patients, Salvan et al. find no correlation between NAA levels measured at short and long TE, whereas they find a nonlinear correlation between Cho levels measured at short and long TE (20). The lack of a simple correlation may indicate that the same metabolite provides different physiologic information at different TE. These differences may account for some of the discrepancies found among MRS studies.

It is also possible that disease affects relaxation rates that would, in turn, affect metabolite measurements. Wilkinson et al found that the T2 of NAA increased in HIV-positive individuals showing diffuse abnormalities on MRI consistent with HIVencephalitis with respect to controls (44). This result would, however, create systematic overestimation of NAA levels in HIV-positive individuals, indicating that drops in NAA levels of patients are actually larger than those reported.

2.1.2. Phosphorus MRS in HIV Infection

Phosphorus MRS is employed to a much lesser degree than proton MRS. The reasons include lower inherent sensitivity of the nucleus, the need for additional hardware such as properly tuned head coils and transceivers, and greater difficulty in spatial localization. Metabolites detected by ^{31}P MRS include ATP, phosphomonoesters (PME), phosphodiesters (PDE), phosphocreatine, and inorganic phosphate (P_i). The P_i resonance is also useful because its chemical shift is pH dependent. Thus, ^{31}P MRS not only gives information about *in vivo* high-energy metabolism, but also provides a noninvasive method for the determination of tissue pH. ^{31}P MRS has not been extensively employed in the study of HIV infection, though investigations have shown that HIV-positive patients have lower ATP/P_i ratios as compared with controls (62, 63). These results suggest that HIV infection alters energy metabolism in the brain, and may contribute to the cognitive impairment in ADC.

3. OPPORTUNISTIC INFECTIONS AND NEOPLASMS IN AIDS

Unusual brain infections in immunocompromised HIV patients are commonly observed with MRI. Bacterial meningitis and pyogenic abscesses are infrequent in patients with AIDS, except in those with a history of being intravenous drug abusers (64). The cellular immunodeficiency in HIV patients may eventually lead to a variety of CNS infections caused by opportunistic organisms such as *Toxoplasma gondii* (65) or viral agents, such as cytomegalovirus, the JC virus (a papovavirus which causes progressive multifocal leukoencephalopathy), and herpes type 1 (66). Also found are fungal infections (cryptococcosis, aspergillosis, candidiasis) (67), mycobacterial infections (68), nocardiosis, and neurosyphilis (69). We will focus on two of the most clinically relevant infections, toxoplasmosis and progressive multifocal leukoencephalopathy.

Toxoplasma encephalitis is the most common cause of intracranial mass lesion in AIDS patients. In AIDS patients, toxoplasma encephalitis is a reactivation infection. The clinical presentation of toxoplasma encephalitis is nonspecific and indistinguishable from manifestations of other CNS infection processes or primary brain lymphoma (e.g., fever, headaches, focal neurologic deficits, seizures). On MRI, toxoplasma encephalitis may appear as a focal or multiple ring and/or nodular enhancing lesion(s) surrounded by variable degrees of vasogenic edema (Fig.5). The lesion(s) tend to be located superficially (at the corticomedullary junction) and/or deeply located (basal ganglia, thalamus). Occasionally, lesions may be encountered in the brainstem or cerebellum.

Progressive multifocal leukoencephalopathy (PML), caused by JC virus infection of oligodendrocytes, leads to demyelination. Once the diagnosis is made, the prognosis is ominous, with death usually occurring within 9 months to 1 year. The incidence of PML in the AIDS population has been estimated to be 1–4%. MR is the modality of choice for evaluating this entity. Characteristic MR findings include areas of decreased T1 and increased T2 signal affecting white matter, usually in asymmetric distribution, but with predilection for the parietal and occipital lobes. The white matter lesion often extends to the subcortical U fibers with characteristic scalloped margins. There is no associated mass effect; indeed, there is often evidence of tissue loss. Contrast enhancement is uncommon. Cerebellum, brainstem, basal ganglia, thalamus, and external capsule involvement by PML, although infrequent, can be seen.

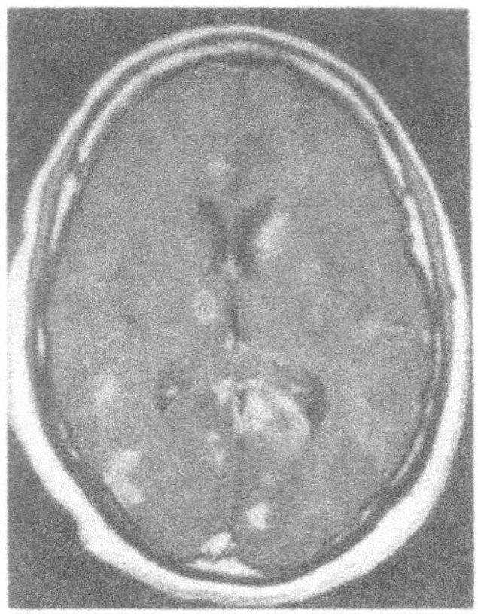

Figure 5. MRI of AIDS patient with toxoplasmosis. T1-weighted image obtained after the administration of intravenous contrast demonstrates numerous enhancing lesions throughout the brain.

The most common malignant neoplasms associated with AIDS are primary CNS lymphoma, which occurs in up to 6% of the neurologically symptomatic AIDS patients; metastatic lymphoma, which may result in leptomeningeal and epidural spread; and Kaposi's sarcoma, which rarely affects the CNS. In AIDS patients, primary CNS lymphoma is the second most common cause of focal CNS mass lesions after toxoplasma encephalitis. Clinically, it is difficult to accurately differentiate between primary brain lymphoma and other intracranial masses, particularly toxoplasma encephalitis. On MRI, lymphoma may appear as focal ring and/or nodular enhancing mass lesion(s) with surrounding edema (Fig.6).

The lesions may be single or multiple, superficial (corticomedullary junction) and/or deep (basal ganglia, thalamus, corpus callosum). The posterior fossa may also be involved (cerebellum, pons, midbrain). Rapid progression of the mass(es) on serial imaging studies also favors lymphoma. However, MRI findings are usually not pathognomonic, and lymphoma cannot be differentiated with certainty from toxoplasma encephalitis or other infectious process. In AIDS patients, primary CNS lymphoma resembles toxoplasmosis in 50–80% of the cases.

MRS may aid in differential diagnosis of brain lesions. MRS measurements of metabolite profiles performed at short and long TE reveal significant differences among toxoplasmosis, PML, lymphomas, and cryptococcomas (70). In fact, MRS alone correctly diagnosed 94% of the lesions. Another study performed only at long TE did not, however, differentiate between toxoplasmosis and lymphoma (71).

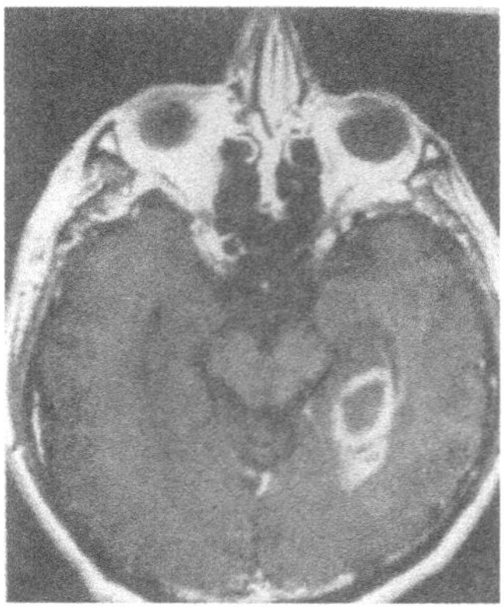

Figure 6. MRI of AIDS patient with primary CNS lymphoma. T1-weighted image obtained after the administration of intravenous contrast demonstrates a ring-enhancing lesion in the medial left temporal lobe.

Functional MRI has also proven useful for differentiating toxoplasmosis and lymphoma. Ernst and colleagues employed dynamic contrast fMRI and found increased rCBV in regions with active lymphoma but reduced rCBV in the central regions of toxoplasmosis lesions (72). rCBV is reduced in the edema surrounding lesions of both types. The rise within active lymphoma probably results from hypervascularity within the tumor, and the drop within toxoplasmosis lesions probably reflects lack of vasculature. The drop within the edema probably reflects vasoconstriction with increased interstitial pressure.

Neuroimaging is a critical diagnostic tool for evaluating the cause of neurologic abnormalities, including those associated with HIV disease. Structural analysis with MRI clearly demonstrates effects of opportunistic infections, lymphoma, and the late sequelae of CNS involvement by HIV. If structural examinations are not illuminating, functional MRI and MRS, among other advanced neuroimaging probes, should be considered. The advanced imaging schemes are less common and are still in the research stage, but show promise for early detection and evaluation of treatment of HIV-associated CNS disease.

REFERENCES

1. Navia BA, Jordan BD, Price RW. The AIDS dementia complex: I. Clinical features. Ann Neurol 1986; 19:517–524.
2. Navia BA, Cho ES, Petito CK, Price RW. The AIDS dementia complex: II. Neuropathology. Ann Neurol 1986; 19:525–535.

3. Price RW, Brew B, Sidtis J, Rosenblum M, Scheck AC, Cleary P. The brain in AIDS: Central nervous system HIV-1 infection and AIDS dementia complex. Science 1988; 239:586–592.
4. Fischer PA, Enzensberger W. Primary and secondary involvement of the CNS in HIV infection. J Neuroimmunol 1988; 20:127–131.
5. Post MJ. Neuroimaging in various stages of human immunodeficiency virus infection. Curr Opin Radiol 1990; 2:73–79.
6. Aronow HA, Brew BJ, Price RW. The management of the neurological complications of HIV infection and AIDS. AIDS 1988; 2:S151–159.
7. Nomenclature and research case definitions for neurologic manifestations of human immunodeficiency virus-type 1 (HIV-1) infection. Report of a Working Group of the American Academy of Neurology AIDS Task Force [see comments]. Neurology 1991; 41:778–785.
8. Post MJ, Sheldon JJ, Hensley GT, et al. Central nervous system disease in acquired immunodeficiency syndrome: Prospective correlation using CT, MR imaging, and pathologic studies. Radiology 1986; 158:141–148.
9. Trotot PM, Gray F. Diagnostic imaging contribution in the early stages of HIV infection of the brain. Neuroimaging Clin North Am 1997; 7:243–260.
10. Tracey I, Hamberg LM, Guimaraes AR, et al. Increased cerebral blood volume in HIV-positive patients detected by functional MRI. Neurology 1998; 50:1821–1826.
11. Achim CL, Wang R, Miners DK, Wiley CA. Brain viral burden in HIV infection. J Neuropathol Exp Neurol 1994; 53:284–294.
12. Masliah E, Ge N, Achim CL, Wiley CA. Differential vulnerability of calbindin-immunoreactive neurons in HIV encephalitis. J Neuropathol Exp Neurol 1995; 54:350–357.
13. Chang L, Speck O, Leonido-Yee M, Ernst T. Cerebral perfusion abnormalities in patients with HIV. 51st Meeting of the American Academy of Neurology, Toronto, 1999.
14. Jarvik JG, Lenkinski RE, Grossman RI, Gomori JM, Schnall MD, Frank I. Proton MR spectroscopy of HIV-infected patients: Characterization of abnormalities with imaging and clinical correlation. Radiology 1993; 186:739–744.
15. Barker PB, Lee RR, McArthur JC. AIDS dementia complex: Evaluation with proton MR spectroscopic imaging. Radiology 1995; 195:58–64.
16. Lu D, Pavlakis SG, Frank Y, et al. Proton MR spectroscopy of the basal ganglia in healthy children and children with AIDS. Radiology 1996; 199:423–428.
17. Moller HE, Vermathen P, Lentschig MG, et al. Metabolic characterization of AIDS dementia complex by spectroscopic imaging. J Magn Reson Imaging 1999; 9:10–18.
18. Salvan AM, Vion-Dury J, Confort-Gouny S, Nicoli F, Lamoureux S, Cozzone PJ. Brain proton magnetic resonance spectroscopy in HIV-related encephalopathy: Identification of evolving metabolic patterns in relation to dementia and therapy. AIDS Res Hum Retroviruses 1997; 13:1055–1066.
19. Salvan AM, Lamoureux S, Michel G, Confort-Gouny S, Cozzone PJ, Vion-Dury J. Localized proton magnetic resonance spectroscopy of the brain in children infected with human immunodeficiency virus with and without encephalopathy. Pediatr Res 1998; 44:755–762.
20. Salvan AM, Vion-Dury J, Confort-Gouny S, Nicoli F, Lamoureux S, Cozzone PJ. Cerebral metabolic alterations in human immunodeficiency virus-related encephalopathy detected by proton magnetic resonance

spectroscopy. Comparison between sequences using short and long echo times. Invest Radiol 1997; 32:485–495.

21. Tracey I, Carr CA, Guimaraes AR, Worth JL, Navia BA, Gonzalez RG. Brain choline-containing compounds are elevated in HIV-positive patients before the onset of AIDS dementia complex: A proton magnetic resonance spectroscopic study [published erratum appears in Neurology 1996; 46(6): 1787]. Neurology 1996; 46:783–788.
22. Vion-Dury J, Confort-Gouny S, Nicoli F, et al. Localized brain proton MRS metabolic patterns in HIV-related encephalopathies. C R Acad Sci III 1994; 317:833–840.
23. Vion-Dury J, Nicoli F, Salvan AM, Confort-Gouny S, Dhiver C, Cozzone PJ. Reversal of brain metabolic alterations with zidovudine detected by proton localised magnetic resonance spectroscopy [letter]. Lancet 1995; 345:60–61.
24. Wilkinson ID, Lunn S, Miszkiel KA, et al. Proton MRS and quantitative MRI assessment of the short term neurological response to antiretroviral therapy in AIDS. J Neurol Neurosurg Psychiatry 1997; 63:477–482.
25. Chang L. Reversal of cerebral metabolite abnormalities with highly active antiretroviral therapy (HAART) in mild HIV-1 dementia. ISMRM, 7th Scientific Meeting. Philadelphia.
26. Rand SD, Prost R, Li SJ. Proton MR spectroscopy of the brain. Neuroimaging Clin North Am 1999; 9:379–395.
27. Chang L, Ernst T, Leonido-Yee M; Walot I, Singer E. Cerebral metabolite abnormalities correlate with clinical severity of HIV-1 cognitive motor complex. Neurology 1999; 52:100–108.
28. Urenjak J, Williams SR, Gadian DG, Noble M. Proton nuclear magnetic resonance spectroscopy in vitro. J Neurochem 1992; 59:55–61.
29. Urenjak J, Williams SR, Gadian DG, Noble M. Proton nuclear magnetic resonance spectroscopy unambiguously identifies different neural cell types. J Neurosci 1993; 13:981–989.
30. Chong WK, Sweeney B, Wilkinson ID, et al. Proton spectroscopy of the brain in HIV infection: Correlation with clinical, immunologic, and MR imaging findings. Radiology 1993; 188:119–124.
31. Chong WK, Paley M, Wilkinson ID, et al. Localized cerebral proton MR spectroscopy in HIV infection and AIDS. Am J Neuroradiol 1994,15:21–25.
32. Gouny S, Vion-Dury J, Nicoli F, et al. A multiparametric data analysis showing the potential of localized proton MR spectroscopy of the brain in the metabolic characterization of neurological diseases. J Neurol Sci 1993; 118:123–133.
33. Cortey A, Jarvik JG, Lenkinski RE, Grossman RI, Frank I, Delivoria-Papadopoulos M. Proton MR spectroscopy of brain abnormalities in neonates born to HIV positive mothers. Am J Neuroradiol 1994; 15:1853–1859.
34. Jarvik JG, Lenkinski RE, Saykin AJ, Jaans A, Frank I. Proton spectroscopy in asymptomatic HIV-infected adults: Initial results in a prospective cohort study. J Acquir Immune Defic Syndr Hum Retrovirol 1996; 13:247–253.
35. Laubenberger J, Haussinger D, Bayer S, et al. HIV-related metabolic abnormalities in the brain: Depiction with proton MR spectroscopy with short echo times. Radiology 1996; 199:805–810.
36. Lopez-Villegas D, Lenkinski RE, Frank I. Biochemical changes in the frontal lobe of HIV-infected individuals detected by magnetic resonance spectroscopy. Proc Natl Acad Sci USA 1997; 94:9854–9859.

37. McConnell JR, Swindells S, Ong CS, et al. Prospective utility of cerebral proton magnetic resonance spectroscopy in monitoring HIV infection and its associated neurological impairment. AIDS Res Hum Retroviruses 1994; 10:977–982.

38. Menon DK, Ainsworth JG, Cox IJ, et al. Proton MR spectroscopy of the brain in AIDS dementia complex. J Comput Assist Tomogr 1992; 16:538–542.

39. Meyerhoff DJ, MacKay S, Bachman L, et al. Reduced brain N-acetylaspartate suggests neuronal loss in cognitively impaired human immunodeficiency virus-seropositive individuals: In vivo 1H magnetic resonance spectroscopic imaging. Neurology 1993; 43:509–515.

40. Meyerhoff DJ, MacKay S, Poole N, Dillon WP, Weiner MW, Fein G. N-acetylaspartate reductions measured by 1H MRSI in cognitively impaired HIV-seropositive individuals. Magn Reson Imaging 1994; 12:653–659.

41. Meyerhoff DJ, Bloomer C, Cardenas V, Norman D, Weiner MW, Fein G. Elevated subcortical choline metabolites in cognitively and clinically asymptomatic HIV+ patients. Neurology 1999; 52:995–1003.

42. Paley M, Wilkinson ID, Hall-Craggs MA, Chong WK, Chinn RJ, Harrison MJ. Short echo time proton spectroscopy of the brain in HIV infection/AIDS. Magn Reson Imaging 1995; 13:871–875.

43. Pavlakis SG, Lu D, Frank Y, et al. Magnetic resonance spectroscopy in childhood AIDS encephalopathy. Pediatr Neurol 1995; 12:277–282.

44. Wilkinson ID, Paley M, Chong WK, et al. Proton spectroscopy in HIV infection: Relaxation times of cerebral metabolites. Magn Reson Imaging 1994; 12:951–957.

45. Wilkinson ID, Miller RF, Miszkiel KA, et al. Cerebral proton magnetic resonance spectroscopy in asymptomatic HIV infection. AIDS 1997; 11:289–295.

46. Miller BL, Chang L, Booth R, et al. In vivo 1H MRS choline: Correlation with in vitro chemistry/histology. Life Sci 1996; 58:1929–1935.

47. Confort-Gouny S, Vion-Dury J, Nicoli F, Dano P, Gastaut JL, Cozzone PJ. Metabolic characterization of neurological diseases by proton localized NMR spectroscopy of the human brain. C R Acad Sci III 1992; 315:287–293.

48. Paley M, Cozzone PJ, Alonso J, et al. A multicenter proton magnetic resonance spectroscopy study of neurological complications of AIDS. AIDS Res Hum Retroviruses 1996; 12:213–222.

49. Vion-Dury J, Salvan AM, Confort-Gouny S, Cozzone PJ. Brain proton magnetic resonance spectroscopy. Indications for diagnosis and follow-up of HIV-related encephalopathy in the adult. Presse Med 1998; 27:1398–1405.

50. Brand A, Richter-Landsberg C, Leibfritz D. Multinuclear NMR studies on the energy metabolism of glial and neuronal cells. Dev Neurosci 1993; 15:289–298.

51. Newsholme P, Gordon S, Newsholme EA. Rates of utilization and fates of glucose, glutamine, pyruvate, fatty acids and ketone bodies by mouse macrophages. Biochem J 1987; 242:631–636.

52. Newsholme P, Newsholme EA. Rates of utilization of glucose, glutamine and oleate and formation of end-products by mouse peritoneal macrophages in culture. Biochem J 1989; 261:211–218.

53. Petroff OA, Graham GD, Blamire AM, et al. Spectroscopic imaging of stroke in humans: Histopathology correlates of spectral changes. Neurology 1992; 42:1349–1354.

54. Graham GD, Blamire AM, Rothman DL, et al. Early temporal variation of cerebral metabolites after human stroke. A proton magnetic resonance spectroscopy study. Stroke 1993; 24:1891–1896.

55. Menon DK, Baudouin CJ, Tomlinson D, Hoyle C. Proton MR spectroscopy and imaging of the brain in AIDS: Evidence of neuronal loss in regions that appear normal with imaging. J Comput Assist Tomogr 1990; 14:882–885.
56. Eng LF, Ghirnikar RS. GFAP and astrogliosis. Brain Pathol 1994; 4:229–237.
57. Everall IP, Heaton RK, Marcotte TD, et al. Cortical synaptic density is reduced in mild to moderate human immunodeficiency virus neurocognitive disorder. HNRC Group. HIV Neurobehavioral Research Center. Brain Pathol 1999; 9:209–217.
58. Gray F, Scaravilli F, Everall I, et al. Neuropathology of early HIV-1 infection. Brain Pathol 1996; 6:1–15.
59. Gonzalez RG, Ruiz A, Tracey I, McConnell J. Structural, functional, and molecular neuroimaging in AIDS. In *The neurology of AIDS*, HE Gendelman, SA Lipton, L Epstein, S Swindells, eds. London: Edwin Arnold, 1998; 333–352.
60. Harrison MJ, Newman SP, Hall-Craggs MA, et al. Evidence of CNS impairment in HIV infection: Clinical, neuropsychological, EEG, and MRI/MRS study. J Neurol Neurosurg Psychiatry 1998; 65:301–307.
61. Wilkinson ID, Miller RF, Paley MN, et al. Cerebral proton magnetic resonance spectroscopy in cytomegalovirus encephalitis and HIV leucoencephalopathy/encephalitis [letter]. AIDS 1996; 10:1443–1444.
62. Deicken RF, Hubesch B, Jensen PC, et al. Alterations in brain phosphate metabolite concentrations in patients with human immunodeficiency virus infection. Arch Neurol 1991; 48:203–209.
63. Bottomley PA, Hardy CJ, Cousins JP, Armstrong M, Wagle WA. AIDS dementia complex: Brain high-energy phosphate metabolite deficits. Radiology 1990; 176:407–411.
64. Cohen WA. Intracranial bacterial infections in patients with AIDS. Neuroimaging Clin North Am 1997; 7:223–229.
65. Ramsey RG, Gean AD. Neuroimaging of AIDS. I. Central nervous system toxoplasmosis. Neuroimaging Clin North Am 1997; 7:171–186.
66. Lizerbram EK, Hesselink JR. Neuroimaging of AIDS. I. Viral infections. Neuroimaging Clin North Am 1997; 7:261–280.
67. Harris DE, Enterline DS. Neuroimaging of AIDS. I. Fungal infections of the central nervous system. Neuroimaging Clin North Am 1997; 7:187–198.
68. Whiteman ML. Neuroimaging of central nervous system tuberculosis in HIV-infected patients. Neuroimaging Clin North Am 1997; 7:199–214.
69. Harris DE, Enterline DS, Tien RD. Neurosyphilis in patients with AIDS. Neuroimaging Clin North Am 1997; 7:215–221.
70. Chang L, Miller BL, McBride D, et al. Brain lesions in patients with AIDS: H-1 MR spectroscopy [published erratum appears in Radiology 1996; 198: 586]. Radiology 1995; 197:525–531.
71. Chinn RJ, Wilkinson ID, Hall-Craggs MA, et al. Toxoplasmosis and primary central nervous system lymphoma in HIV infection: Diagnosis with MR spectroscopy. Radiology 1995; 197:649–654.
72. Ernst TM, Chang L, Witt MD, et al. Cerebral toxoplasmosis and lymphoma in AIDS: Perfusion MR imaging experience in 13 patients. Radiology 1998; 208:663–669.

Chapter 8

CONGENITAL INFECTIONS

Korgun Koral, M. Albert Thomas, and John G. Curran

Certain infectious agents (viruses, protozoa, and bacteria) may infect the fetus and damage the developing central nervous system (CNS) (1). The nature of the infective agent and the resulting neuropathology are, however, not the only determinants of the clinical presentation and disease course in congenital infections of the pediatric CNS (2). The stage of development and the maturation of the host CNS and host immune system play additional critical roles (2). The immature fetal brain may show deficient responses such as an inability to remove damaged cells or to repair damaged tissue (3).

Since the early 1970s, TORCH has been a useful acronym to refer to the most important congenital CNS infections: *Toxoplasma gondii*, other microorganisms, Rubella virus, Cytomegalovirus, and Herpesviruses (4).

Spontaneous abortion or fetal death, intrauterine growth retardation, developmental delay, seizures, hypotonia, increased intracranial pressure, multiple congenital anomalies, microcephaly, ocular abnormalities (e.g., cataracts, glaucoma, chorioretinitis, microphthalmia), other congenital defects (e.g., congenital heart disease), skin rash, hepatosplenomegaly, lymphadenopathy, and thrombocytopenia may be the clinical manifestations of congenital infections (1, 5). The presentation of infection in the neonatal period may be with sepsis and meningoencephalitis, hyperthermia or hypothermia,

Korgun Koral, M.Albert Thomas and John G. Curran • Department of Radiological Sciences, UCLA School of Medicine, Los Angeles, California 90095.

MR Imaging and Spectroscopy of Central Nervous System Infection, edited by Gupta and Lufkin. Kluwer Academic / Plenum Publishers, New York, 2001.

distress, feeding difficulties, or lethargy (1, 5). Developmental delay with mental retardation, motor deficits, visual impairment, deafness, seizures, microcephaly, or hydrocephalus may constitute the long-term sequelae of the congenital or neonatal CNS infections (2).

1. BACTERIAL MENINGITIS

Though bacterial meningitis has been discussed in detail in an earlier chapter, here we will highlight some of the points specifically related to neonatal bacterial meningitis.

Neonatal meningitis is seen more commonly in premature infants, male infants and infants born to mothers with complicated pregnancies or deliveries (6). The major organisms responsible for neonatal meningitis are Group B beta-hemolytic streptococci and *E. coli* accounting for approximately 75% of all cases. The next most common agent is *Listeria monocytogenes* (7). Meningitis is closely related to sepsis. Septic dissemination may result from pneumonia, gastroenteritis, umbilical or skin infections, and other systemic infections. Infection from a contiguous focus is rare. Two patterns of neonatal meningitis are seen. Early onset meningitis begins in the first week postpartum, usually in newborns with a history of complications during labor and delivery. The organisms are Group B streptococci and *E. coli*. Late-onset meningitis is due to postnatal contamination and caused by a variety of organisms.

The acute changes of bacterial meningitis are dramatic and include arachnoiditis, ventriculitis, vasculitis, cerebral edema, infarction, and associated encephalopathy (8). Various degrees of phlebitis and arteritis of intracranial vessels can be found in all infants (6). Thrombosis with infarction may occur as the early as the first week (Fig.1). The lesions are usually related to venous occlusions and may be hemorrhagic (9). In the second and third week, an active ependymitis occurs and glial bridges may develop within the ventricles and subarachnoid spaces. The complications in the acute phase may be communicating or noncommunicating hydrocephalus, subdural effusions (Fig.1), subdural empyema (Fig.2), and ventriculitis.

The major neuropathologic sequelae are hydrocephalus, multicystic encephalomalacia, and cerebral cortical and white matter atrophy (8). Dense fibrosis in the leptomeninges and ventricular system obstruct CSF flow and may cause hydrocephalus. The imaging features are similar to what has been described in Chapter 2 on pyogenic infections.

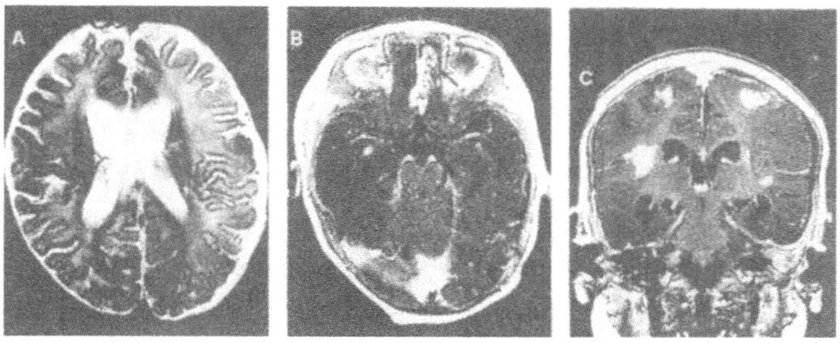

Figure 1. Gram-negative bacterial meningitis in an infant with complications. Axial T2 image (a) demonstrating abnormal hyperintensity in the distribution of the left middle cerebral artery suggestive of infarction. Postcontrast axial T1 image (b) through the inferior frontal lobes shows tentorial and falcine enhancement associated with meningitis. The enhancement in the anterior interhemispheric fissure is seen to involve the adjacent left frontal lobe, a manifestation of local cerebritis (arrow). There are foci of enhancement in the right and left temporal lobe, also attributed to cerebritis. Coronal T1 postgadolinium image (c) demonstrating focal enhancement at the base of individual sulci. Enhancement extends on the left into the basal ganglia. On the right, there is enhancement in the putamen. These enhancing lesions could be due to either ischemia or focal cerebritis. A small interhemispheric subdural effusion is present (arrow).

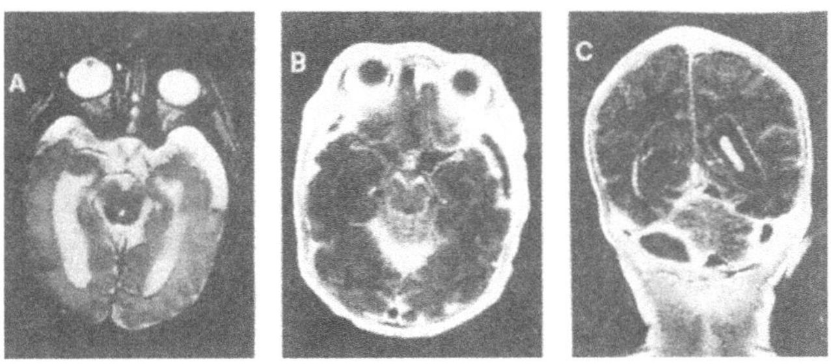

Figure 2. Gram-negative meningitis with subdural empyema. Axial T2-weighted image (a) demonstrating bilateral hyperintense collections in the temporal regions. Axial postgadolinium T1-weighted image (b) at the same level shows bilateral subdural collections, with thick pial enhancement suggesting empyema. Also note pial enhancement surrounding the midbrain and in the depths of the Sylvian fissures suggestive of meningitis. Coronal postgadolinium T1-weighted image (c) demonstrates the bilateral posterior fossa empyemas along with tentorial enhancement. Note the enhancing walls of the ventricles susggetive of ependymitis.

2. TOXOPLASMOSIS

T. gondii is an intracellular parasite that may proliferate asexually in numerous vertebrate species, including humans, but undergoes sexual reproduction only in the cat; therefore, it is found wherever cats are present (10–12). Congenital toxoplasmosis occurs when maternal infection is acquired during pregnancy (12). The oocyst stage of *T. gondii* passed in cat feces becomes infective after 24 to 48 hours under optimal conditions (12) and may survive longer than 1 year in soil (13,14) or circulate in contaminated water (15). Ingestion of oocysts in either contaminated water or soil or dormant forms encysted in muscle results in human infection (12). Pork and beef are most often implicated when transmission is through ingestion of undercooked meat (16).

Almost exclusively, congenital toxoplasmosis occurs when maternal infection is acquired during pregnancy (12). Clinically evident disease does not occur in infants whose immunocompetent mothers have chronic infection resulting from acquisition of toxoplasma infection before conception (17). Approximately 40% of infants born to untreated, acutely infected mothers develop congenital infection (12). Transmission of infection to the fetus is more likely to occur when infection is acquired late in pregnancy (12). However, the risk of spontaneous abortion, stillbirth, or severe disease in the neonate is much higher when infection is acquired in the first two trimesters. Approximately 3000 infants with congenital toxoplasmosis are born annually in the United States (18). Mortality rates for infants with congenital toxoplasmosis are between 1 and 6% (18, 19). The majority of surviving infants who have neurologic symptoms at birth have significant long-term disabilities (1). Characteristic histopathologic findings include a focal leptomeningitis with adjacent necrotic lesions in the cerebral cortex, subcortical white matter, and especially in the basal ganglia; focal cerebral calcifications; and scattered glial nodules (12). Periaqueductal or periventricular vasculitis is diagnostic (20–23).

Ultrasonography, CT, and MR may demonstrate calcifications in the basal ganglia, periventricular white matter, and peripheral cerebral tissue (24, 25). Aqueductal stenosis resulting in hydrocephalus is usually an outcome of earlier intrauterine infection (2). There may also be atrophy, encephalomalacia, hydranencephaly, or microcephaly (2).

3. SYPHILIS

Treponema pallidum is the organism that causes syphilis. Congenital infection occurs by vertical transmission as a result of hematogenous dissemination across the placenta (26).

The incidence of syphilis varies substantially worldwide (1). A dramatic increase was reported in syphilis cases among young women in the late 1980s and early 1990s (26). The recent rise in syphilis is linked to illegal drug use, particularly crack cocaine usage (26). In 1990, nearly 3000 cases were reported to the Centers for Disease Control (26, 27). For reasons that remain unclear, the incidence of syphilis saw a decline in the late 1990s (26).

Syphilis may have a profound effect on the health of a pregnancy (26). Untreated or inadequately treated syphilis can result in spontaneous abortion, stillbirth, nonimmune hydrops, premature delivery, and perinatal death (28–35). New guidelines that were established in 1989 for the diagnosis of congenital syphilis include not only all infants with clinical evidence of active syphilis but also asymptomatic infants and stillbirths born to women with untreated or inadequately treated syphilis (26).

Early congenital syphilis is a term used to refer to disease where symptoms appear within the first 2 years of life (26). The CNS abnormalities in early congenital syphilis may include acute leptomeningitis, chronic meningovasculitis, hydrocephalus, cranial nerve palsies, cerebral infarction, seizures, and hypopituitarism.

Neuroimaging may show enhancement of the leptomeninges. This may extend through the meningeal vessels and the cortical perforators to involve brain parenchyma, and appear on imaging as an enhancing parenchymal mass. Infarction may also occur, attributed to the accompanying arteritis, or less commonly, phlebitis (28).

4. RUBELLA

Rubella virus is a non-arthropod-borne togavirus that infects humans endemically during winter and spring (36, 37). Since the licensing of rubella vaccine in 1969, the incidence has shown a dramatic decrease (38). The number of reported cases in 1988 was 225 (39). In the early 1990s a significant increase in the incidence of rubella was encountered with coincident increase in the incidence of congenital rubella syndrome (CRS) (39).

The most serious consequence of rubella infection is CRS in the offspring of mothers who contracted rubella during pregnancy (40–42). The risk of fetal infection is more than 80% during the first trimester and then decreases significantly thereafter (1). After the 16th week, even if the fetal infection occurs, it is usually without symptoms (37, 42).

Growth retardation, ocular abnormalities (41) (cataracts and pigmentary retinopathy), congenital heart disease (43) (usually patent ductus arteriosus or pulmonary artery stenosis), hepatosplenomegaly, jaundice, purpuric rash, sensorineural hearing loss (44), or signs of meningoencephalitis constitute the typical features of CRS.

Inhibition of mitosis by the cytolytic rubella virus may result in microcephaly (2). Meningoencephalitis with ventriculitis, calcifications, and delayed myelination may be encountered in the neonate (2). Deafness is the most common feature of congenital rubella that often remains undetected until the infant is older (45). Vasculopathy is also present with relatively little cellular necrosis (2, 46). Ishikawa et al. described CT findings in five patients with CRS (47). Four patients had low-density areas in the centrum semiovale and periventricular white matter, and two had calcified nodules. Sugita et al. have described MRI findings in two cases of CRS. These include periventricular/subcortical hyperintensities and delayed myelination (48)

5. CYTOMEGALOVIRUS

Human cytomegalovirus (HCMV) a DNA virus, is the most common congenital viral infection in the world (49–51). In United States, approximately 37,000 newborns are affected annually (50). Transplacental infection or breastfeeding are the causes of congenital HCMV infection (1). The risk of transmission is approximately 40% during primary maternal HCMV infection. As with other congenital infections, earlier contraction with the virus in gestation results in a greater likelihood of severe infection (52). The HCMV invades the immature cells of the germinal matrix in early gestational infections (49,50,53,54). This phenomenon may result in microcephaly, lissencephaly, heterotopias, or polymicrogyria or periventricular calcifications, hydranencephaly, porencephaly (26,55). Cerebellar hypoplasia, cerebral atrophy, and subependymal cysts are other pathologic features of congenital HCMV (53). Recurrent maternal HCMV infections do not present significant risk to the fetus and newborn (50). CNS involvement rarely occurs with parturitional or neonatal infection, and in these cases, characteristic calcifications are not seen (2). Only about 5 to 10% of the infected newborns show signs of HCMV in the neonatal period (1).

CT and MRI are used to determine the extent of disease. Calcifications are better assessed with CT, the parenchymal abnormalities being more accurately defined on MR. Intracranial calcifications are present in up to 40% of the affected individuals and are typically located in the periventricular areas (Fig.3). However, they can be seen in the basal ganglia, subcortical and cortical regions (53). MRI findings regarding cytomegalovirus have included cerebellar hypoplasia, cerebral atrophy, cortical gyral anomalies, dilated ventricles, large subarachnoid spaces, T1 and T2 prolongation of the white matter and delayed myelination, pachygyria, and paraventricular cysts. Hemorrhage, focal white matter injury, and hippocampal abnormalities have also been reported (48, 55–59).

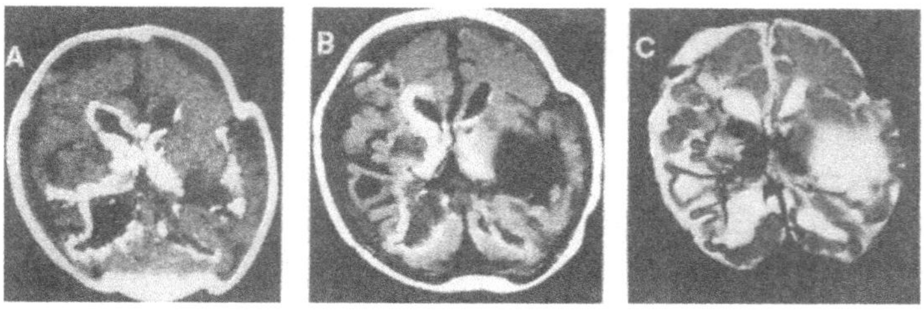

Figure 3. Congenital cytomegalovirus infection. CT axial section (a) at midventricular level in a neonate shows severe brain atrophy accompanied by extensive calcification. Axial T1-weighted image (b) through the ventricles demonstrate atrophy; calcification is seen as hyperintensity, mainly in a periventricular distribution. Axial T2-weighted image (c) with findings similar to T1. Here the denser portions of calcification are producing low signal on T2.

6. HERPES SIMPLEX VIRUSES

Herpes simplex virus (HSV) infection is prevalent worldwide, particularly among women of childbearing age (60). Herpes labialis, caused by HSV-1, is nonsexually transmitted and commonly produces skin lesions and encephalitis (2, 60). HSV-2 is the agent for herpes vulvovaginitis and is sexually transmitted. Both viruses can infect the neonate (61–63).

The incidence of neonatal infection varies between 1 per 2500 and 1 per 20,000 live births (64–68). Two-thirds of neonatal HSV infections result from HSV-2 (63, 64, 68), and intrapartum or postnatal acquisition of the virus account for a majority of the cases (1).

Clinically, congenital herpes is divided into three categories (1, 56):

(1) skin, eye, and mouth;
(2) disseminated;
(3) CNS infections.

Although the mucocutaneous disease is the mildest form, if untreated it may progress to disseminated or CNS disease in three-fourths of the infected children (56). The disseminated form is characterized by multiple organ involvement and CNS manifestations are present in approximately 50% of the cases (56).

Isolated CNS herpesvirus infection account for approximately 35% of the congenital herpes infections (69, 70). The disease presents 2–4 weeks postnatally with fever, lethargy, and seizures (71). All cells of the CNS can be infected, however, the propensity of endothelial cell involvement results in vascular thrombosis and hemorrhage (Fig.4) (72). On CT and MRI, diffuse brain swelling may be demonstrated (2). The most frequent initial finding on CT is patchy areas of low attenuation in the cerebral cortex and white matter (73). On MR, increased T2 signal is noted in cerebral hemispheres, basal ganglia, thalamus, and cerebellum (56). In contrast to older children or adult herpes encephalitis, there is no temporal lobe predilection in congenital herpes CNS infection (74). Encephalomalacia may be apparent (Fig.5) within 2–3 weeks. On contrast-enhanced CT studies, a linear gyriform enhancement has been reported (73, 74). On MRI, linear gyriform hyperintensities on T2-weighted images have also been described (55). These abnormal densities and signal abnormalities have been attributed to faint calcification or cortical laminar necrosis (56, 73, 75).

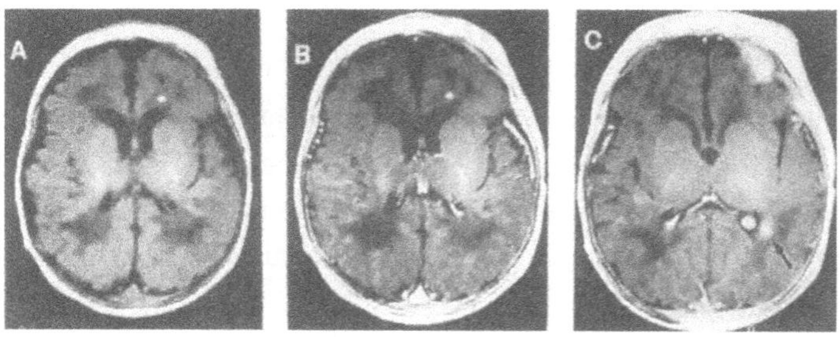

Figure 4. Congenital herpes simplex encephalitis. Axial T1-weighted (a) noncontrast image at the level of the frontal horns of the lateral ventricles shows a small hyperintensity in the right frontal white matter consistent with focal hemorrhage. Axial T1-weighted image (b) postgadolinium through the frontal horns demonstrating the focus of hemorrhage, without any enhancement. Axial T1-weighted image (c) through atria of ventricles demonstrating focal hyperintensity (arrow) adjacent to the atrium of the left lateral ventricle, another focal hemorrhage.

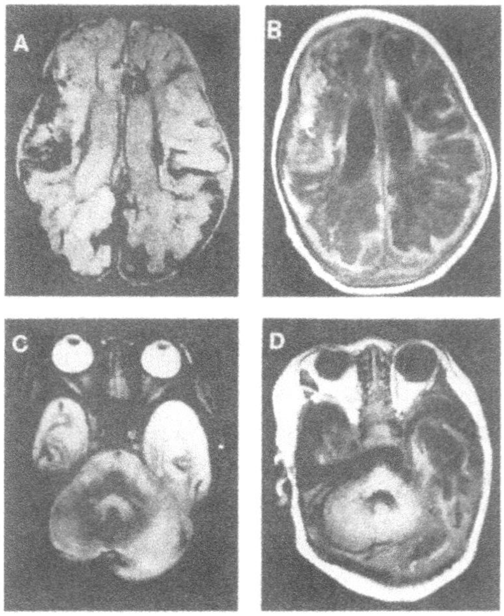

Figure 5. Congenital herpes simplex encephalitis. Axial T2- (a) and T1-weighted (b) images demonstrating extensive cystic encephalomalacic changes in bilateral cerebral hemispheres in a neonate. Axial T2- (c) and T1-weighted (d) images through the posterior fossa demonstrate relative sparing of the cerebellum and brainstem.

7. MR SPECTROSCOPY OF CNS INFECTIONS

Since 1983, CNS involvement in AIDS has been well recognized. Viral infections of the brain and spinal cord cause significant morbidity and mortality in patients afflicted with AIDS. Debate continues over the specific mechanisms and pathways of how HIV-1 manifests itself in the brain and spinal cord. Attempts to predict which seropositive patients develop cognitive deficits caused by HIV-1 and how soon these deficits will occur in the course of disease have had limited success. The neuropathologic changes of HIV-1 must be distinguished from other viral infections, such as CMV, JC papovavirus (progressive multifocal leukoencephalopathy), herpes simplex virus type 1, and varicella zoster virus.

Toxoplasmosis is the most common cerebral mass lesion encountered in HIV-positive patients. Previously uncommon, this disease has increased markedly since the AIDS epidemic. Imaging methods are important in diagnosing AIDS-related primary and secondary CNS processes as HIV-1-associated encephalopathy, cerebral toxoplasmosis, primary CNS lymphoma, CMV encephalitis, progressive multifocal leukoencephalopathy, and infectious spinal cord granulomas. There are occasionally unusual appearances of CNS toxoplasmosis that make diagnosis by standard imaging techniques difficult or impossible. Recent advances in MR, including MRS and diffusion-weighted imaging, may provide additional information in the evaluation of these diseases. MRS of various AIDS brain lesions shows distinct chemical profiles (76–81).

Twenty-seven lesions were studied (18 *T. gondii* lesions, 9 lymphoma lesions) at 1.5 T (74). Spectra were acquired at an echo time of 135 ms from voxels centered on the lesions. Both visual analysis and spectral fitting were used to obtain metabolite ratios for choline (Cho), creatine (Cr), *N*-acetylaspartate (NAA), lactate, and lipids. Three spectral categories were seen. One had large lipid peaks with suppression of other metabolites. Another had an elevated Cho/Cr ratio with relatively diminished NAA. The third had features of the other two. Examples of each spectrum type were acquired from both *T. gondii* and lymphoma lesions. Neither method of analysis allowed differentiation between lesion types. MRS showed an overlap of spectral pattern.

Hitosugi and co-workers have used localized MRS (81) in two patients with HSV-1 encephalitis (HSE). MR spectra of bilateral temporal lobes were acquired by the single-voxel method using a 1.5 T MRI/MRS scanner. Peaks indicating NAA, Cho, and creatine including phosphocreatine (Cr) were identified and ratios of NAA/Cr and Cho/Cr were calculated. These ratios were compared with those of the contralateral side characterized by normal MRI findings and with the control spectra obtained from normal volunteers. Three abnormal findings were observed in the spectra of the patients suffering from HSE: (1) significant reduction of the NAA/Cr ratio at the involved temporal lobe, (2) mild reduction of the NAA/Cr ratio at the normal temporal lobe, and (3) elevation of the Cho/Cr ratio at the bilateral lobes, but more significant on the involved sides. These results indicated that neural loss and gliosis occurred in the contralateral area with normal MR images as well as the involved hemisphere.

REFERENCES

1. Bale JF, Murph JR. Congenital infections and the central nervous system. Pediatr Clin North Am 1992; 39:669–690.
2. Barnes PD, Poussaint TY, Burrows PE. Imaging of pediatric central nervous system infections. Neuroimaging Clin North Am 1994; 2:367–391.
3. Weil ML, Levin M. Infections of the nervous system. In *Textbook of child neurology*, JH Menkes, ed. Baltimore: Williams & Wilkins, 1995; 411–418.
4. Nahmias AJ. The TORCH complex. Hosp Pract (Off) 1974; 65.

5. Klein JO, Remington JS. Current concepts of infections of the fetus and newborn infant. In *Infectious disease of the fetus and newborn infant*, JS Remington, JO Klein, eds. Philadelphia: Saunders, 1990; 1–14.
6. Friej BJ, McCracken GH Jr. Acute infections. In *Neonatology. Pathophysiology & management of the newborn*, GB Avery, MA Fletcher, MG MacDonald, eds. Philadelphia: Lippincott/Williams & Wilkins, 1999; 1207–1211.
7. Smith AL. Neonatal bacterial meningitis. In *Infections of the central nervous system*, WM Scheld, RJ Whitley, OT Durrack, eds. Philadelphia: Lippincott–Raven, 1997; 313.
8. Volpe JJ. *Neurology of the newborn*. Philadelphia: Saunders 1998; 730–766.
9. Kinney HC, Armstrong DD. Perinatal neuropathology. In *Greenfields neuropathology*, DI Graham, RL Lantos eds. London: Arnold, 1997; 561–563.
10. Feldman HA. Toxoplasmosis. (Two parts). N Engl J Med 1968; 279:1370–1375, 1431–1437.
11. Freij BJ, Sever JL. Toxoplasmosis. Pediatr Rev 1991; 12:227–236.
12. Lowichik A, Siegel JD. Parasitic infections of the central nervous system in children. Part I: Congenital infections and meningoencephalitis. J Child Neurol 1995; 10:4–17.
13. Dubey JP, Miller NL, Frenkel JK. Characterization of the new fecal form of Toxoplasma gondii. J Parasitol 1970; 56:447–456.
14. Frenkel JK, Ruiz A, Chinchilla M. Soil survival of Toxoplasma oocysts in Kansas and Costa Rica. Am J Trop Med Hyg 1975; 24:439–443.
15. Benenson MW, Takafuji ET, Lemon SM, et al. Oocyst-transmitted toxoplasmosis associated with ingestion of contaminated water. N Engl J Med 1982; 307:666–669.
16. Kean BH, Kimball AC, Christenson WN. An epidemic of acute toxoplasmosis. J Am Med Assoc 1969; 208:1002–1004.
17. Desmonts G, Couvreur J. Congenital toxoplasmosis: A prospective study of 378 pregnancies. N Engl J Med 1974; 290:1110–1116.
18. Koskiniemi M, Lappalainen M, Hedman K. Toxoplasmosis needs evaluation. An overview and proposals. Am J Dis Child 1989; 143:724.
19. Couvreur J, Desmonts G. Congenital and maternal toxoplasmosis. A review of 300 congenital cases. Dev Med Child Neurol 1962; 4:519.
20. Remington JS, Desmonts G. Toxoplasmosis. In *Infectious disease of the fetus and newborn infant*, JS Remington, JO Klein, eds. Philadelphia: Saunders, 1990; 89–195.
21. Rodney MB, Mitchell N, Redner B, Turin R. Infantile toxoplasmosis: Report of a case with autopsy. Pediatrics 1950; 5:649–663.
22. Wolf A, Cowan D, Paige B. Human toxoplasmosis: Occurrence in infants as an encephalomyelitis, verification by transmission to animals. Science 1939; 89:226–227.
23. Frenkel JK. Toxoplasmosis. Pediatr Clin North Am 1985; 32:917–1085.
24. Diebler C, Dusser A, Dulac O. Congenital toxoplasmosis: Clinical and neuroradiological evaluation of the cerebral lesions. Neuroradiology 1985; 27:125–130.
25. Fitz CR. Inflammatory diseases of the brain in childhood. Am J Neuroradiol 1992; 13:551–568.
26. Hollier LM, Cox SM. Syphilis. Semin in Perinatol 1998; 22:323–331.
27. Centers for Disease Control. Summary of notifiable diseases, United States, 1990.MMWR 1991; 39(53).
28. Weil ML, Levin M. Infections of the nervous system. In *Textbook of child neurology*, JH Menkes, (ed). Baltimore: Williams & Wilkins, 1995; 457–459.

29. Rolfs RT, Nakashima AK. Epidemiology of primary and secondary syphilis in the United States, 1981–1989. J Am Med Assoc 1990; 264:1432.
30. Wendel GD. Gestational and congenital syphilis. Clin Perinatol 1988; 15:287–303.
31. Barton JR, Thorpe EM, Shaver DC, et al. Nonimmune hydrops fetalis associated with maternal infection with syphilis. Am J Obstet Gynecol 1992; 167:56–58.
32. Fiumara NJ, Fleming WL, Downing JG, et al. The incidence of prenatal syphilis at the Boston City Hospital. N Engl J Med 1952; 247:48–52.
33. Hallak M, Peipert JF, Ludomirsky A, et al. Nonimmune hydrops fetalis and fetal congenital syphilis: A case report. J Reprod Med 1992; 37:173–176.
34. Ingraham NR. The value of penicillin alone in the prevention and treatment of congenital syphilis. Acta Dermatol Venereol Suppl 1951; 31:60–65.
35. Mascola L, Pelosi R, Blount JH, et al. Congenital syphilis revisited. Am J Dis Child 1985; 139:575-580.
36. Bakshi SS, Cooper LZ. Rubella. Clin Dermatol 1989; 7: 8–18.
37. Miller E. Rubella in the United Kingdom. Epidemiol Infect 1991; 107: 31–42.
38. Lindergren ML, Fehrs LJ, Hadler SC, Hinman AR. Update: Rubella and congenital rubella syndrome, 1980–1990. Epidemiol Rev 1991; 13: 341–348.
39. Increase in rubella and congenital rubella syndrome in United States. 1988–1990. MMWR 1991; 40: 93–99.
40. Cradock-Watson JE, Miller E, Ridehalgh MKS, et al. Detection of rubella virus in fetal and placental tissues and in the throats of neonates after serologically confirmed rubella in pregnancy. Prenatal Diagn 1989; 9: 91.
41. Dudgeon JA. Congenital rubella. J Pediatr 1975; 87:1078–1086.
42. Ueda K, Nishida Y, Oshima K, et al. Congenital rubella syndrome: Correlation of gestational age at time of maternal rubella with type of defect. J Pediatr 1979; 94:763–765.
43. Hastreiter AR, Joorabchi B, Pujatti G, et al. Cardiovascular lesions associated with congenital rubella. J Pediatr 1967; 71:59–65.
44. Wild NJ, Sheppard S, Smithells RW, et al. Onset and severity of hearing loss due to congenital rubella infection. Arch Dis Child 1989; 65:1280–1283.
45. Preblud SR, Alford CA. Rubella. In *Infectious disease of the fetus and newborn infant*, JS Remington, JO Klein, eds. Philadelphia: Saunders, 1990; 196–224.
46. Rorke LB, Spiro AJ. Cerebral lesions in congenital rubella syndrome. J Pediatr 1957; 70:243–255.
47. Ishikawa A, Murayama T, Sakuma N, et al. Computed cranial tomography in congenital rubella syndrome. Arch Neurol 1982; 39:420–421.
48. Sugita K, Ando M, Makino M, Takanashi J, Fujimoto N, Niimi H. Magnetic resonance imaging of the brain in congenital rubella virus and cytomegalovirus infections. Neuroradiology 1991; 33: 239–242.
49. Alford CA, Britt WJ. Cytomegalovirus. In *Virology*, BN Fields, DM Knipe, eds. New York: Raven Press, 1990; 1981.
50. Demmler GJ. Summary of a workshop on surveillance for congenital cytomegalovirus disease. Rev Infect Dis 1991; 13:315–329.
51. Stinski MF. Cytomegalovirus and its replication. In *Virology*, BN Fields, DM Knipe, eds. New York: Raven Press, 1990; 1959.
52. Stagno S, Pass RF, Cloud G, et al. Primary cytomegalovirus infection in pregnancy. Incidence, transmission to fetus, and clinical outcome. J Am Med Assoc 1986; 256:1904–1908.

53. Bale JF. Human cytomegalovirus infection and disorders of the nervous system. Arch Neurol 1984; 41:310–320.
54. Becker LE. Infections of the developing brain. Am J Neuroradiol 1992; 13:537–550.
55. Hayward JC, Titelbaum DS, Clancy RR, et al. Lissencephaly-pachygyria associated with congenital cytomegalovirus infection. J Child Neurol 1991; 6:109–114.
56. Shaw DWW, Cohen WA. Viral infections of the CNS in children: Imaging features. Am J Roentgenol 1993; 160:125–133.
57. Boesch CH, Issakainen J, Kewitz G, et al. Magnetic resonance imaging of the brain in congenital cytomegalovirus infection. Pediatr Radiol 1989; 19:91–93.
58. Bignalli A, Appicciutolo L. Micropolygyria and cerebral calcification in cytomegalic inclusion disease. Acta Neuropathol 1964; 4:127–137.
59. Barkovich AJ, Lindan CE. Congenital cytomegalovirus infection of the brain: Imaging analysis and embryologic considerations. Am J Neuroradiol 1994; 15: 703–715.
60. Riley LE. Herpes simplex virus. Semin Perinatol 1998; 22:284–292.
61. Cuthbertson G, Weiner CP, Giller RH, et al. Prenatal diagnosis of second-trimester congenital varicella syndrome by virus-specific immunoglobulin M. J Pediatr 1987; 111:592–595.
62. Whitley RJ, Arvin A, Prober C, et al. Predictors of morbidity and mortality in neonates with herpes simplex virus infections. N Engl J Med 1991;324:450.
63. Whitley RJ, Nahmias AJ, Visintine AM, et al. The natural history of herpes simplex virus infection of mother and newborn. Pediatrics 1980; 66:489.
64. Baldwin S, Whitley RJ. Teratogen update: Intrauterine herpes simplex virus infection. Teratology 1989; 39:1.
65. Corey L, Spear PG. Infections with herpes simplex viruses (two parts). N Engl J Med 1986; 314: 686, 749.
66. Koskiniemi M, Happonen J-M, Jarvenpaa A-L, et al. Neonatal herpes simplex virus infection: A report of 43 patients. Pediatr Infect Dis J 1989; 8:30.
67. Stone KM, Brooks CA, Guinan ME, et al. National surveillance for neonatal herpes simplex virus infections. Sex Transm Dis 1989; 16:152.
68. Sullivan-Bolyai J, Hull HF, Wilson C, et al. Neonatal herpes simplex virus infection in King County, Washington. Increasing incidence and epidemiologic correlates. J Am Med Assoc 1983; 250:3059.
69. Whitley RJ, Arvin A, Prober C, et al. A controlled trial comparing vidarabine with acyclovir in neonatal herpes simplex virus infection. N Engl J Med 1991;324:450.
70. Whitley RJ, Corey L, Arvin A, et al. Changing presentation of herpes simplex virus infection in neonates. J Infect Dis 1988; 158:109–116.
71. Herman TE, Cleveland RH, Kushner DC, Taveras JM. CT of neonatal herpes encephalitis. Am J Neuroradiol 1985; 6:773–775.
72. Whitley RJ. Herpes simplex virus infections. In *Infectious disease of the fetus and newborn infant*, JS Remington, JO Klein, eds. Philadelphia: Saunders, 1990; 82–301.
73. Noorbehesht B, Enzmann DR, Sullender W, et al. Neonatal herpes simplex encephalitis: Correlation of clinical and CT findings. Radiology 1987; 162:813–819.
74. Benator HM, Magill HL, Gerald B, et al. Herpes simplex encephalitis: CT findings in the neonate and young infant. Am J Neuroradiol 1985; 6:539–543.
75. Taccone A, Gambaro G, Chiorzi M. CT in children with herpes simplex encephalitis. Pediatr Radiol 1988; 19: 9–12.

76. Chinn RJ, Wilkinson ID, Hall-Craggs MA, Paley MN, Miller RF, Kendall BE, Newman SP, Harrison MJ. Toxoplasmosis and primary central nervous system lymphoma in HIV infection: Diagnosis with MR spectroscopy. Radiology 1995; 197: 649–654.
77. Ramsey RG, Gean AD. Neuroimaging of AIDS. I. Central nervous system toxoplasmosis. Neuroimaging Clin North Am 1997; 7: 171–186.
78. Chang L, Ernst T. MR spectroscopy and diffusion-weighted MR imaging in focal brain lesions in AIDS. Neuroimaging Clin North Am 1997; 7: 409–426.
79. Arendt G. Imaging methods as a diagnostic tool in neuro-AIDS. A review. Bildgebung 1995; 62: 310–319.
80. Lizerbram EK, Hesselink JR. Neuroimaging of AIDS. I. Viral infections. Neuroimaging Clin North Am 1997; 7: 261–280.
81. Hitosugi M, Ichijo M, Matsuoka Y, Takenaka N, Fujii H. [Proton MR spectroscopy findings in herpes simplex encephalitis]. Rinsho Shinkeigaku Clin Neurol 1996; 36: 839–843.

Chapter 9

PARAINFECTIOUS AND OTHER INFECTIOUS DISEASES

Sunil Pradhan and Rakesh K. Gupta

Several rare infections occasionally involve the central nervous system (CNS). In addition, various demyelinating diseases of the brain and spinal cord are of unknown etiology but infections and immunizations have been strongly implicated in their pathogenesis. These disorders are generally grouped under the syndrome "acute disseminated encephalomyelitis" and are supposed to have a close relationship with experimental allergic encephalomyelitis (EAE). It has been suggested that some decapeptides found in certain viruses have homology to myelin proteins and there could be a cross-reacting hypersensitivity reaction induced by the viral infection or immunization. Apart from parainfectious and immuno-allergic disorders, there are several inflammatory diseases of the CNS where infections, though yet unknown, have been strongly implicated. In this chapter, we will discuss clinical and imaging features of some of these conditions in detail.

1. PARAINFECTIOUS MYELITIS

Parainfectious myelitis (PIM) is commonly known as acute transverse myelitis.

Sunil Pradhan • Department of Neurology, Sanjay Gandhi Post-Graduate Institute of Medical Sciences, Lucknow 226014, India. Rakesh K. Gupta • Department of Radiodiagnosis, Sanjay Gandhi Post-Graduate Institute of Medical Sciences, Lucknow 226014, India.

MR Imaging and Spectroscopy of Central Nervous System Infection, edited by Gupta and Lufkin. Kluwer Academic / Plenum Publishers, New York, 2001.

However, the latter term denotes diverse etiologies (1,2) including the infection related to immune-mediated disorders. The initial reports suggested acute transverse myelitis as a single homogeneous entity with uniform characteristics (1,2). Subsequent attempts have been made to differentiate PIM from other varieties of acute transverse myelitis such as multiple sclerosis and vascular insufficiency (3,4). Acute or subacute inflammatory myelopathy of unknown etiology, in which most authors implicate a postinfection autoimmune process (5–8), is now categorized in the parainfectious group. In multiple sclerosis and ischemic myelopathy, the lesion in the spinal cord is usually confined longitudinally to a few segments only (1,2,9) and this justifies the term "acute transverse myelitis." In postinfectious inflammatory myelopathy, a long vertical segment of the spinal cord is often affected and the term " parainfectious myelitis" is preferred (7). The clinical diagnosis of PIM is essentially based on a rapid onset of spinal motor and sensory deficits with a transverse-level, nonprogressive course stabilizing within a few weeks and the absence of cord compression or any other known neurologic disease related to trauma, malignancy, encephalitis, granulomatous lesion, and so on (7,10,11).

It is important to understand the differentiating features of different types of acute transverse myelitis. PIM is usually a uniphasic illness but recurrences have been reported. On the other hand, recurrence or rarely a chronic progressive course suggests multiple sclerosis or the vasculitis-related ischemic myelopathy. Appearance of cranial signs in patients with acute transverse myelitis may indicate more extensive central demyelination that is usually seen in acute disseminated encephalomyelitis (12). However, the involvement of specific tracts of the brain as evidenced by evoked potential and MRI studies suggests multiple sclerosis. Multiple ischemic lesions in the brain or spinal cord would suggest vasculitis-related ischemic encephalomyelopathy. PIM generally causes spastic paraplegia with a thoracic transverse sensory-motor level. On rare occasions, cervical lesion may produce spastic quadriparesis. On MRI, involvement of more than one or two spinal cord segments indicates idiopathic or parainfectious myelitis and helps differentiate it from multiple sclerosis and ischemic myelopathy (13,14). In the axial plane, partial cord involvement with lesions confined only to a few of the long tracts usually suggests multiple sclerosis (15) while symmetric central or whole cord involvement in the transverse plane may suggest PIM (13). On the basis of the vertical extent of the inflammatory lesions as observed on MRI, the PIM has been classified into four groups (Fig.1) (7,16): parainfectious focal segmental myelitis (PIFSM), parainfectious ascending myelitis (PIAM), parainfectious disseminated myelitis (PIDM), and parainfectious conus myelitis (PICM).

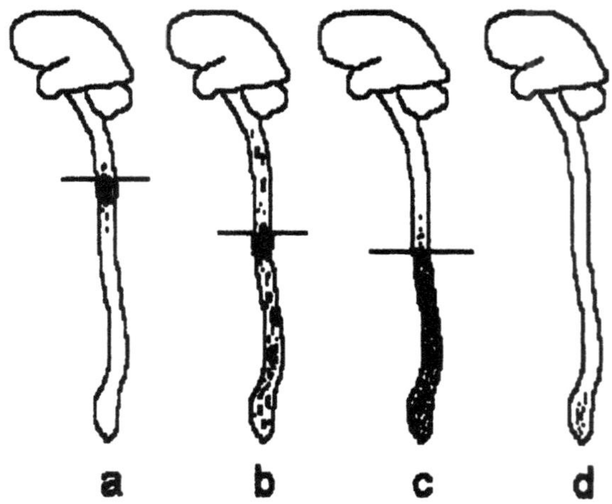

Figure 1. Line diagram showing the distribution of lesions in four groups of myelitis. (Reprinted from Pradhan S, et al. J Neurol Sci 1998; 161:156–162.)

1.1. Parainfectious Focal Segmental Myelitis

PIFSM patients have a focal demyelinating inflammatory spinal cord lesion somewhere in the middle of the cord extending to only a few spinal segments in the longitudinal plane. On MRI, clinically insignificant small discrete lesions may be seen scattered above and below the main lesion that appear hyperintense on T2- and iso- to hypointense on T1-weighted images, the latter being responsible for the rapidly evolving clinical transverse sensory-motor level (Fig.2). Spastic paralysis is observed either from the beginning or after a brief period of spinal shock. Autonomic dysfunction in these patients is minimal and is confined to bladder and bowel involvement.

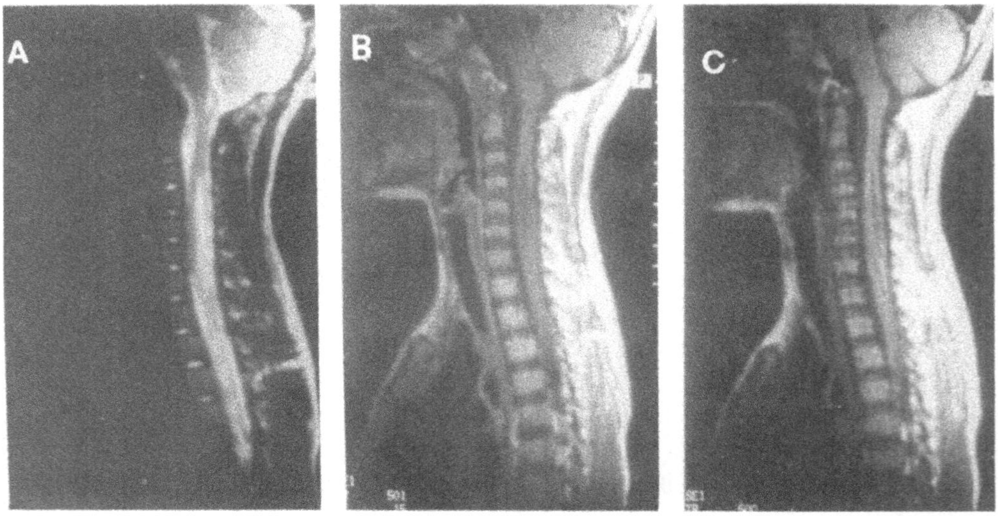

Figure 2. Parainfectious focal segmental myelitis. T2-weighted sagittal image (a) of the cervical cord shows focal hyperintensity of the cord from fourth cervical vertebra to first dorsal vertebra. It appears fusiform and slightly hypointense on T1-weighted image (b). Postcontrast T1-weighted image (c) shows doubtful enhancement.

1.2. Parainfectious Ascending Myelitis

Most PIAM patients reveal a smoothly ascending sensory-motor transverse level. MRI shows a dense longitudinal lesion continuously from conus medullaris to dorsal (or rarely cervical) spinal cord (Fig.3). The clinical transverse level corresponds to the uppermost segment of the involved spinal cord visible on MRI. Due to long and dense inflammatory lesion, the intermediolateral columns and anterior horn cells are also involved. This results in severe autonomic dysfunction in the form of loss of skin temperature, hair loss, pedal edema and bladder/bowel involvement, and wasting in the lower limbs due to anterior horn cell involvement. Electromyography shows denervation in the paraspinal and lower limb muscles. In these patients, the initial "spinal shock" never reverts to spasticity and hypotonia persists due to the neuronal cell damage caused by the longitudinally extended inflammation of the spinal cord.

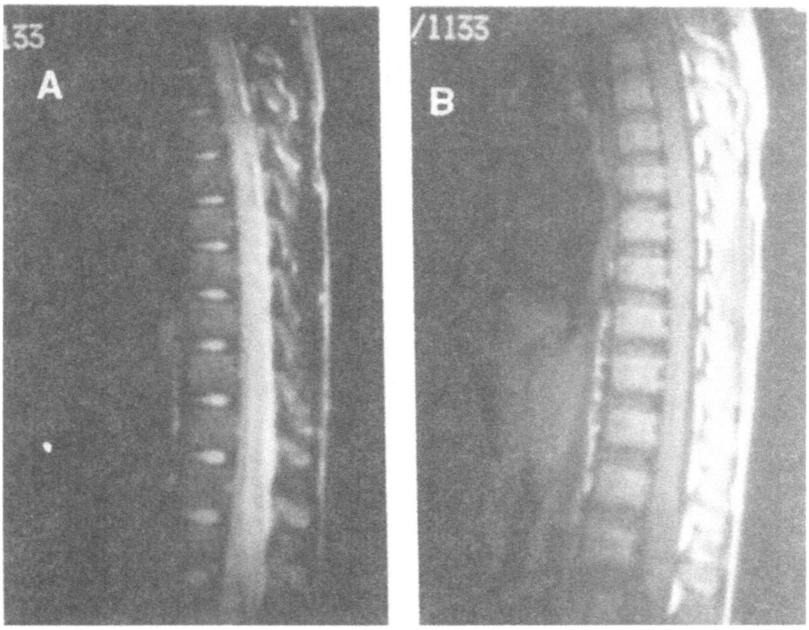

Figure 3. Parainfectious ascending myelitis. T2-weighted image (a) of the dorsal spine shows diffuse hyperintensity of the spinal cord from conus to upper dorsal region. Postcontrast T1-weighted (b) image shows minimal swelling of the cord with no abnormal enhancement.

1.3. Parainfectious Disseminated Myelitis

These patients often present with stepladder ascent of sensory-motor level with or without initial spinal shock, mild to moderate autonomic dysfunction, and occasional abnormalities in electromyography. MRI shows multiple dense T2 hyperintense lesions scattered over the length of the spinal cord (Fig.4), the sensory-motor transverse level corresponding to the uppermost dense inflammatory lesion. A careful clinical examination of these patients may reveal segmental sensory-motor deficits above and below the transverse level that are the result of multiple discrete lesions scattered in the spinal cord. Some of these patients may show a unique phenomenon in which the transverse sensory level is different for pinprick and fine touch. This is perhaps due to separate inflammatory lesions in posterior and lateral columns of the spinal cord to account for the upper transverse level.

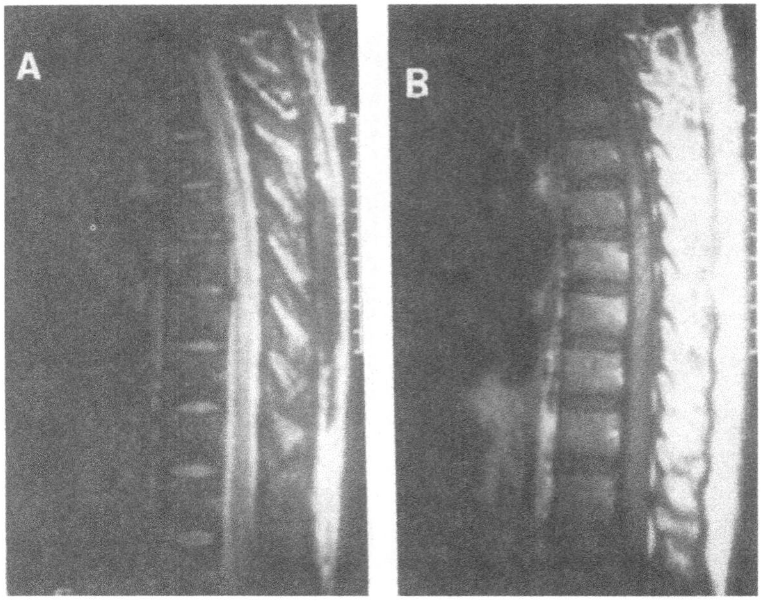

Figure 4. Parainfectious disseminated myelitis. T2-weighted image (a) of the dorsal spine shows areas of hyperintensities in the cord with partial sparing in between the lesions. T1-weighted image (b) shows foci of hyperintensities in the middorsal cord suggesting small areas of hemorrhage.

1.4. Parainfectious Conus Myelitis

This a new clinical entity described recently (16). These patients have a unique clinical presentation with urinary symptoms for which they often have initial consultation in urology outpatient clinic. Careful examination reveals mild sensory-motor segmental dysfunction in the sacral and lower lumbar segments. A few patients may not have any sensory-motor deficit and the sole manifestation is the bladder/bowel involvement. The clinical diagnosis of myelitis is difficult as there is no transverse sensory level around the trunk and the segmental neurologic deficit, if present, is confined to the leg or perineal region only. Motor paralysis is rare and, whenever present, is mild and flaccid. Some of these patients may give a history of preceding viral or bacterial infection. MRI reveals inflammatory lesion in a the conus–epiconus region of the spinal cord, visible as hyperintense on T2 and isointense on T1, with often no enhancement after contrast administration. In some patients, sagittal MRI may not reveal any lesion. Hence, in suspected cases, MRI in the axial plane is mandatory (Fig.5); it may show symmetric or asymmetric inflammatory lesions predominantly in the central gray with spread to adjoining white matter.

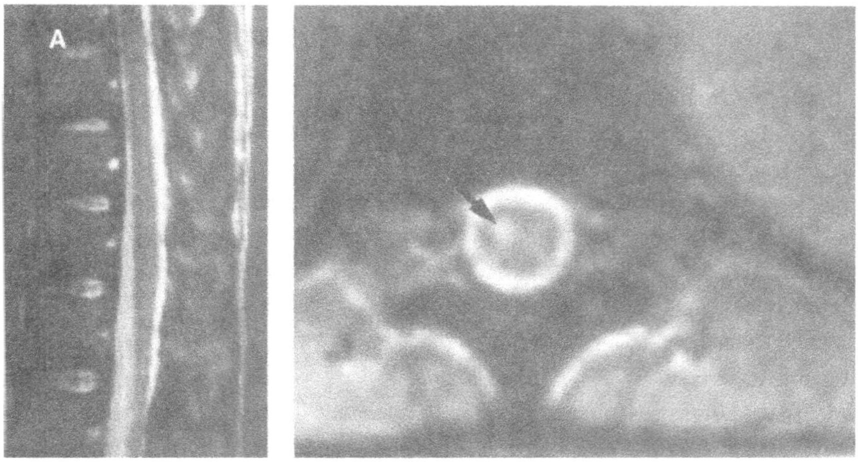

Figure 5. Conus myelitis. T2-weighted sagittal image (a) of the conus does not show any abnormality. T2-weighted axial image (b) at the level of dorsal 12 vertebra shows a hyperintense lesion (arrow) anteriorly and to the right. (Reprinted from Pradhan S, et al. J Neurol Sci 1998; 161:156–162.)

This anatomic classification of PIM into four groups on the basis of clinical, electrophysiologic, and MRI findings has significant prognostic implications. While PIFSM has good prognosis, PIDM has moderate and PIAM has poorest prognosis. Severe denervation potentials observed in PIAM also contribute to poor prognosis. The only exception to this rule is the presence of large dense inflammatory lesion with or without hemorrhagic component; these patients have absent somatosensory evoked potentials and central motor conductions and have a poor prognosis even if they fall in the PIFSM group. PICM has variable outcome; though the sensory-motor deficit invariably improves, some patients continue to use intermittent self-catheterization for their persistent urinary complaints.

2. ACUTE DISSEMINATED ENCEPHALOMYELITIS

ADEM is an acute focal or multifocal (disseminated) inflammatory disorder of the brain and/or spinal cord in which the white matter is predominantly involved through an unknown but most probably a parainfectious autoimmune process. Due to its close resemblance with experimental allergic encephalomyelitis, the condition is thought to be precipitated by certain antigenic decapeptides present in viruses and vaccines that cross-react with myelin basic and other proteins present mainly in the CNS (17).

ADEM is often preceded by certain viral infections and vaccinations (18–21). The viral prodrome may be nonspecific such as a common cold, flulike illness, pharyngitis, bronchitis, or gastroenteritis or else there may be specific viral infections such as measles, German measles, chicken pox, influenza, or infectious mononucleosis (18,22). The onset of the disease is usually acute or subacute and the progression of symptoms may continue for a few days to 6 weeks. It is a uniphasic illness but recurrence may be observed in a few patients (19,20). Children and young adults are most commonly affected. The initial symptoms could be weakness, numbness, hemiplegia, paraplegia, ataxia, hemianopia, cortical blindness, and so on. Although the lesions in the brain are multiple, the clinical signs are often localized to one or a few of the lesions (23). If the disease is more extensive, the clinical picture is that of generalized encephalopathy with multifocal neurodeficits. After a few days of stationary course, the symptoms usually resolve largely but some patients have relentlessly progressive course resulting in death. At autopsy, typically there is perivenous lymphocytic infiltration with patches of demyelination. Depending on the severity and rapidity of disease progression, several variations have been noted in ADEM. If the disease progression is very rapid with fever, neurologic deterioration, raised intracranial pressure, and coma, the condition is called acute hemorrhagic leukoencephalitis (AHL) of Weston Hurst. AHL is characterized by perivenous mononuclear infiltration, demyelination with significant hemorrhages, and necrosis. Some autopsies may reveal microglial proliferation (24). In some rare cases, ADEM may show chronicity with recurrence. Although there is controversy about the recurrent form of ADEM and its overlap with multiple sclerosis, viral prodrome or a febrile illness preceding each episode favors the existence of the former. Another group of patients may present with ascending type of acute transverse myelitis with minimal or no cranial signs but with MRI evidence of small focal demyelinating areas in the brain (25). On the other hand, some patients may have clinical and MRI evidence of predominant involvement of the brain but on MRI, a few lesions may be observed in the spinal cord as well (26).

MRI typically shows T2 hyperintense lesions of varying size predominantly in the white matter of the brain (Fig.6) and spinal cord. These lesions are hypointense on T1-weighted image. Hemorrhages in different stages may be observed in a subtle manner in ADEM and in an extensive manner in its AHL variant. The clinical improvement often correlates with the resolution of lesions on MRI but sometimes, fresh lesions may be observed at the stage when the old lesions were resolving (Fig. 6) (25,26). Apart from brain and spinal cord, a few rare patients may have optic nerve involvement as well (26). In these patients, the differentiation from multiple sclerosis is difficult. However, the presence of associated large cortical lesions on MRI in the acute phase of illness with complete resolution in the follow-up, favors the diagnosis of ADEM. Often the MRI findings correlate well with the clinical features (27) but at times, the MRI shows many more lesions than what are expected on clinical grounds alone (23,28–30). The MRI findings in ADEM reflect the morphologic changes seen on histopathology. As happens in multiple sclerosis, the classical pathologic lesion in ADEM is perivenous inflammation and demyelination (31,32). On MRI, the lesions of ADEM are primarily located at subcortical white matter particularly at centrum semiovale and cortico-subcortical junctions (33–35). If the lesions are multiple, these are larger and more asymmetric in ADEM than in MS (35). Rarely basal ganglia (36,37) and thalamus (28,36,37) may be involved in ADEM. Demyelinating lesions with destructive or hemorrhagic components, favor the diagnosis of ADEM over multiple sclerosis (38). Gadolinium enhancement may be observed in some (Fig.7) but not all patients (26); at times, the enhancement may have a ring shape (39).

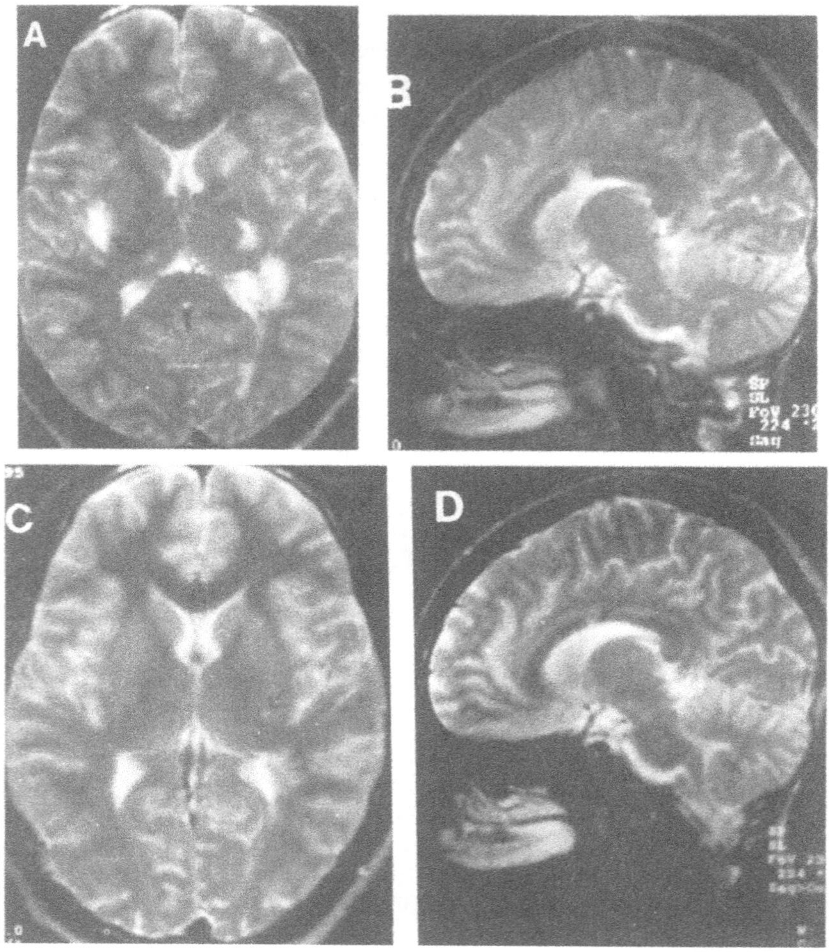

Figure 6. Acute disseminated encephalomyelitis. T2-weighted axial (a) and sagittal (b) images through the lateral ventricle show multiple hyperintensities in the white matter, basalganglia, and body of the corpus callosum (arrows). Follow-up T2-weighted axial (c) and sagittal (d) images after the treatment show regression of the lesions (arrow).

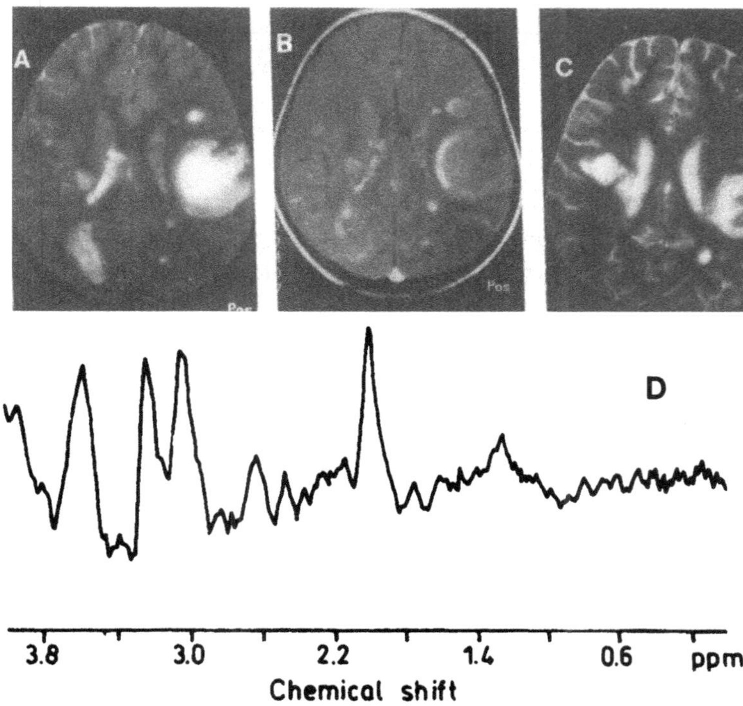

Figure 7. Acute disseminated encephalomyelitis: T2-weighted axial image (a) shows multiple hyperintense lesions in periventricular white matter. The large lesion on the left side is extending to the cortical gray matter. Postcontrast (b) study shows rim enhancement of the lesions. *In vivo* single voxel proton MR spectroscopy (d) (STEAM, TR/TE/TM/n=3000/20/30/128) from the largest lesion on the left side shows a decrease in NAA (1). Repeat T2-weighted image (c) after 6 weeks of treatment shows regression /disappearance of some lesions while other lesions not seen earlier have appeared. Patient showed improvement even when the new lesions were seen on imaging.

3. SARCOIDOSIS

Sarcoidosis is a systemic, noncaseating granulomatous disease of idiopathic nature (40). The diagnosis is suspected on clinical and radiologic grounds and confirmed by histopathology. Around 5% patients with systemic sarcoidosis have clinically evident neurologic involvement that is essentially nonspecific. In an autopsy series, however, subclinical neurologic involvement has been observed 3–5 times more often (41,42). Patients who develop CNS sarcoidosis do so within 2 years of initial diagnosis (42). Primary CNS involvement without systemic manifestations is extremely rare. The neurosarcoidosis has variable clinical presentations, which mimic other chronic granulomatous infections of the CNS. Hence, the clinical diagnosis is essentially based on documentation of systemic sarcoidosis in the absence of other neurological disorder (42).

The clinical symptoms of neurosarcoidosis are quite variable. Some patients have a single episode while others show chronic relapsing disease. Several patients present with headache in the initial phase of illness. Most of the clinical symptoms are due to cranial nerves or leptomeningeal involvement. Some patients have sarcoid granulomas within the brain parenchyma. Among the cranial nerves, the seventh and lower four are most often involved. The cranial nerve involvement is either due to direct sarcoid infiltration or due to adjacent leptomeningeal disease and is observed in 50–75% of patients with neurosarcoidosis (43). Predominant leptomeningeal disease is observed in around 20% of patients (43). Some patients develop communicating hydrocephalus due to leptomeningeal disease (44). The intra-axial sarcoidosis presents as a focal granulomatous disease or as diffuse sarcoid infiltration (45). The focal lesions may be observed in any part of the brain but have a specific predilection for the hypothalamic–pituitary axis. If this site is involved, endocrine abnormalities, especially the inappropriate secretion of antidiuretic hormone, are common (46). The initial extra-axial infiltration of meninges and subarachnoid space may extend through perivascular (Virchow–Robin) spaces and gives rise to lesions in the perforating vessel-supplied areas such as basal ganglia and brain stem. If basal ganglia are primarily involved, extrapyramidal symptoms predominate (43).

Histopathology of the sarcoid tissue shows epitheloid granulomas without caseation. The staining for infectious agents is negative. The granulomas predominantly contain lymphocytes and multinucleated giant cells (47–49). Compared with sarcoidosis granulomas elsewhere in the body, CNS sarcoid granulomas have high vascularity and fewer multinucleated giant cells (50,51). The new and old granulomas are seen simultaneously, suggesting an ongoing disease process with a waxing and waning course (41).

On MRI, sarcoid granulomas appear hypointense on T2- and isointense in T1-weighted images and show enhancement on contrast administration (52). Patients who present with headache or cranial nerve involvement, show often diffuse but sometimes-focal meningeal enhancement predominantly in the basal region. In addition, patients with cranial nerve palsy often show contrast enhancement of the involved cranial nerve (52). The contrast enhancement is not a universal phenomenon and at times it is difficult to differentiate nonenhancing sarcoid lesions from demyelinating disorders when the lesions are intra-axial. It is important to remember that both sarcoidosis and multiple sclerosis not only occur in similar locations, but also more commonly affect young and middle aged women. Intra-axial lesions may be solitary or multiple and may be located in the subcortical white matter or at more peripheral cortex. In the former location and when these are multiple, they are again indistinguishable from multiple sclerosis. Gd-enhanced MRI is considered the method of choice for the imaging of meningeal sarcoidosis (53). In one study, contrast-enhanced MRI could detect meningeal lesions in 75% of those patients

(Fig.8) who were suspected of having neurosarcoidosis (46). Large focal sarcoid collections in the meningeal spaces may be indistinguishable on imaging from other inflammatory disorders such as tuberculosis and hypertrophic pachymeningitis, primary meningeal tumors such as meningiomas, or metastasis including lymphoma and plasmacytoma (54). It is important for the evaluation of meningeal sarcoidosis to pay more attention to falx, tentorium, basal cisterns (particularly suprasellar and cerebellopontine angle cisterns), and communicating hydrocephalus (44,55). The meningeal involvement in sarcoidosis appears as intermediate signal intensity thickening of the subarachnoid space on T1- as well as T2-weighted images and on T2, it is clearly distinguishable from high signal intensity of normal CSF. This feature could be a reflection of cellular compactness as it enhances homogeneously on Gd-DTPA administration (56). MRI has a role not only in the diagnosis of sarcoidosis but also in the follow-up of these patients either on or off steroid therapy.

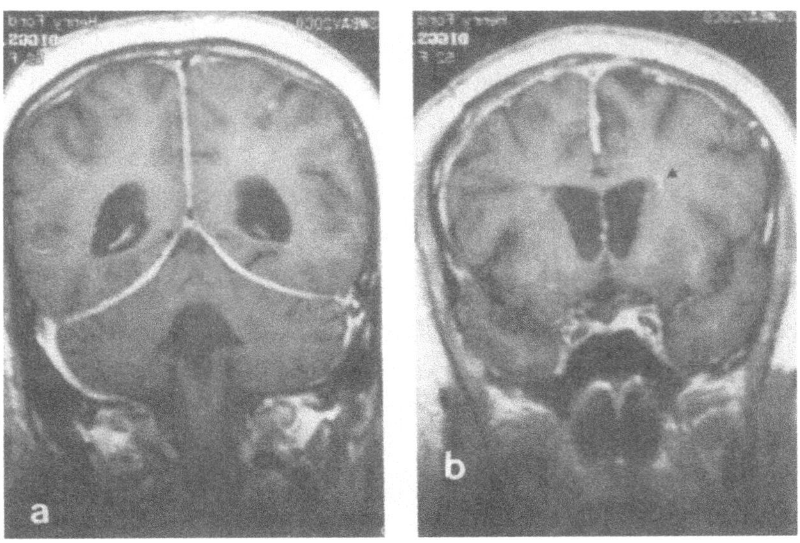

Figure 8. Meningeal sarcoidosis. Postcontrast T1-weighted coronal images (a, b) show enhancement of the pachymeninges. Note small enhancing lesions in the left periventricular region (arrowhead) suggestive of sarcoid granulomas. Suprasellar cisterns do not show any abnormal enhancement. (Courtsey Dr. Suresh Patel, Henry Ford Hospital, Detroit, Michigan.)

4. WEGENER'S GRANULOMATOSIS

Wegener's granulomatosis is a systemic necrotizing vasculitis characteristically involving the respiratory tracts and kidneys. The CNS involvement may occur in association with other organs or sometimes as a sole manifestation (57). The diagnosis can be made by the typical respiratory signs and symptoms, characteristic histologic findings in the nasal mucosa, and positive circulating antineutrophil cytoplasmic antibodies in the serum. The typical histopathologic features include granulomatous changes in the upper respiratory organs and nephritis. There are areas of necrosis as well as hemorrhages due to generalized necrotizing vasculitis. Wegener's granulomatosis often responds to cyclophosphamide and prednisolone with maintenance treatment recommended for more than 2 years as the recurrences are common (58). The CNS involvement in Wegener's granulomatosis has been reported in about 4–8 % of patients (59–61). Granulomatous thickening of the meninges and cerebrovascular events are the most common CNS manifestations. The cerebrovascular events include intracerebral hemorrhage, subarachnoid hemorrhage, arterial thrombosis, and venous thrombosis (60). As it is a small vessel vasculitis involving blood vessels with a diameter between 50 and 300 µm, it is seldom revealed by cerebral angiography (62,63). Some of the neurologic manifestations, however, are due to ischemic peripheral nerve damage and should not be confused with CNS damage. In young patients without any history of head trauma, subdural hematomas have been reported with Wegener's granulomatosis (57,58,64). These hematomas are primarily due to dural vasculitis and should be differentiated from systemic lupus erythematosus and polyarteritis nodosa that are known to cause dural vasculitis.

On cranial MRI, attention must be paid to paranasal cavities where, instead of typical sinusitis, granulomatous lesions may be seen. The meninges may show thickening and enhancement. The latter may extend to the adjoining subdural or subarachnoid space. Depending on the site of involvement and the kind of pathology, MRI may reveal multiple ischemic or hemorrhagic areas in the brain parenchyma. More often, the granulomatous lesions in the meninges show up only as meningeal enhancements. Large areas of hemorrhages in the adjoining subdural or the subarachnoid space surround these meningeal lesions.

5. RASMUSSEN'S ENCEPHALITIS

Rasmussen et al. first described Rasmussen's encephalitis in 1958 (65). Patients have intractable seizure activity caused by a kind of chronic encephalitis. In the original description of 27 children by Rasmussen, there was severe and progressive focal epilepsy with frequent episodes of epilepsia partialis continua. Rarely, generalized status epilepticus had occurred. The disease had a devastating course with progressive neurologic deficit. There were no clinical findings of encephalitis, but the histology of the brain revealed viral encephalitis. The primary presentation in patients with Rasmussen's encephalitis is partial motor seizures, although the other seizure types can also be observed. Most of the patients have progressive motor deficit as well as mental deterioration. Rasmussen's encephalitis is primarily an event of childhood with 85% of children having onset before 10 years of age. In Rasmussen's own series, the median age of onset was 5 years. It has been suggested that initially there may be bilateral cerebral hemispheric involvement but very soon the disease tends to localize to one cerebral hemisphere (66). The disease appears to be self-limiting in only 10% of patients. A majority of patients have varying degree of permanent neurologic deficits including hemiparesis, dysphasia, hemianopia, and mental deterioration (66). In spite of devastating course, the disease is rarely fatal (66).

The etiology of Rasmussen's encephalitis is not certain. Possibly, it is slow virus encephalitis, the virus being similar to the etiologic agent found in subacute sclerosing panencephalitis (66). In more recent studies, inflammatory or infectious episode has been observed in two-thirds of patients or in a close relative, immediately before the onset of seizures (66,67). In some cases, cytomegalovirus and Epstein–Barr virus have been suggested as the possible cause (68, 69).

The important histopathologic features in Rasmussen's encephalitis are perivascular lymphocytic cuffing, discrete glial nodules, and diffuse proliferation of microglia in the cortical gray matter as well as in the white matter. These changes are consistent with viral infections although rickettsial infections may also produce similar changes. In the chronic stage of disease, there is no evidence of inflammation and only cortical atrophy is visible with scattered areas of spongy degeneration and neuronal loss.

The CT scan findings in the initial stage of the disease are often normal. In one study, even the MRI done in the initial stage of the disease with and without contrast enhancement, was normal. In later stages, there were T2 hyperintense lesions suggestive of inflammation and edema in the left temporal as well as left parietal lobes (70). In another report, xenon CT cerebral blood flow study showed selectively decreased blood flow to the affected hemisphere despite rather unremarkable CT studies. In other patients, CT and MRI showed atrophy of one hemisphere with ipsilateral ventricular dilatation and decreased tracer uptake in the atrophic area (Fig.9) (71). Rarely there may be gliosis in the basal ganglia and periventricular region (71). *In vitro* MR spectroscopy has shown a decrease in NAA concentration in proportion to the severity and extent of the histologic abnormalities in excised neocortical tissue from a patient with Rasmussen's encephalitis (72). There are reports of *in vivo* proton MR spectroscopy showing changes in the ratios of different metabolites. Most reports reveal a decrease in NAA in the hemisphere showing abnormality on imaging (69,73–75). Few studies have also shown a decrease in NAA even in regions where imaging was normal (69,75). Serial spectroscopy in a child with Rasmussen's encephalitis revealed no descending trend in NAA while the patient had shown clinical progression (69). Presence of lactate has also been demonstrated along with decrease in NAA in patients who had epilepsia partialis continua at the time of MR spectroscopy (74).

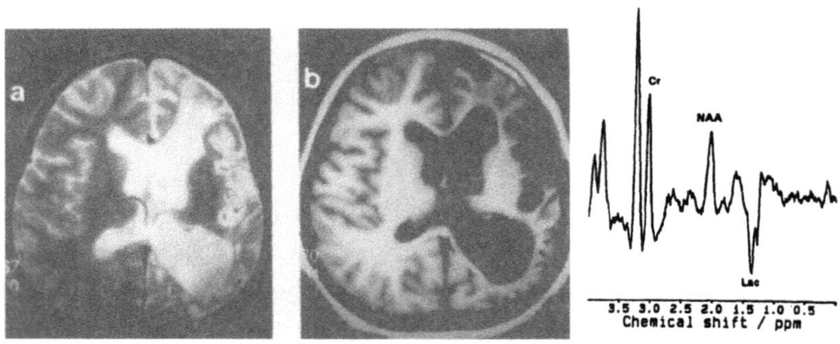

Figure 9. Rasmussen's encephalitis. T2- (a) and T1-weighted (b) axial images through the lateral ventricles show atrophy of the left lobe with ipsilateral ventricular dilatation. *In vivo* single voxel proton MR spectroscopy (SE, TR/TE/n = 3000/135/128) from the left frontoparietal lobe (c) shows a decrease in NAA and presence of lactate (Lac). Spectroscopy was done while patient was having epilepsia partialis continua.

6. WHIPPLE'S DISEASE

Whipple's disease is a relatively rare multisystem disorder that sometimes also has a neurologic presentation. The actual incidence of the CNS-restricted form of the disease is not clearly known, but it is undoubtedly low (76). However, some authors believe that all patients with Whipple's disease may have CNS involvement, although only 10–20% of them will exhibit clinical, laboratory, and radiologic features consistent with such involvement (77). The disease is now known to be a chronic granulomatous disorder due to *Tropheryma whippelli* bacillus (78). Cerebral manifestations of Whipple's disease are mainly observed during relapse of the disease, particularly when initial antibiotic treatment has not been adequate (79). Whipple's disease confined to CNS in the absence of jejunal infiltration by PAS-positive sickleparticle-containing cells is rare (80). However, PAS-positive macrophages have rarely been observed in the CSF (79). The most common neurologic manifestations are organic psychosis, dementia, deterioration of consciousness, ataxia, nystagmus, and supranuclear gaze palsy (81). Usually the foci are microscopic granulomas and are well beyond the resolution of MRI. There are few reports describing the MRI features of Whipple's disease. MRI features are nonspecific and are mainly confined to cortical and subcortical gray matter; rarely white matter is also involved (79). The lesions appear hyperintense on T2-weighted images and may rarely show rim enhancement on postcontrast study (80).

7. PROGRESSIVE MULTIFOCAL LEUKOENCEPHALOPATHY (PML)

PML is a subacute demyelinating disease originally described as a complication of chronic lymphocytic leukemia and Hodgkin's disease (82). As the disease is known to be have associated with illnesses that suppress cell-mediated immunity, an opportunistic noninflammatory viral infection of the nervous system was postulated (83,84). The viral theory is also supported by the histopathologic finding of inclusion bodies in large distorted oligodendrocyte nuclei surrounding the demyelinated foci and bizarre changes in the astrocytes within the areas of demyelination. The viral origin of the disease was established in the 1960s and 1970s with demonstration of virions within the lesions and consistent recovery of JC papovavirus from the lesion.

The neurologic symptoms start any time during the course of the underlying immunodeficiency disorder. The onset is insidious with symptoms and signs suggesting multifocal disorder. Mental decline, visual symptoms, paralysis, and sensory abnormalities are common. Cerebellar, brainstem, and spinal cord involvement is relatively rare. There is definite upsurge of the disease after the eruption of HIV infection. CSF may show JC virus DNA on polymerase chain reaction in about 80–90% of patients (85,86).

The MRI shows characteristic multifocal lesions involving subcortical white matter without enhancement. On T2-weighted images, lesions demonstrate increased signal intensity in the periventricular and/or subcortical white matter (Fig.10a). Lesions may be initially small, but usually progress to larger areas of involvement. A multifocal distribution pattern is seen, which may be unilateral but is more often bilateral and asymmetric (87). There is an absence of mass effect and lesions rarely enhance. When enhancement is present, it is faint and peripheral (87). The subcortical lesions follow the gray–white interface because of involvement of subcortical U-fibers (87). On T1-weighted images, the lesions are usually hypointense to parenchyma, although early lesions may be isointense.

Any lobe may be involved by PML, but the frontal and parieto-occipital locations are most common. Posterior fossa may be affected in about one-third of cases either as a part of the generalized parenchymal disease or in isolation (87,88). PML may also appear to involve the deep gray structures because of involvement of small myelinated fibers that course through the basal ganglia and adjacent parenchyma (87).

In vivo proton MR spectroscopy has been used for the differentiation of PML from other focal lesions such as lymphoma and toxoplasmosis. The usual spectroscopic findings are a decrease of NAA/Cr ratio (Fig.10b) and an increase of lactate/Cr, Cho/Cr, and lipids/Cr ratios compared with normal healthy control subjects (89–91). Creatine concentration may decrease in PML (89,90). Presence of lactate is related to cellular hypoxia and decreased mitochondrial ATP synthesis (92). Increase in choline and lipids may reflect the accumulation of myelin breakdown products and rapid cell membrane synthesis (92).

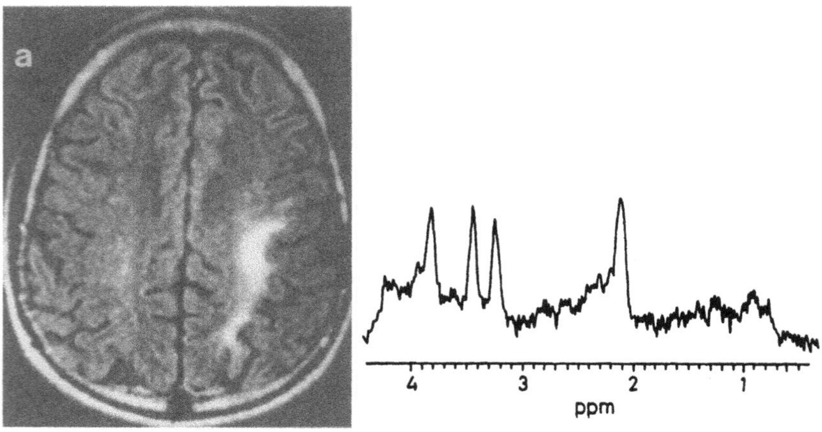

Figure 10. Progressive multifocal leukoencephalopathy. FLAIR image (a) through the supraventricular region shows hyperintensity in the white matter on the left side. *In vivo* single-voxel proton MR spectroscopy (b) from the hyperintense lesion shows a decrease in NAA (1). (Courtesy Drs. L. Chang and T. Ernst, UCLA–Harbour Medical Center, Los Angeles.)

REFERENCES

1. Walton J. *Brain's diseases of the nervous system*, 9 ed. London: Oxford University Press, 1985; 420–422.
2. Byrne TN. Non compressive myelopathies simulating spinal cord compression. Contemp Neurol Sci 1990; 33:229–264.
3. Ropper AH, Poskanzer DC. The prognosis of acute and subacute transverse myelopathy based on early signs and symptoms. Ann Neurol 1978; 4:51–59.

4. Jeffery DR, Mandler RN, Davis LE. Transverse myelitis: Retrospective analysis of 33 cases with differentiation of cases associated with multiple sclerosis and parainfectious events. Arch Neurol 1993; 50:532–535.
5. Abramsky O, Teitelbaum D. The autoimmune features of acute transverse myelopathy. Ann Neurol 1977; 2:36–40.
6. Al Deeb SM,Yaqub BA, Bruyn GW, Biary NM. Acute transverse myelitis—A localised form of post-infectious encephalomyelitis. Brain 1997; 120:1115–1122.
7. Pradhan S,Gupta RK, Ghosh D. Parainfectious myelitis: Three distinct clinico-imageological patterns with prognostic implications. Acta Neurol Scand 1997; 95:241–247.
8. Reik L. Disseminated vasculo-myelopathy: An immune complex disease. Ann Neurol 1980; 7:291–296.
9. Sanders KA, Khandji AG, Mohr JP. Gadolinium MRI in acute transverse myelopathy. Neurology 1990; 40:1614–1616.
10. Kincaid JC. Myelitis and myelopathy. In *Clinical neurology*, RJ Joynt, ed. Philadelphia: Lippincott, 1989; 1–32.
11. Berman M, Geldman S, Alter M, Zilber N, Kahana E. Acute transverse myelitis: Incidence and etiologic considerations. Neurology 1981; 31:966–971.
12. Al Deeb SM, Yaqub BA, Bruyn GN, Biary NM. Acute transverse myelitis—A localised form of postinfectious encephalomyelitis. Brain 1997; 120:1115–1122.
13. Campi A, Fillippi M, Comi G, et al. Acute transverse myelopathy: Spinal cord and cranial MR study with clinical follow up. Am J Neuroradiol 1995; 16:115–123.
14. Tartaglino LM, Liem M, Friedman DP, Flanders AE. Differentiation of idiopathic acute transverse myelitis from multiple sclerosis by MR (abstr). In: Proceedings of the American Society of Neuroradiology. Oak Brook, IL: American Society of Neuroradiology, 1993:10.
15. Miller DH, McDonald WI, Blumhardt LD, et al. Magnetic resonance imaging in isolated noncompressive spinal cord syndromes. Ann Neurol 1987; 22:714–723.
16. Pradhan S, Gupta RK, Shashank S, Kapoor R, Kathuria MK. Parainfectious conus myelitis. J Neurol Sci 1998; 161:156–162.
17. Jahnke U, Fischer EH, Alvord (Jr) EC. Hypothesis—Certain viral proteins contain encephalitogenic and/or neuritogenic sequences. J Neuropathol Exp Neurol 1985; 44: 320.
18. Miller HG, Stanton JB, Gibbons JL. Parainfectious encephalomyelitis and related syndromes. Q J Med 1956; 25: 427–505.
19. Yahr MD, Lobo-Antunes J. Relapsing encephalomyelitis following the use of influenza vaccine. Arch Neurol 1972; 27: 182–183.
20. Saito H, Endo M, Takase S, Itahara K. Acute disseminated encephalomyelitis after influenza vaccination. Arch Neurol 1980; 37: 564–566.
21. Fukuda H, Umehara F, Kawahigashi N, Suehara M, Osame M. Acute disseminated myelitis after Japanese B encephalitis vaccination. J Neurol Sci 1997; 148: 113–115.
22. Ambler M, Stoll J, Tzamaloukas A, Albala MM. Focal encephalomyelitis in infectious mononucleosis—A report with pathological description. Ann Intern Med 1971; 75: 579–583.
23. Pradhan S, Pandey N. Acute disseminated encephalomyelitis presenting as ataxic hemiparesis. Neurol India 1998; 46: 156–158.
24. Hurst EW. Acute haemorrhagic leukoencephalitis—A previously undefined entity. Med J Aust 1941; 28: 1–6.

25. Pradhan S, Gupta RP, Shashank S, Pandey N. Intravenous immunoglobulin therapy in acute disseminated encephalomyelitis. J Neurol Sci 1999; 165: 56–61.
26. Caldemeyer KS, Smith RR, Harris TM, Edwards MK. MRI in acute disseminated encephalomyelitis. Neuroradiology 1994; 36: 216–220.
27. Murthy JMK, Yangala R, Meena AK, Reddy JJ. Acute disseminated encephalomyelitis: Clinical and MRI study from south India. J Neurol Sci 1999; 165: 133–138.
28. Kesselring J, Miller DH, Robb SA, et al. Acute disseminated encephalomyelitis. Brain 1990; 113: 291–302.
29. Miller DH, McDonald WI, Blumhardt LD, et al. Magnetic resonance imaging in isolated noncompressive spinal cord syndromes. Ann Neurol 1987; 22: 714–723.
30. Poser CM. Magnetic resonance imaging in asymptomatic disseminated vasculomyelinopathy. J Neurol Sci 1989; 94: 69–77.
31. Swamy HS, Shanker SK, Satishchandra P, Aroor SR, Sivaramkrishna A, Kaliperumal VG. Neurological complications due to betapropiolactone (BPL) inactivated anti-rabies vaccination. J Neurol Sci 1984; 63: 111–128.
32. Sriram S, Steinman L. Post-infectious and post-vaccinal encephalomyelitis. Neurol Clin 1987; 2: 341–353.
33. Atlas SW, Grossman RI, Goldberg HI, et al. MR diagnosis in acute disseminated encephalomyelitis. J Comput Assist Tomogr 1986; 10: 798–801.
34. Epperson LW, Whiteker JN, Kapila A. Cranial MRI in acute disseminated encephalomyelitis. Neurology 1988; 38: 332–333.
35. Andreula CF, Rechia Luciani ANM, Millela D. Magnetic resonance imaging in the diagnosis of acute disseminated encephalomyelitis (ADEM). Int J Neuroradiol 1997; 3: 21–34.
36. Baum PA, Barkovich JA, Koch TK, Berg BO. Deep gray matter involvement in children with acute disseminated encephalomyelitis. Am J Neuroradiol 1994; 15:1275–1283.
37. Okumura A, Hayakawa M, Watanabe K, et al. Two cases of acute disseminated encephalomyelitis with lesion in the thalamus or basal ganglia on MRI. No To Hattatsu 1992; 24: 278–282.
38. Youl BR, Kermode AG, Thompson AJ, et al. Destructive lesions in demyelinating disease. J Neurol Neurosurg Psychiatry 1991; 54: 288–292.
39. van der Meyden, de Villiers JFK, Middlecote BD, Terblanche J. Gadolinium ring enhancement and mass effect in acute disseminated encephalomyelitis. Neuroradiology 1994; 36: 221–223.
40. Dantzker DR, Tobin MJ, Whatley RE. Respiratory diseases. In *Cecil essentials of medicine*, TE Andreoli, CCJ Carpenter, F Plum, LH Smith, eds. Philadelphia: Saunders, 1986;149.
41. Oksanen VE. Neurosarcoidosis. In *Sarcoidosis and other granulomatous disorders*, DG James, ed.. New York: Dekker, 1994; 285–309.
42. Stern BJ, Krumholz A, Johns C, Scott P, Nissan J. Sarcoidosis and its neurological manifestations. Arch Neurol 1985; 42: 909–917.
43. Ulmer JL, Elster AD. Sarcoidosis and the central nervous system. Neuroimag Clin North Am 1991; 1: 141–158.
44. Delaney P. Neurologic manifestations in sarcoidosis: Review of the literature, with a report of 23 cases. Ann Intern Med 1987; 87: 336-345.
45. Sherman JL, Stern BJ. Sarcoidosis and the CNS: Comparison of unenhanced and enhanced MR images. Am J Neuroradiol 1990; 11: 915–923.

46. Fick RB, Hunnininghake GW. Sarcoidosis. In *Immunological diseases*, 4th ed. M. Samter, ed. Boston: Little Brown; 1985.
47. Urich H. Neurological manifestations of sarcoidosis. Practitioner 1969; 202: 632–636.
48. Waxman JS, Sher JH. The spectrum of central nervous system sarcoidosis: A clinical and pathologic study. Mt Senai J Med 1979; 46:309–317.
49. Burger PC, Scheithauer BW. Reactive and inflammatory masses simulating neoplasia. In *Tumours of the central nervous system: Atlas of tumour pathology. Third series*. Washington, DC: Armed Forces Institute of Pathology, 1994; 391–413.
50. Nathan MPR, Chase PH, Elguezabel A, Weinstein M. Spinal cord sarcoidosis. N Y State J Med 1976; 76: 748—752.
51. Schmidt RE. Demyelinating diseases. In *Principles and practice of neuropathology*, JS Nelson, JE Parisi, SS Schochet, eds. St. Louis. Mosby Year Book, 1993; 361–397.
52. Christopheridis GA, Spickler EM, Recio MV, Mehta BM. MR of CNS sarcoidosis: Correlation of imaging features to clinical symptoms and response to treatment. Am J Neuroradiol 1999; 20: 655–669.
53. Williams DW, Elster AD, Kramer SI. Neurosarcoidosis: Gadolinium-enhanced MR imaging. J Comput Assist Tomogr 1990; 14: 704–707.
54. Brooks ML, Wang A, Black P, et al. Subdural mass lesion secondary to sarcoid granuloma. MR and CT findings and differential diagnosis. Comput Med Imaging Graph 1989; 13: 199–205.
55. Post MJD, Quencer RM, Tabei S. CT demonstration of sarcoidosis of the optic nerve, frontal lobes, and falx cerebri. Case report and literature review. Am J Neuroradiol 1982; 3: 523–526.
56. Wilson JD, Castillo M. Magnetic resonance imaging of granulomatous infiltrations: Sarcoidosis and tuberculosis. Top Magn Reson Imaging 1994; 6: 32–40.
57. Yokote H, Terada T, Nakai K, Itakura T. Subdural and meningeal involvement related to Wegener's granulomatosis: Case report. Neurosurgery 1997; 40: 1071–1074.
58. Hoffman GS, Kerr GS, Leavitt RY, Hallahan CW, Lebovics RS, Travis WD, Rottem M, Fauci AS. Wegener's granulomatosis: An analysis of 158 patients. Ann Intern Med 1992; 116: 488–498.
59. Anderson JM, Jamieson DG, Jefferson JM: Non-healing granuloma and the nervous system. Q J Med 1975; 44: 309–323.
60. Drachman DD. Neurological complications of Wegener's granulomatosis. Arch Neurol 1963; 8: 145–155.
61. Nishino H, Rubino FA, DeRemee RA, Swanson JW, Parisi JE. Neurological involvement in Wegener's granulomatosis. An analysis of 324 consecutive patients at Mayo Clinic. Ann Neurol 1993; 33: 4–9.
62. Satoh J, Miyasaka N, Yamada T, Nishido T, Okuda M, Kuroiwa T, Shimokawa R. Extensive cerebral infarction due to involvement of both anterior cerebral arteries by Wegener's granulomatosis. Ann Rheum Dis 1988; 47: 606–611.
63. Yamashita Y, Takahashi M, Bussaka H, Miyawaki M, Tosak K. Cerebral vasculitis secondary to Wegener's granulomatosis: Computed tomographic and angiographic findings. J Comput Assist Tomogr 1986; 10: 115–120.
64. Futran J, Shore A, Urowitz MB, Grossman H. Subdural hematoma in systemic lupus erythematosus: Report and review of literature. J Rheumatol 1987; 14: 378–381.
65. Rasmussen T, Olszweski J, Lloyd-Smith D. Focal seizures due to chronic localised encephalitis. Neurology 1958; 8: 435–455.

66. Rasmussen T, Andermann F. Update on the syndrome of "chronic encephalitis" and epilepsy. Cleveland Clin J Med 1989; 56 [suppl]: 181–184.
67. Oguni H, Andermann F, Rasmussen T. The natural history of the syndrome of chronic encephalitis and epilepsy: A study of the MRI series of 48 cases. Neurology 1989; 39 [suppl 1]: 155.
68. Power C, Poland SD, Blume WT, Girvin JP, Rice GPA. Cytomegalovirus and Rasmussen's encephalitis. Lancet 1990; 336: 1282–1284.
69. Sundgren PC, Burtscher IM, Lundgren J, Geijer B, Holtas S. MRI and proton spectroscopy in a child with Rasmussen's encephalitis—Case report. Neuroradiology 1999; 41: 935–940.
70. Zupank ML, Handler EG, Levine RL, et al. Rasmussen's encephalitis: Epilepsia partialis continua secondary to chronic encephalitis. Pediatr Neurol 1990; 6: 397–401.
71. Tien RD, Ashdown BC, Lewis DV, Atkins MR, Burger PC. Rasmussen's encephalitis: Neuroimaging findings in four patients. Am J Radiol 1992; 158: 1329–1332.
72. Peeling J, Sutherland G. 1H magnetic resonance spectroscopy of extracts of human epileptic neocortex and hippocampus. Neurology 1993; 43: 589–594.
73. Geller E, Faerber EN, Legido A, et al. Rasmussen encephalitis: Complementary role of multitechnique neuroimaging. Am J Neuroradiol 1998; 19: 445–449.
74. Matthews PM, Andermann F, Arnold DL. A proton magnetic resonance spectroscopy study of focal epilepsy in humans. Neurology 1990; 40: 985–989.
75. Cendes F, Andermann F, Silver K, Arnold DL. Imaging of axonal damage in vivo in Rasmussen's syndrome. Brain 1995; 118: 753–758.
76. Adams M, Rhyner PA, Day J, DeArmond S, Smuckler EA. Whipple's disease confined to the central nervous system. Ann Neurol 1987; 21: 104–108.
77. Dobbins WO. Whipple's disease (editorial). Mayo Clin Proc 1988; 63: 623–624.
78. Relman D, Schmidt T, McDermott R, Falkow S. Identification of the uncultured bacillus of Whipple's disease. N Engl J Med 1992; 327: 293–301.
79. Davion T, Rosat P, Sevestre H, et al. MR imaging of CNS relapse of Whipple's disease. J Comput Assist Tomogr 1990; 14: 815–817.
80. Wroe S, Pires M, Harding B, Youl B, Shorvon S. Whipple's disease confined to the CNS presenting with multiple intracranial mass lesions. J Neurol Neurosurg Psychiatry 1991; 54: 989–992.
81. Feldman M. Whipple's disease. Am J Med Sci 1986; 291: 56–67.
82. Astrom KE, Mancall EL, Richardson EP Jr. Progressive multifocal leucoencephalopathy. Brain 1958; 81: 93–111.
83. Cavanagh JB, Greenbaum D, Marshall AHE, Rubinstein LJ. Cerebral demyelination associated with disorders of the reticulo-endothelial system. Lancet 1959; 2: 525–529.
84. Richardson EP Jr. Progressive multifocal leucoencephalopathy. N Engl J Med 1961; 265: 815–823.
85. Weber T, Turner RW, Frye S, et al. Specific diagnosis of progressive multifocal leukoencephalopathy by polymerase chain reaction. J Infect Dis 1994; 169: 1138–1141.
86. McGuire D, Barhite S, Hollander H, Miles M. JC virus DNA in cerebrospinal fluid of human immunodeficiency virus infected patients: Predictive value of progressive multifocal leukoencephalopathy. Ann Neurol 1995; 37: 395–399.
87. Whiteman MLH, Post MJD, Berger JR, et al. Progressive multifocal leukoencephalopathy in 47 HIV-seropositive patients: Neuroimaging with clinical and pathologic correlation. Radiology 1993; 187:233–240.

88. Sze G, Brant-Zawadzki MN, Norman D, Newton HT. The neuroradiology of AIDS. Semin Roentgenol 1987; 22:42–53.

89. Chang L, Miller BL, McBride D, et al. Brain lesions in patients with AIDS: H-1 MR spectroscopy. Radiology 1995; 197: 525–531.

90. Chang L, Ernst T, Tornatore C, et al. Metabolic abnormalities in progressive multifocal leukoencephalopathy by proton magnetic resonance spectroscopy. Neurology 1997; 48: 836–845.

91. Iranzo A, Moreno A, Pujol J, et al. Proton magnetic resonance spectroscopy pattern of progressive multifocal leukoencephalopathy in AIDS. J Neurol Neurosurg Psychiatry 1999; 66: 520–523.

92. Ross BD, Michaelis T. Clinical applications of magnetic resonance spectroscopy. Magn Reson Q 1994; 10: 191–247.

Chapter 10

IMAGING OF POSTOPERATIVE INFECTIONS

Gregory Christoforidis, Suresh C. Patel, and Mark Rosenblum

The efficient recognition of infection is essential to the correct management of any postoperative patient. Although there are signs, that may lead a clinician to suspect postoperative infection, the radiologist is often called on to help confirm or identify the source of the infection.

In assessing a patient's imaging studies for postoperative infection, one faces the dilemma of differentiation of normal postsurgical repair from infection. It is therefore important to be familiar with the appearance of an uncomplicated postoperative site. Furthermore, the radiologist needs to be familiar with the risk factors and expected course and site of infectious complications. Finally, one needs to scrutinize imaging studies for radiographic signs of an infectious process.

According to the National Academy of Sciences–National Research Council (1), postoperative wounds are classified into four categories: clean surgery, clean–contaminated

---

Gregory Christoforidis • Department of Radiology, Ohio State University School of Medicine, Columbus, Ohio 43210-1228. Suresh C. Patel • Section of Neuroradiology, Henry Ford Hospital, Detroit, Michigan 48202. Mark Rosenblum • Department of Neurosurgery, Henry Ford Hospital, Detroit, Michigan 48202.

MR Imaging and Spectroscopy of Central Nervous System Infection, edited by Gupta and Lufkin. Kluwer Academic / Plenum Publishers, New York, 2001.

surgery, contaminated surgery, and dirty surgery. Accordingly, the risk of infection varies (Table 1).

Table 1. Classification of Postoperative Wounds

Wound class	Rate of infection	Description
Clean surgery	<5%	Clean elective neurosurgical procedure
Clean-contaminated	5–15%	Operative site in which the bronchus, gastrointestinal tract, paranasal sinuses, or oropharyngeal cavities are entered but without unusual contamination
Contaminated	10–25%	Contamination is known to have occurred, as in compound skull fracture, gunshot wound, and break in sterile technique
Dirty surgery	30–80%	There is already a source of infection present in the wound, such as a brain abscess, subdural empyema, meningitis, osteomyelitis, or established wound infection

Risk factors for infection following surgery include surgical length of stay, reoperation, use of postoperative drains, entry into paranasal sinuses, craniofacial procedures, compound fracture, CSF leak, retained fragments following trauma, immunocompromised patient, diabetes, use of corticosteroids, prior radiation therapy, concomitant systemic infection, and clean surgery with foreign body. In assessing a patient for a wound infection, one needs to take into consideration such risk factors.

In this chapter imaging of postcraniotomy infections will be reviewed separately from post-spinal surgery infections.

1. IMAGING OF POSTCRANIOTOMY INFECTIOUS COMPLICATIONS

The overall incidence for a postcraniotomy wound infection ranges from 2 to 8%. Potential sequelae include osteomyelitis, meningitis, subdural empyema, sinus thrombosis, and brain abscess. One out of 26 postcraniotomy extradural infections resulted in neurologic sequelae. Twenty one percent of the patients with intradural postcraniotomy infections die or are neurologically crippled (2).

Assessment of the cranial postoperative site is typically performed on the basis of CT in order to assess for hemorrhagic complication, infarction, or hydrocephalus. In general, this is sufficient as a screening tool that does not incur the expense and time of MR examination. MR, however, has been shown to be more sensitive than CT in delineating intracranial infections (3–5). This is preferably performed with the benefit of gadolinium. MR has advantages over CT. Its multiplanar capability and improved soft tissue contrast resolution allow for more accurate differentiation between intra- and extra-axial fluid collections, and more sensitive characterization of meningeal lesions. Furthermore, MR lacks the ionizing radiation that accompanies CT (6,7). CT with contrast, however, can provide an adequate evaluation for complications of meningitis such as hydrocephalus, cerebritis, abscess, subdural empyema or effusion, ventriculitis, and infarction and in predicting neurologic outcome (4).

1.1. Postcraniotomy MRI Appearance

Before assessing a patient's imaging studies for postcraniotomy infection, one needs to be familiar with the appearance of the normal postoperative site (3). In the first few postoperative days, expected findings include intracranial air, parenchymal and intraventricular hemorrhage, extra-axial fluid, edema and minor hemorrhages along the margins of resection, and mass effect (8). In general, these findings are expected to resolve in the weeks following the operation. A relatively thin and smoothly enhancing meningogaleal complex may also be present (8). A sterile fluid collection may sometimes accompany dural enhancement following surgery. This collection is usually but not always nonhemorrhagic (3,9). MR enhancement of pia or brain at the encephalotomy site is often seen but usually disappears by 1 year (8). This probably represents normal enhancement of brain along the resection site (blood–brain barrier damage) and is usually not seen on CT (8).

Patterns of meningeal enhancement following surgery include (1) no dural enhancement, (2) pachymeningeal enhancement at the postoperative site (Fig.1), or (3) mild smooth diffuse enhancement over the convexities. In a study of 20 patients, Burke et al. (10) found that 80% of postcraniotomy patients had nonneoplastic enhancement, 9/20 generalized and 7/20 localized. In addition, the meningogaleal complex may thicken following surgery. Meningeal enhancement as noted on MR can be found as early as 18 hours following surgery (8). Elster and DiPersio have shown that meningeal enhancement may persist for years. Over time this meningogaleal enhancement and thickening decreases over the first year and in most cases disappears after the first year (8). A mild degree of thickening (less than 3 mm) may persist for decades. This meningeal time course of enhancement differs from CT. Meningeal enhancement on CT starts later and disappears early (11,12). Neither a history of prior radiation nor the type of operation has a recognizable relationship to the presence or degree of dural enhancement (8). This postoperative enhancement may be explained on the basis of extrusion of contrast media, dysautoregulation (13), granulation tissue, or neovascularity (11). Furthermore, chemical meningitis may also be an underlying cause of meningeal enhancement seen following surgery (10). The difference in the time course of contrast enhancement seen on CT is probably related to the fact that mole per kilogram, MR is 20 times more sensitive in detecting Gd-DTPA than CT in detecting iodinated contrast (14). In addition, CT will have greater difficulty in detecting enhancement located adjacent to bone due to volume averaging (8).

In addition to normal postoperative finding, one must consider the presence of residual tumor or infection for which the surgery was originally performed. Leptomeningeal and ependymal enhancement may be related to recurrent tumor or infection. Nodular enhancement is more typical of recurrent or residual exophytic tumor (3). Patients with normal dura, pia, and arachnoid or ependyma on gadolinium-enhanced MR have been shown not to have evidence of leptomeningeal or ependymal disease based on clinical or laboratory data (3).

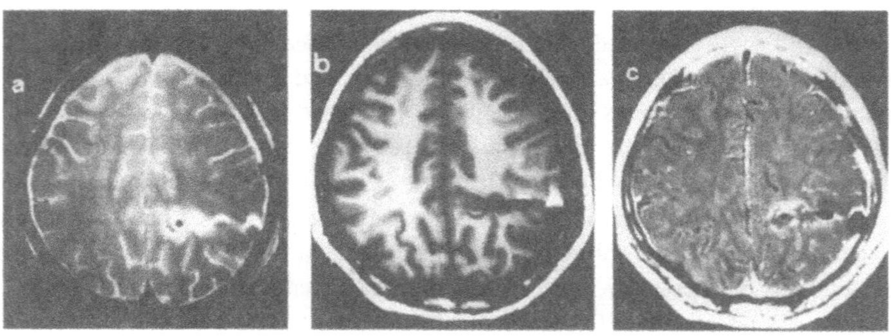

Figure 1. Normal postoperative appearance: There is evidence of surgery in the left parietal region seen as hyperintense on T2- (a) and hypointense on T1-weighted image (b). Note the hyperintensity seen on the lateral margin of the linear surgical cavity on T1-weighted image, consistent with subacute bleed. Postcontrast T1-weighted image (c) shows meningeal enhancement around the operated region on the left side.

Overall postsurgical infection rates of 6.3% for supratentorial, 5.4% for infratentorial, and 1.4% for extradural procedures have been described in the literature dating back to the 1960s; however, a more recent review indicates that these numbers are more likely under 5% (15). Predisposing factors for postcraniotomy infections include obesity, diabetes mellitus, malnutrition, malignancy, immunosuppression, and advanced age. Use of steroids and advanced age did not lead to increase in wound infection (15). Prolonged preoperative hospitalization, shaving the operative site on the day before surgery, infection in a remote location of the body, or a lengthy operation, roughly doubles the rate of infection (15). Prophylactic antibiotics recently have been shown to help (15).

Wound infections are rarely seen in the first 48 hours following surgery. Signs include tenderness, erythema, purulence, wound dehiscence, and warmth. Fever, leukocytosis, lymphadenopathy, and widespread edema are systemic signs of infection (16). Postoperative infections often begin at the suture line or among the several layers of superficial closure. Proposed mechanisms for intracranial involvement include direct extension of a superficial infection and intraoperative contamination (17). In general, early signs of postoperative infection are a change in the appearance of the meningogaleal complex. This is seen as thickening and enhancement on CT and MRI. The normal dark dural line seen on MRI becomes poorly defined and has mixed signal because of edema or granulation tissue. A ring-enhancing fluid collection can signify the development of an abscess. These findings in and by themselves, however, are not specific for infection. The development of a sterile granuloma following craniotomy has been described. Because it can have a ring-enhancing appearance, it can even mimic a postoperative abscess. In addition, cystic tumors can manifest at the operative site and can mimic an abscess.

1.2. Subdural Empyema

Isolated subdural empyema without accompanying osteomyelitis following a craniotomy has been previously reported (18,19). Unlike a subdural empyema unrelated to craniotomy, postoperative subdural empyema can present without superficial signs of wound infection and as such seen to have a more indolent course with a time lag ranging from 5 weeks to over a year following the operation and little or no fever (18,20). If the patient presents with signs of wound infection, however, the time lag may be shorter (19). The causative organism for postoperative subdural empyema is usually *Staphylococcus* (*S. epidermidis* or *S. aureus*). This is in contrast to the typical causative agent of most subdural empyemas — *Streptococcus*. Unlike subdural empyemas from other causes which have been reported to have a mortality rate as high as 15 to 40% (21), patients with subdural empyemas following craniotomy have a much better outcome and typically do not have significant neurologic sequelae (18,20). Because of multiplanar imaging and greater contrast resolution, MRI is more sensitive than CT to the detection of small extra-axial fluid collections such as an empyema (22). On MRI, a subdural empyema tends to be isointense on the T1-weighted sequence and hyperintense on the T2-weighted sequence. The adjacent sulci are frequently effaced and there is often mass effect. In addition, the adjacent pachymeninges and leptomeninges are enhanced. On noncontrast imaging, it may be easy to confuse a subacute subdural hematoma with an empyema; however, the associated inflammatory changes seen with infection may help differentiate the two. In addition to evaluating the extent of the empyema and the associated mass effect, one should also scrutinize the dural venous sinuses for signs of thrombosis.

1.3. Osteomyelitis

Osteomyelitis of the skull accompanies approximately 14% of superficial wound infection (2). The patient typically will present with local signs of a scalp defect or swelling with headache but without systemic signs and symptoms occurring between 2 months and 32 years following the initial inoculation (1). Cranial nerve palsies may accompany skull base osteomyelitis. MRI is useful in the detection and evaluation of the extent of involvement of osteomyelitis in regions of the skull and skull base containing fatty marrow. In the presence of postoperative changes, the sensitivity of MRI in detecting osteomyelitis is lower (23). MRI can also assess for adjacent sinus thrombosis, dural involvement, cavernous sinus involvement, and presence of dural venous sinus thrombosis. Furthermore, in regions of the skull where there is no bone marrow present, CT and radionuclide imaging are more sensitive modalities in evaluating for osteomyelitis (1,23). Both clinical and radiographic signs of osteomyelitis may be mimicked by recurrent tumor. As a result, biopsy may still be necessary to establish the diagnosis (23,24).

Both CT and MRI may take months to years to return to normal following the resolution of osteomyelitis (1,23,25). Resolution of osteomyelitis on CT is based on remineralization of bone following the complete resolution of infection. Thus, a large time interval must pass before this can occur. On the other hand, indium-labeled WBC scan have been found to be very useful in this regard returning to normal within a month following successful treatment (23,26). Similarly, serial gallium scans have been found to be sensitive to the presence of recurrent or continuing infection, although sometimes it may take up to 6 months to return to normal (1,25–27). More recently, combined use of indium-111 WBC and Tc-99m MDP bone SPECT scans has been shown to be superior to planar radionuclide imaging in differentiating osteomyelitis from adjacent soft tissue infection (23). Color images and region of interest quantitative data may assess whether the degree of uptake is likely to be of clinical importance (23). Follow-up assessment with imaging is best performed with In-111 WBC and Tc-99m MDP bone imaging since this imaging technique reverts to normal sooner than CT, MR, and gallium-67 imaging (23).

1.4. Ventriculitis/Meningitis

Postoperative meningitis and ventriculitis have been reported to occur after 0.34% (28) to 2% (29) of all neurosurgical operations and constitute up to 46% of all postoperative infections. The most frequent pathogens are reported to be *S. aureus* and *S. epidermidis* with the latter organism being the most frequent pathogen involved in shunt-related infections (30). Other pathogens include gram-negative aerobic bacilli (15). A normal enhanced examination does not absolutely exclude meningeal or ventricular disease. Imaging complements clinical examination and CSF findings. Typical findings associated with meningitis or ventriculitis include leptomeningeal or ependymal enhancement. MRI can also be useful in assessing complications of meningitis such as hydrocephalus, infarction, dural sinus thrombosis, or subdural empyema (22,31).

1.5. Brain Abscess

Since the advent of CT, the mortality from brain abscess has decreased (32). This has been attributed to earlier detection and more accurate localization. Serial CT scans allow for the determination of optimal timing for surgical intervention and more rapid detection and treatment of postoperative complication (32). However, the mortality of a brain abscess remains high at 14%. It is found to occur in 0.1% of postoperative patients (2). Typical pathogens include *S. aureus*, streptococci, and enterobacter (15). The appearance of a brain abscess on MRI depends on the stage of development (33). As a cerebritis it is seen as a low-signal area on T1-weighted imaging, a high-signal focus on T2-weighted imaging, and patchy enhancement following contrast administration. With time a ring-enhancing fluid collection can develop (Fig.2).

1.6. CSF Shunt Infections

Following obstruction, infection is the second most common complication affecting CSF shunts (34). Factors that place patients at risk include length of time from the most recent shunt revision and patient age. Most (80–95%) shunt-related infections occur within 6 months of shunt manipulation (35,36). Infection rates in children under 1 year of age are higher (36,37).

A glycoprotein film covers a shunt following its implantation. This film can serve as a receptor site for organisms as well as host tissue cells. If bacterial layering precedes host tissue cell layering, an infection can follow which may be difficult for antibiotics to effectively reach. Typical pathogens include *S. epidermidis* and less commonly *S. aureus* (34). Although diagnosis is based on neurologic and laboratory examination, diagnostic

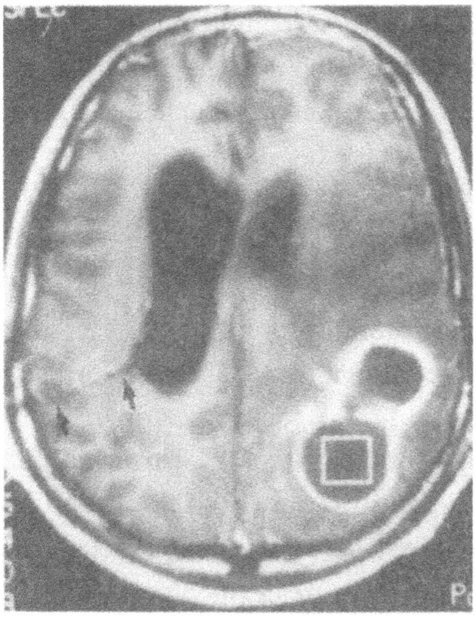

Figure 2. Brain abscess developing in a postoperative case of ventriculoperitoneal shunt. Contrast-enhanced T1-weighted axial section showing evidence of ring-enhancing areas in the left parietal region. Note the shunt tract on the right side (arrows) and the persisting hydrocephalus. The probable mechanism of abscess formation on the other side is septic embolism. The culture from the aspirated pus revealed anaerobic streptococci and bacteroids.

imaging including MRI can serve to identify focally affected meninges, subdural fluid collections, and peritoneal loculations as described above (34).

2. POSTOPERATIVE SPINAL INFECTIOUS COMPLICATIONS

Spinal infections can be described in terms of vertebral osteomyelitis, diskitis, or epidural abscess (38,39). Infection rate following lumbar diskectomy with antibiotic prophylaxis has been estimated to be 0.7–0.8% (40,41). With fusion and instrumentation the infection rate is higher (40). Risk factors predisposing to infection include age, poorly controlled diabetes, prior radiation, and presence of tumor. MRI is the most accepted imaging modality for the initial evaluation of patients with suspected spinal infection (42).

2.1. Postoperative Diskitis

The appearance of the normal and infected postoperative lumbar spine following discectomy procedures overlaps. The classic signs of discitis on MR include decreased disc space signal on T1, increased disc space signal on T2, possibly with effacement of the internuclear cleft and enhancement of the disc and endplates. Unfortunately, these changes normally occur following diskectomy surgery especially within the first 6 months. This most likely reflects granulation or scar tissue formation (43–45). Intervertebral disk enhancement has been found to occur more frequently following surgery for larger disk herniation at 3 months. Increased signal following disk surgery has been demonstrated in all patients 2 days following diskectomy and most patients at 3 months (46).

Smooth linear bands of enhancement have been described as benign postoperative findings following diskectomy, often associated with increased signal on T2 and decreased signal on T1 imaging similar to early degenerative change. This pattern has been found to occur in approximately 20% of postdiskectomy MR of the lumbar spine at 3 months (43). In general, the enhancement pattern seen with diskitis is amorphous rather than bandlike (43).

2.2. Postoperative Osteomyelitis

The diagnosis of vertebral osteomyelitis is often delayed. The results from clinical studies, laboratory tests, and radiographic studies are usually nonspecific. In general, MRI is the most sensitive and specific examination for the purpose of detecting vertebral osteomyelitis. It can be diagnostic in 55% of cases in the first 2 weeks of symptom onset and 76% after 2 weeks (47). The diagnosis can be suggested in 90% of cases before or 2 weeks after symptom onset. Fine needle aspiration or blood cultures can confirm the diagnosis in over 70% of cases (47); however, open or core biopsy may be required to confirm or make the diagnosis.

On CT examination, the diagnosis can be suggested based on new bone formation and paravertebral mass, which may be difficult to distinguish from postoperative change. On MRI, the marrow signal is abnormally low on T1-weighted imaging and high on T2-weighted imaging (Fig.3). The usual demarcation between disk and vertebral body is often

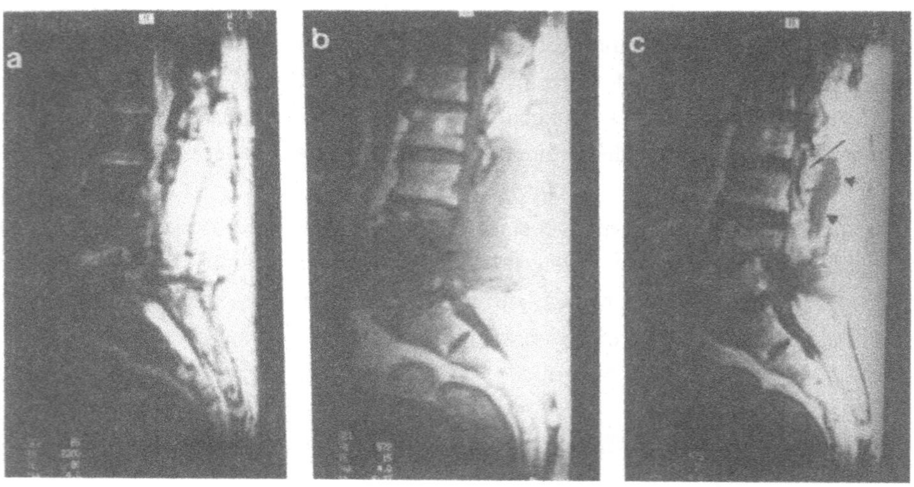

Figure 3. Postoperative spine infection. T2-weighted sagittal image (a) of lumbosacral spine shows evidence of hyperintense area in the operated site. Note the irregularity of the L4–L5 disk space. T1-weighted image (b) shows loss of signal from the L4 and L5 vertebrae and the intervening disk is not separately visualized. Postcontrast T1-weighted image (c) shows enhancement of the operated site along with a ring-enhancing area in the epidural location (arrow) opposite the L2 and L3 vertebrae suggestive of epidural abscess formation. Note the enhancement of the L4 and L5 vertebrae also.

lost in the face of vertebral osteomyelitis (48). In addition, associated disk space abnormalities and changes within the paravertebral musculature are common. Typically, this paravertebral muscular involvement is in the form of an enhancing cuff around the vertebral body (47,49). The decrease in paravertebral inflammatory response is thought to represent the earliest sign resolution following treatment; vertebral body enhancement, on the other hand, can persist in the face of clinical improvement (47).

2.3. Postoperative Infected Fluid Collections

Fluid collections can occur following surgeries such as laminectomy, corpectomy, and diskectomy. These fluid collections can become secondarily infected and result in an abscess. Such collections can be anterior or posterior in location (Figs.3,4). These infections can spread into the spinal canal depending on the operative tract and result in an epidural abscess or arachnoiditis. This needs to be differentiated from postoperative edema that is less hyperintense on T2-weighted images compared with collections. The management of collection is drainage whereas edema resolves with time.

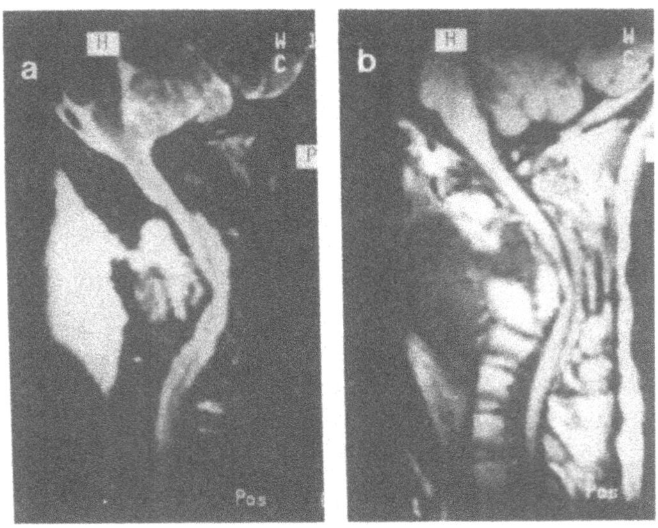

Figure 4. Prevertebral collection in a postoperative case of corpectomy. T2-weighted sagittal image (a) shows evidence of a hyperintense area in the prevertebral region continuous with a small pocket in the operative site through a tract. There is evidence of angulation deformity of the spine with compression of the cord. On T1-weighted image (b) the collection appears hypointense and is seen to extend up to the epidural location.

2.4. Postoperative Epidural Abscess

Epidural abscess is regarded as a medical and surgical emergency (40). Early clinical signs include pain over the suspected area of the spine and paresthesia (50). In general, it may be difficult to detect on plain radiographs, myelography, or radionuclide examination. MR with gadolinium is very sensitive to the detection of epidural abscess (Fig.5) (50). Because both fat and gadolinium have high signal intensity on T1-weighted imaging, fat suppression techniques increase the sensitivity of MR examinations (50,51). It appears as high-signal-intensity fluid collection on T2-weighted imaging but may be confused with normal CSF. It is frequently accompanied by spondylodiskitis (Fig.3), although it can also occur in isolation (50). Gadolinium-enhanced MRI may show a ring-enhancing fluid collection peripherally within the spinal canal often with associated enhancement of the adjacent meninges. Another pattern is that of a homogeneously enhancing epidural phlegmon. The signal from an epidural abscess may be heterogeneous (48). Because of its

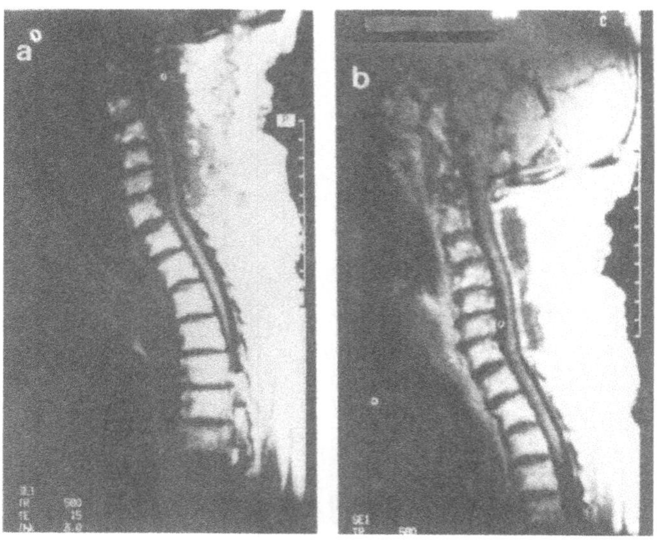

Figure 5. Postoperative epidural collection. This patient underwent laminectomy for compressive myelopathy. T1-weighted sagittal image (a) shows evidence of hypointense region posterior to the thecal sac and extending from craniovertebral junction to sixth cervical vertebra. Postcontrast T1-weighted image (b) shows rim enhancement of the hypointense region consistent with an extradural collection. Fine needle aspiration from the area revealed frank pus.

location, it can be confused with the epidural venous plexus (50). Awareness of the normal anatomy may help avoid this pitfall. The epidural venous plexus is more prominent in the cervical spine and can be followed into the intervertebral foramina (50).

2.5. Arachnoiditis

Arachnoidal scarring has been documented as a rare complication of spinal meningitis. Typical pathogens include tuberculosis, listeria, cryptococcus, syphilis, and rarely pyogenic meningitis (34). T2-weighted MRI in the axial plane can display abnormal-appearing thecal sac with clumping of nerve roots (Fig.6) either centrally or eccentrically (52). Myelography may be more sensitive in detecting arachnoiditis in patients with limited areas of involvement (53).

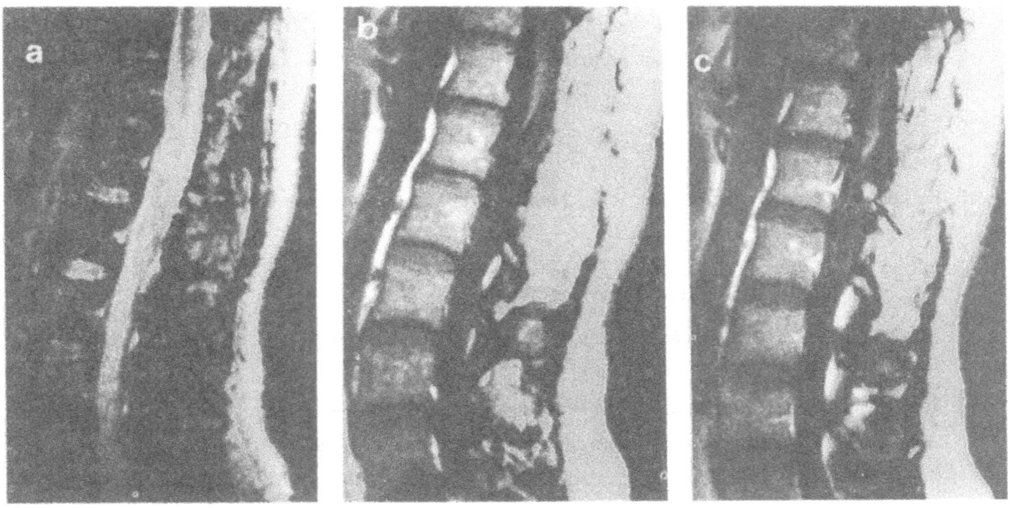

Figure 6. Postoperative arachnoiditis in a case of laminectomy. T2-weighted sagittal image (a) shows evidence of hyperintensity in the lower cord and conus region and the nerve roots are not separately visualized. T1-weighted image (b) shows evidence of isointense nodule in relation to the roots below the conus and enhances strongly in the postcontrast image (c). Note also the enhancing nerve roots anteriorly (arrow).

REFERENCES

1. Olson JJ, Bingaman KD. Cranial bone flap infections and osteomyelitis of the skull. In *Infections in neurological surgery – Diagnosis and management*, RK Osenbach, SM Zeidman, eds. Philadelphia: Lippincott–Raven, 1999; 65–84.
2. Balch RE: Wound infection complicating neurosurgical procedures. J Neurosurg 1967; 26:41–45.
3. Hudgins PA, Davis PC, Hoffman JC. Gadopentetate dimeglumine-enhanced MR imaging in children following surgery for brain tumor: Spectrum of meningeal findings. Am J Neuroradiol 1991; 12:301–307.
4. Matthews VP, Kuharik M, Edwards MK, D'Amour PG, Azzarelli B, Dreesen RG. Gd-DTPA-enhanced MR imaging of experimental bacterial meningitis: Evaluation and comparison with CT. Am J Neuroradiol 1988; 9:1045–1050.
5. Chang KH, Han M, Jae KR, Kimj IO, Han MC, Kim CW. Gd-DTPA-enhanced MR imaging of the brain in patients with meningitis: Comparison with CT. Am J Neuroradiol 1990; 11: 69–76.
6. Sze G, Soletsky S, Bronen R, Krol G. MR imaging of cranial meninges with emphasis on contrast enhancement and meningeal carcinomatosis. Am J Neuroradiol 1989; 10: 965–975.
7. Hudgins PA, Davis PC, Hoffman JC. Gadopentate dimeglumine-enhanced MR imaging in children following surgery for brain tumor: Spectrum of meningeal findings. Am J Neuroradiol 1991; 12: 301-307.
8. Elster A, DiPersio DA. Cranial postoperative site: Assessment with contrast enhanced MR images. Radiology 1990; 174: 93-98.
9. Lanzieri CF, Larkins M, Mancall A, et al. Cranial postoperative site: MR imaging appearance. Am J Neuroradiol 1988; 9:27–34.
10. Burke JW, Podrasky AE, Bradley WG. Meninges: Benign postoperative enhancement on MR images. Radiology 1990; 174:99–102.
11. Jeffries BF, Kishore PRS, Singh KS, Ghatak NR, Krempa J. Contrast enhancement in the postoperative brain. Radiology 1981; 139:409–413.
12. Lanzieri CF, Som PM, Sacher M, Solodrik P, Moore F. The postcraniotomy site: CT appearance. Radiology 1986; 159:165–170.
13. Penn RD, Walser R, Kurtz D, et al. Tumor volume, luxury perfusion, and regional blood volume changes in man visualized by subtraction computerized tomography. J Neurosurg 1976; 44:449–457.
14. Runge VM, Clanton JA, Price AC, et al. Evaluation of contrast-enhanced MR imaging in a brain-abscess model. Am J Neuroradiol 1985; 6:137–147.
15. Krauss WE, Post KD. General complication in neurologic surgery. In *Postoperative complications in intracranial Neurosurgery*, KD Post, ED Friedman, P McCormick, eds. New York: Thieme, 1993; 2–27.
16. Bullit E, Lehman RAW. Osteomyelitis of the skull. Surg Neurol 1979; 11:163–166.
17. Garvey G. Current concepts of bacterial infections of the central nervous system: Bacterial meningitis and bacterial brain abscess. J Neurosurg 1983; 59:735–744.
18. Post EM, Modesti LM. Subacute postoperative subdural empyema. J Neurosurg 1981; 55:761–765.
19. Luken MG III, Wealen MA. Recent diagnostic experience with subdural empyema. J Neurosurg 1980; 52:764–771.
20. Renaudin JW, Frazee J. Subdural empyema—Importance of early diagnosis. Neurosurgery 1980; 7:477–479.

21. Stern WE, Boldery E. Subdural purulent collections. Surg Gynecol Obstet 1952; 95:623–630.
22. Meltzer CC, Fukui MB, Kanal E, Simirniatopoulos JG. MR imaging of the meninges: Part I. Normal anatomic features, and non-neoplastic disease. Radiology 1996; 201:297–308.
23. Seabold JE, Simonosob TM, Weber PC, et al. Cranial osteomyelitis: Diagnosis and follow-up with In 111 white blood cell and Tc-99m methylene diphosphonate bone SPECT, CT and MR imaging. Radiology 1995; 196: 779–788.
24. Gherini SG, Brackmann DE, Bradley WG. Magnetic resonance imaging and computerized tomography in malignant external otitis. Laryngoscope 1986; 96:542–548.
25. Malone DG, O'Boynick PL, Ziegler DK, Batnizky S, Hubble JP, Holaday FP. Osteomyelitis of the skull base. Neurosurgery 1992; 30:426–431.
26. Epstein JS, Ganz WI, Lizak M, Grobman L, Goodwin WJ, Dewanjee MK. Indium 111-labelled leukocyte scintigraphy in evaluating head and neck infections. Ann Otol Rhinol Laryngol 1992; 101:961–968.
27. Noyek AM, Kirsch JC, Grayson ND, et al. The clinical significance of radionuclide bone and gallium scanning in osteomyelitis of the head and neck. Laryngoscope 1984; 94:1–21.
28. Buckwold FJ, Hand R, Hansebout RR. Hospital-acquired bacterial meningitis in neurosurgical patients. J Neurosurg 1977; 46:494–500.
29. Balch RE. Wound infection complicating neurosurgical procedures. J Neurosurg 1967; 26:41–45.
30. Chang KH, Han M, Roh J, Kim I, Han M, Kimm C. Gd-DTPA-enhanced MR imaging of the brain in patients with meningitis: Comparison with CT. Am J Neuroradiol 1990; 11:69–76.
31. Rosenblum ML, Hoff JT, Norman D, Weinstein PR, Pitts L. Decreased mortality from brain abscesses since the advent of computerized tomography. J Neurosurg 1978; 49:658–668.
32. Haimes AB, Zimmerman RD, Morgello S, et al. MR imaging of brain abscesses. Am J Neuroradiol 1989; 10:279–291.
33. LaRocca H. Spinal sepsis. In *The spine*, RH Rothman, FA Simeonse, eds. Philadelphia: Saunders, 1982; 757–774.
34. Moores LE, Elenbogen RG. Cerebrospinal fluid shunt infections. In *Infections in neurosurgery. AANS Publications Committee*, WA Hall, IE McCutcheon, eds. Parkridge, IL: American Association of Neurologic Surgeons, 2000; 141–153.
35. Borgbjerg BM, Gjerris F, Albeck MJ, et al. Risk of infection after cerebrospinal fluid shunt: Analysis of 884 first time shunts. Acta Neurochir 1995; 136:1–7.
36. Ammitati M, Raimondi AJ. Cerebrospinal fluid shunt infections in children. A study on the relationship between the etiology of hydrocephalus, age at the time of shunt placement, and infection rate. Childs Nerv Syst 1987; 3:106–109.
37. Piatt JH Jr, Carlson CV. A search for determinants of cerebrospinal fluid shunt survival: Retrospective analysis of 14-year institutional experience. Pediatr Neurosurg 1993; 19:233–242.
38. Parker LM, McAfee PC, Fedder IL, Weis JC, Geis WP. Minimally invasive surgical techniques to treat spine infections. Orthop Clin North Am 1996; 27:183–199.
39. Ozuna RM, Delamarter RB. Pyogenic vertebral osteomyelitis and post-surgical disc space infections. Orthop Clin North Am 1996; 27:87–94.
40. El-Glindi S, Aref S, Salama M, Andrew J. Infection of intervertebral discs after operation. J Bone Joint Surg 1976; 58B: 114–119.

41. Post MJD, Quencer RM, Montalvo BM, et al. Spinal infection: Evaluation with MR imaging and intraoperative US. Radiology 1988; 169:765–771.

42. Ross JS, Zepp R, Modic MT. The postoperative lumbar spine: Enhanced MR evaluation of the intervertebral disc. Am J Neuroradiol 1996; 17: 323–331.

43. Heuftle MG, Modic MT, Ross JS, et al. Lumbar spine: Postoperative MR imaging with Gd-DTPA. Radiology 1988; 167:817–824.

44. Ross JS, Modic MT, Masaryk TJ. Tears of the annulus fibrosus: Assessment with Gd-DTPA enhanced MR imaging. Am J Neuroradiol 1989; 10:1251–1254.

45. Belagura S, Neumann JF. Magnetic resonance imaging of the postoperative intervertebral disk: The first eight months. Clinical and legal complications. J Spinal Disorders 1993; 6:212–217.

46. Carragee EJ. The clinical use of magnetic resonance imaging in pyogenic vertebral osteomyelitis. Spine 1997; 22:780–785.

47. Djukie S, Lang P, Morris J, Hoaglund F, Genant HK. The postoperative spine: Magnetic resonance imaging. Orthop Clin North Am 1990; 21:603–624.

48. Thrush A, Enzman D. MR imaging of infectious spondylitis. Am J Neuroradiol 1990; 11:1171–1180.

49. Kuker W, Mull M, Mayfrank L, Topper R, Thron A. Epidural spinal infection: Variability of clinical and magnetic resonance imaging findings. Spine 1997; 22:544–551.

50. Georgy BA, Hesselink JR. MR imaging of the spine: Recent advances in pulse sequences and special techniques. Am J Roentgenol 1994; 162: 923–934.

51. Venes JL. Control of shunt related infection: Report of 150 consecutive cases. J.Neurosurg 1976; 45: 311–314.

52. Ross J, Masaryk T, Modic MT, et al. MR imaging of lumbar arachnoiditis. Am J Roentgenol 1987; 149:1025–1032.

53. Heuftle MG, Modic MT, Ross JS, et al. Lumbar spine: Postoperative MR imaging with Gd-DTPA. Radiology 1988; 167: 817–824.

Chapter 11

DIAGNOSTIC ALGORITHM FOR A T2 HYPOINTENSE LESION

Sunil Kumar and Rakesh K. Gupta

Presently MRI is the primary investigating modality for diseases involving the CNS. The past few years have seen the development of newer imaging sequences to resolve the wide range of diseases that produce almost similar signal characteristics on conventional sequences. The majority of pathologic processes in the brain parenchyma produce hyperintensity on T2-weighted images (WI); however, similar- appearing hypointense lesions on T2-weighted images are also seen in a wide range of pathologic processes. The spectrum of diseases that appear hypointense on T2-weighted images has enlarged continually with the increase in the use of MRI. Explanation of this signal has produced many theories and some of these are substantiated by the pathologic correlation. In this chapter, we have attempted to minimize the differential diagnosis of similar-appearing T2 hypointense lesions by using various pulse sequences that are presently available on most MR scanners.

1. DEFINITION OF T2 HYPOINTENSE LESION

When the entire lesion or at least a part of the lesion appears darker than the adjoining normal brain structure on a long TR and TE spin-echo (SE) sequence, it is considered as T2 hypointense or T2 black lesion (Fig.1). This may at times include isointense lesions on T2

Sunil Kumar and Rakesh K. Gupta • Department of Radiodiagnosis, Sanjay Gandhi Post-Graduate Institute of Medical Sciences, Lucknow 226014, India.

MR Imaging and Spectroscopy of Central Nervous System Infection, edited by Gupta and Lufkin. Kluwer Academic / Plenum Publishers, New York, 2001.

images that are accompanied by hyperintense perifocal edema, and therefore appear relatively hypointense. Flow void signals are not included in this discussion because they can be easily suspected and confirmed on the basis of their structure, usually linear or round on all SE sequences and with flow-related enhancement on gradient echo (GRE) sequences. T2 hypointense lesions in the spinal cord can also be analyzed in a similar manner, although there are very few spinal conditions that are hypointense on T2 images.

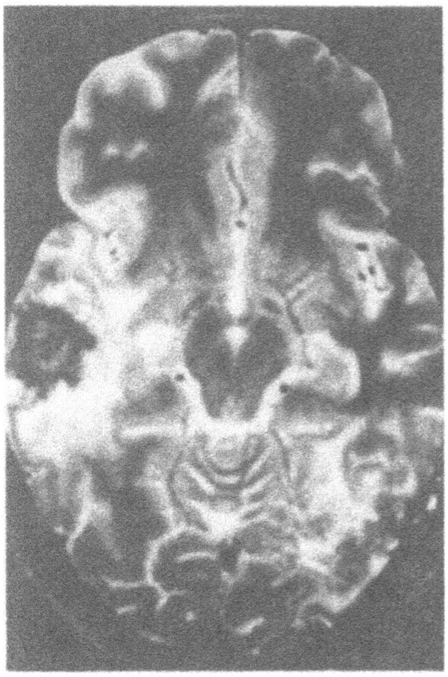

Figure 1. A typical example of a T2 hypointense lesion seen in the right temporal region with perifocal edema.

2. TECHNIQUES OF EXAMINATION

Alhough these techniques were described in detail in Chapter 1, they are summarized below as a ready reference for the reader.

- Spin echo (SE). (1) T2 WI: This has been the mainstay of the imaging protocol in the brain examination ever since MRI entered the clinical arena. In this pulse sequence, a 180° refocusing pulse follows an initial 90° pulse before reading the signal. The typical pulse sequence consists of a long repetition time (TR), usually more than 2 seconds in order to minimize the T1 effects, and two echo times (TE), one short, giving mainly proton-density weighted images (PDWI), and the second long, producing T2 WI. The imaging time ranges from 9 to 12 minutes, as each of the phase-encoding steps is applied in one TR interval, and the usual phase encoding steps vary from 192 to 256. (2) T1WI: Using a similar pulse sequence with a short TR and a short TE produces a predominantly T1 WI. Imaging times typically vary from 5 to 7 minutes for a matrix size of 256 × 192/256 and 2–3 averages. T1 WI produces an excellent anatomical display of structures.

- Gradient echo (GRE). This family of pulse sequences utilizes shorter flip angles than the initial 90° pulse used in SE and uses gradients to refocus the signal instead of the 180° radio frequency (RF) pulse used in SE sequences. This helps in reducing the TR to milliseconds and thus allows much faster acquisitions. The GRE can be of steady state type that produces images of a higher SNR, greater contrast, and reduced phase artifacts or of the spoiled gradient type in which T2 dependence is eliminated and the images are proton density and/or T1 WI.
- Fast spin echo (FSE). In FSE imaging sequence, multiple phase-encoding steps are applied in each TR interval, and multiple TEs are acquired in each TR interval. The initial pulse and refocusing pulses are the same as used in SE. The number of echoes acquired per TR is termed the echo train length and the time between each echo is termed the echo space. The result is that imaging time is shorter than SE by a factor in direct proportion to the number of echoes collected. Here all the echoes in each TR interval contribute to the signal instead of one echo, and this is termed effective echo and is determined by the echo that has the lowest phase-encoding gradient (center of *k*-space).
- Magnetization transfer (MT). This imaging sequence exploits the interchange of bound protons present in macromolecules such as proteins or myelin and free protons as in water. An additional off-resonance saturation pulse is applied just before an imaging sequence. This reduces the signal intensity of the structures containing macromolecules and the degree of reduction is variable in different tissues depending on their composition. The reduction is quantifiable and consistent results are obtained in the normal tissues in the brain.
- Contrast-enhanced imaging (CE). Gadolinium compounds are used as injectable contrast media that act by shortening the T1 of the tissues in which this contrast enters. In brain, the effects are seen in lesions that have an associated breakdown of the blood–brain barrier (BBB) and/or increased vascularity. The usual dose for intravenous administration is 0.1mmol/kg body weight, but currently double or even triple dose injections are being increasingly used to increase the sensitivity of the CE studies.
- Diffusion-weighted imaging (DWI). This technique utilizes the microvascular water milieu and is dependent on the movement of water molecules over short distances. This movement depends on the presence of permeable and impermeable barriers. A pair of gradient pulses is placed symmetrically around the 180° refocusing pulse, to dephase and rephase water protons. Diffusing water protons are dephased and produce signal loss. By changing the time and amplitude of gradient pulse (*b*-value), an apparent diffusion coefficient (ADC) can be calculated and differences in various pathologic

processes can be examined. Inflammation may decrease ADC with little loss of anisotropy while axonal loss leads to loss of anisotropy with an increase in ADC (1).

- Magnetic resonance spectroscopy (MRS). Proton MRS brings out the metabolic information present in the tissue under examination. Prominent peaks that are identified in normal and abnormal brain are *N*-acetylaspartate (NAA), choline-containing compounds (Cho), creatine/phosphocreatine (Cr), and lactate at long TE and methyl groups of lipids, and myoinositol at short TE sequences. NAA is a neuronal marker found in mature brain and any loss of neuronal tissue results in decreased NAA. Increase in Cho peak may result from release of choline containing membrane lipids during active myelin breakdown or from inflammation (2). Lactate peak is normally not visible in normal brain. Similarly, presence of other peaks and their analysis may lead to an insight of the pathologic process, may help in evaluation of the therapy, and in some cases may provide a characteristic MR spectrum unique to the specific disease.
- Perfusion imaging. Perfusion describes the rate at which the circulating blood perfuses the tissue (3). There are many methods of obtaining perfusion imaging. These include contrast bolus tracking, use of intravoxel incoherent motion, and techniques that magnetically label the proximal endogenous water in arterial blood (4–6). Some of these have been clinically applied (7).

3. CAUSES OF T2 HYPOINTENSE LESIONS

Some of the commonly seen conditions that may show a hypointensity on T2-weighted images include:

- Calcification
- Degenerative conditions such as Shy–Drager syndrome and Hallervorden–Spatz disease
- Hematoma or hemorrhage
- Vascular malformations
- Tumors such as lymphoma, some gliomas, and metastases especially from melanoma and colonic carcinoma
- Infections such as tuberculosis, some fungal infections, healing abscesses, and healing neurocysticercosis

In the following discussion, the above-mentioned conditions are discussed briefly including the proposed reasons for the T2 hypointensity, and most of these can be read in much detail elsewhere in this book.

3.1. Calcification

Calcification occurs in the normal as well as abnormal brain. It is well seen in computerized tomography (CT) as a bright density. On MRI, the intensity pattern is variable (Fig.2). It is expected that since there are no mobile protons, calcified areas will show hypointensity on both T2 and T1 images (Fig.2a). However, this is not always the case. At times, T1 images may show a hyperintense picture, presumably due to reduction of T1 relaxation time by surface relaxation mechanism, especially when the concentration of particulate calcium by weight is up to 30% (8). T2 images in such cases may be normal or may show a hyperintense rim around a hypointense center (9). The use of GRE imaging has been advocated in which calcification is seen as a marked hypointensity (Fig.2b) (10). This hypointensity is pronounced on a GRE sequence with long TR and TE and low flip angle. This phenomenon was ascribed to T2* shortening due to static local magnetic field gradients at interfaces of regions differing in magnetic susceptibility. In addition, paramagnetic ions, which at times are present with diamagnetic calcium salts, may also contribute to this effect. However, small calcifications may not be picked up on routine imaging sequences (11). Concentrated efforts to solve this problem using MR technology are going on. In a recent study, GRE phase imaging was used successfully to differentiate calcification, which showed a positive phase shift as opposed to a negative phase shift in chronic bleed (Fig.2c) (12). CT may be used as a complementary study to identify calcification where its presence may significantly alter the diagnosis and thus management where MRI is showing equivocal signals.

3.2. Degenerative

Certain degenerative conditions show hypointense signals in a normal brain structure. Some of these are as follows.

3.2.1. Shy–Drager Syndrome (SDS)

In SDS, abnormal hypointense signals on T2-weighted MR images are symmetrically demonstrated in bilateral putamen, where a remarkable positive Prussian blue stain reaction was revealed on pathologic examinations related to non-heme iron deposition (13).

3.2.2. Wallerian Degeneration

At 4 weeks, a well-defined band of hypointense signal appears onT2-weighted images in the topographic distribution of the corticospinal tract, which gradually changes into a hyperintense signal by 10–14 weeks. The dark signal intensity observed onT2-weighted images between 4 and 14 weeks is believed to result primarily from transitory increased lipid–protein ratio (14). This hypointensity is usually seen along the whole tract and does not show any T2* effect.

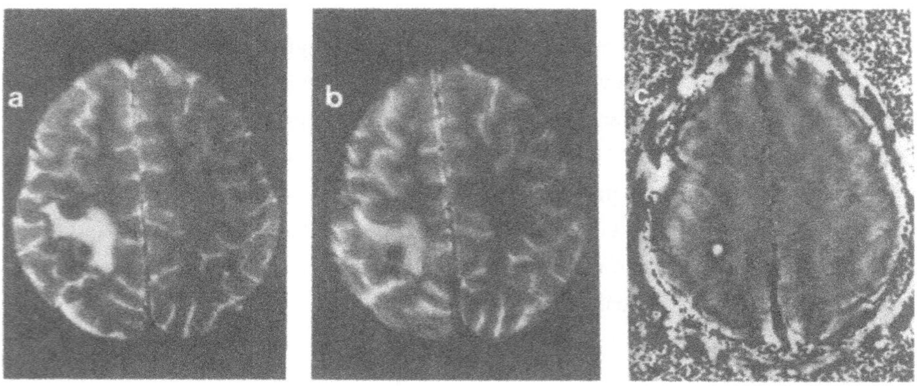

Figure 2. Calcification in a degenerating neurocysticercosis. Axial T2-weighted image (a) showing a hypointense, rounded lesion with perifocal edema in the right parietal region. On GRE sequence (b) the lesion shows T2* effect. Phase image (c) shows bright signal from the lesion consistent with positive phase.

3.2.3. Hallovarden–Spatz Disease

This is a disorder that affects children between 7 and 15 years of age. Non-heme iron deposition in the globus pallidus and less commonly in the substantia nigra causes the T2 hypointensity. Occasionally, early in the disease, the iron deposition is seen only in the periphery (Fig.3) with the central area showing hyperintensity on T2 images (15).

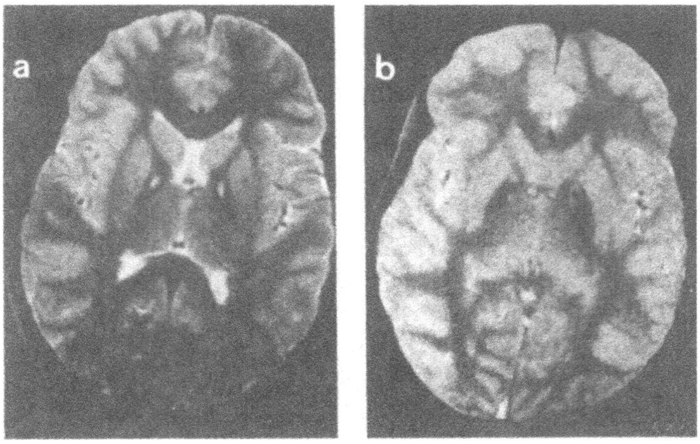

Figure 3. Hallervorden–Spatz disease. T2-weighted (a) and GRE (b) images at the level of basal ganglia showing hypointensity in the globus pallidus with a central hyperintensity, which is seen only on T2-weighted image.

3.3. Hematoma

Different stages of hematoma (evolution with time) appear to have different signals on MRI.

3.3.1. Hyperacute Hematoma

In the initial few hours of an intracranial bleed, a hematoma appears iso- to slightly hypointense on short TR/TE images and shows mixed signal intensity on long TR images (16,17). Features that suggest a hyperacute hematoma include a hypointense rim on long TR/TE images and to a lesser extent on T1 images. It has been found histologically that the hyperacute clot shows interdigitation of intact erythrocytes without significant hemosiderin, ferritin, or phagocytic activity at the hematoma–tissue interface. Biochemically, the iron from the extravasated blood was present only as heme proteins within the first 24 hours (18). The signal intensity pattern seen is consistent with magnetic susceptibility variations of paramagnetic deoxygenated hemoglobin within intact erythrocytes at a microscopically irregular tissue–clot interface (18). The intensity pattern seen on lower field strengths may vary from that seen at high field strength, but most hyperacute lesions would show some hypointensity on T1 GRE images (19). GRE scans that reveal characteristic peripheral hypointensity are critical to the detection and delineation of hyperacute hematoma (20).

3.3.2. *Acute Hematoma*

After the initial few hours, acute hematoma appears as markedly hypointense on long TR/TE images due to the presence of intracellular deoxyhemoglobin (21,22). Deoxyhemoglobin is paramagnetic and is packed in the red blood cells creating susceptibility that manifests as a marked hypointensity on long TR/TE images. On short TR/TE images, the lesion is either isointense or slightly hypointense. This occurs due to lack of dipole–dipole interaction as the distance between water protons and the unpaired electrons of deoxyhemoglobin is relatively large in the cell (23). This is supported by a study using magnetization transfer technique which concluded that hypointensity on T2 images is mainly a magnetic susceptibility effect (24). Another contributor to the hypointensity seen at high field strengths is that of hemoconcentration (25). The signal intensity changes on long TR/TE images are dependent on the type of pulse sequence and its parameters, especially echo time (TE), field strength, and internal milieu of the lesion. Among the newer imaging sequences, a recent study has shown a dual echo combined gradient and spin-echo sequence (GRASE) superior to FSE in delineating the dark lesions (26). At low field strengths, it is believed that the retraction of clot is responsible for T2 hypointensity rather than the intracellular deoxyhemoglobin (27).

In a recent study, although the ADC was decreased in patients with hemorrhagic as well as nonhemorrhagic strokes, T2 images (echo-planar imaging) in patients with acute hemorrhagic stroke showed hypointensities compared to patients with nonhemorrhagic strokes in which T2 images showed hyperintense pattern (28).

3.3.3. *Subacute Hematoma*

After a few days, further changes take place in the evolution of the hematoma, resulting in different MR signals. Hemoglobin is oxidized to methemoglobin. There is also an inflammatory response of the macrophages and glial cells changing the integrity of the red blood cell and producing consequent changes in its enzymatic function (29). Changes also occur in the globin moiety and the distance between the water and the electrons decreases resulting in pronounced relaxivity effects due to dipole–dipole interactions. This accounts for the hyperintensity on T1-weighted images (Fig.4). In the early subacute stage, that is, within 2 to 7 days following the hemorrhage, marked hypointensity still can be seen on T2-weighted images. Intracellular methemoglobin has been suggested as the reason for this susceptibility effect (21). This effect disappears later as the red blood cell lyses, probably due to further T1 shortening of extracellular methemoglobin as well as to the lesion behaving like a very high proton density solution. Free methemoglobin formation starts from the periphery and extends toward the center. On T2 images, undiluted free methemoglobin appears iso- to slightly hypointense and dilute free methemoglobin appears hyperintense (30).

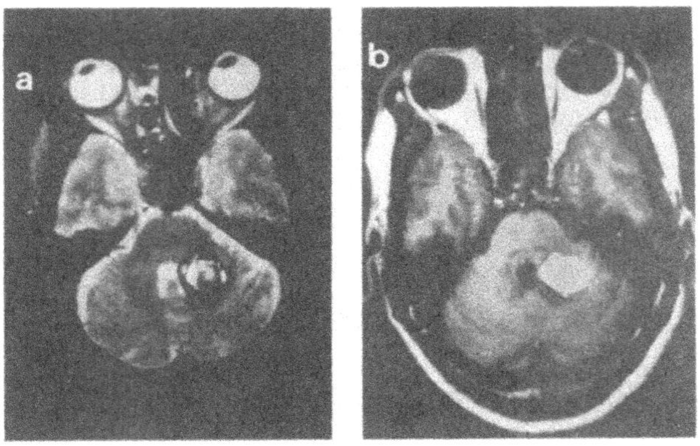

Figure 4. Subacute hematoma. T2 axial image (a) shows evidence of a mixed intensity lesion in the left cerebellar peduncle indenting on the fourth ventricle. The lesion appears hyperintense on T1-weighted image (b).

3.3.4. Chronic Hematoma

As the hemorrhage evolves in the following weeks to months, heme proteins are degraded by the lysosomes,and the released iron is stored in the hydrophobic center of the major iron storage protein called ferritin (31). Excess iron is stored as hemosiderin (31). These large storage compounds are paramagnetic at body temperature, but the iron is not accessible to water, so relaxivity effects are not seen. Instead, magnetic susceptibility variations are pronounced and can be seen as T2* effects. On long TR/TE images, a hypointense rim appears, initially thin, which later thickens as the hematoma resolves (21). This hypointensity can persist for up to many years. As methemoglobin breaks down, its hyperintense signal on T1-weighted images is gradually lost, turning into a low-intensity area decreasing with time. Chronic bleed can be distinguished from calcification on GRE phase imaging where it has a negative phase shift (Fig.5) (12).

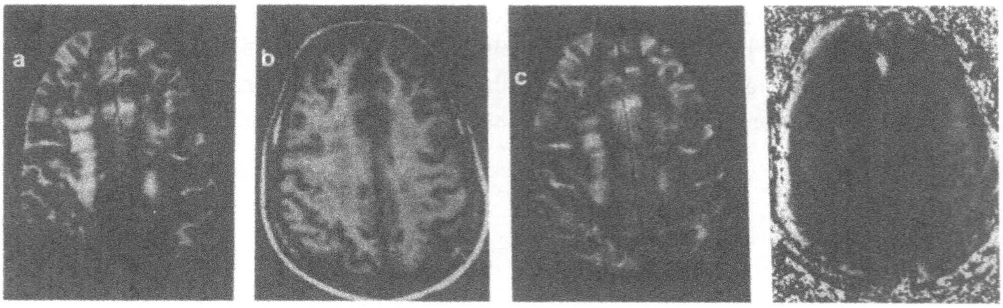

Figure 5. Chronic hematoma. T2-weighted axial image (a) at supraventricular level showing small well-defined hypointense areas in bilateral frontal region. T1-weighted image (b) is unremarkable while GRE image (c) shows lesions greater in number and prominence due to T2* effect. The phase image (d) shows low signal consistent with negative phase in all of these lesions.

3.4. Hemorrhagic Malignancies

Neoplasm can be the cause of an intracranial hemorrhage in about 10% of cases (32). Though the incidence reported is less in primary lesions such as glioblastoma multiforme, oligodendrogliomas, and ependymomas (5%), bleed is more commonly reported in metastases (14%), especially melanoma, renal cell carcinoma, bronchogenic carcinoma, and thyroid carcinoma (32–34). The hemorrhagic neoplastic lesions are more heterogeneous as many different stages of hematoma can be seen simultaneously compared to benign hematoma. The temporal evolution is different and delayed in such cases (35). Peripheral hemosiderin may not form. Hypoxia and recurrent bleeding are proposed as the reasons for this different behavior (35,36). Tumor tissue may be separately visible and may show variable enhancement. Presence of edema long after the episode also helps in differentiation.

3.5. Vascular Malformations

Apart from hematoma, various vascular malformations appear as T2 hypointense. Arteriovenous malformations and venous malformations having typical flow void of vessels are not considered here.

3.6. Cavernous Angioma

Histologically, cavernous angiomas are the most common form of occult vascular malformation (37). These lesions have a characteristic MR pattern (Fig.6) (38). The heterogeneous central areas, containing methemoglobin, are surrounded by a hypointense peripheral rim representing hemosiderin or other iron storage products. There is no edema or any mass effect. There are no feeding arteries or draining veins. These lesions may be enhanced on contrast MR studies. Typically, these lesions are seen in the supratentorial region in close proximity to a CSF space, although they may be seen in the brainstem and cerebellum. Angiography in these cases is normal. In an event of an acute bleed, these characteristic features may not be seen. Multiplicity, seen in about 20–30% of cases, may then lead to the correct diagnosis. The venous angioma may be associated with cavernous angioma (39).

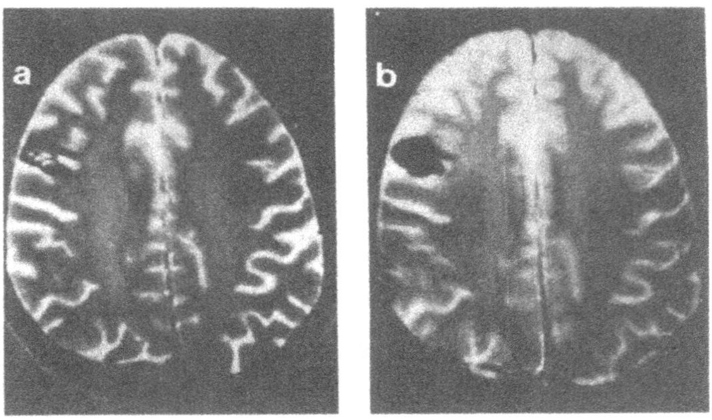

Figure 6. Cavernous angioma. T2-weighted axial image (a) showing a heterogeneous area with a peripheral hypointense rim in the right frontal region. On GRE image (b) the lesion shows T2* effect.

3.7. Miscellaneous Conditions

Entities leading to multiple areas of bleed may be considered in the differential diagnosis of cavernous angioma. These include disseminated intravascular coagulopathy (DIC), amyloid angiopathy, and old hypertensive hemorrhages. Age and clinical history helps in the differentiation of these conditions. Also, the characteristic pattern described for cavernous angioma may be absent. Bleed can be confirmed on GRE sequences, if SE images are equivocal.

Capillary telangiectasia may be encountered occasionally as isointense lesions on T2-weighted images (40). These lesions are seen in the pons and on intravenous administration of Gd-DTPA, enhance as a discrete focus with irregular or brushlike borders. Many of these may have an anomalous draining vein to the surface of the pons. GRE T2-weighted images show marked hypointensity, which is suspected to represent elevated intravascular deoxyhemoglobin from stagnant blood flow (41).

In patients with chronic hypertension, there may be multiple areas of hypointensities on T2*-weighted images, especially in the basal ganglia, thalamus, and other deep cerebral structures (42).

Focal hypointense lesions may also develop after radiation therapy and may appear similar to a cryptic or occult vascular malformation. They represent radiation-induced telangiectasia and may bleed. Pathologic features include ectatic thin-walled vessels surrounded by hemosiderin and gliosis, with minimal necrosis (43).

3.8. Tumors

These consist of mainly metastases and a few primary brain tumors.

3.8.1. Metastases

Unlike most metastases, some lesions are iso- or hypointense on long TR/TE images. Varying signal intensities may be seen due to necrosis and hemorrhage. In general, there is a greater degree of edema around the lesions compared to benign entities.

Melanotic metastases are reported to be isointense on long TR/TE images and hyperintense on short TR/TE images, ascribed to the free radical content of melanin (44). In another study it was shown that the melanotic pattern, that is, hyperintensity on T1 and hypointensity on T2 images, was relatively less frequent in melanoma, but when seen, was usually associated with a higher percentage of melanin-containing cells (45). However, in a recent study, there was no clear association between signal intensity and melanin content on theT2-weighted sequence. There was also no significant association between the signal intensities on the MR images and the iron content, tumor size, or tumor cell type of these lesions (46). The hemorrhage as a cause for this hypointensity has been excluded on pathologic examination, even though there is a high incidence of hemorrhage in metastases of melanoma (47). The intracranial metastases of melanoma may be solitary and may be the only metastatic site in the body (48).

Metastases from mucinous cystadenocarcinoma of the colon (Fig. 7) are hypointense on T2-weighted images (49). The hypointense area corresponds to peripheral necrosis and probably some viable tumor (50). No correlation has been shown with tumor histology or the presence of mucin, blood products, iron, or calcium and it is postulated that this phenomenon most likely reflects the relaxation parameters of the tissue from which the metastases arise (51).

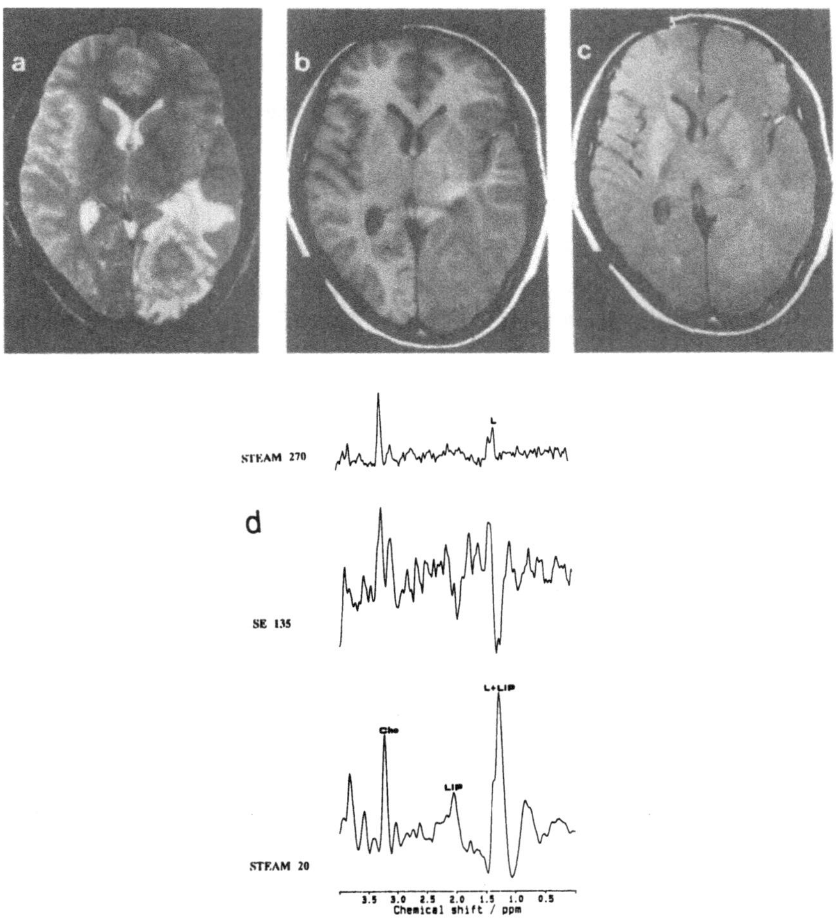

Figure 7. Metastases from a colonic adenocarcinoma. T2 axial image (a) showing a large heterogeneous, predominantly hypointense lesion with perifocal edema and mass effect in the parieto-occipital region. T1-weighted image (b) does not reveal presence of any hemorrhage. MT T1-weighted image (c) shows nondefinition of the rim with a variable MT ratio in the lesion. MR spectroscopy shows presence of large concentration of lipid and lactate at STEAM 20 along with high choline resonance. Note the phase reversal of the lactate (SE 135 ms) along with marked reduction in the signal at 1.3 ppm (STEAM 270 ms).

MR spectroscopy of metastases shows an increase in choline compounds and a decrease in NAA levels (52). Presence of lactate and lipids in most or all lesions has been reported in some studies (52,53). Similar MRS characteristics were reported in metastases from different primary malignancies, but a correlation between specific metabolites and stage of the lesions was noted in this series (54). Early stages of metastases showed elevated Cho, intermediate stages showed lipids and elevated Cho, and late stages showed presence of lactate and lower Cho levels.

3.8.2. *Primary Intracranial Tumors*

Some primary tumors may appear iso- to hypointense on long TR/TE sequences. In most instances, this appearance is due to hemorrhage in the lesion. Features such as heterogeneity, fluid levels, and areas of tumor tissue with edema and enhancement, delayed evolution of hematoma, and frequent absence of hemosiderin help to differentiate these lesions from benign hemorrhage (35). Pigmentation of melanin may be seen in medulloblastoma and neuroectodermal tumor of infancy. Markedly hypercellular tumors with scant cytoplasm also show iso- to hypointensity on T2 images. Undifferentiated small round cell tumors such as medulloblastoma, pinealoblastoma and neuroblastoma fall into this category. Age and location will usually help to differentiate these lesions. Other tumors that display similar signal characteristics include meningioma, lymphoma, and colloid cyst. Signal characteristics on T2 images matched well with histopathologic findings in a series where meningiomas markedly hypointense to cortex onT2-weighted images were composed predominantly of fibroblastic or transitional elements (55).

Primary intracranial lymphoma may rarely develop in a nonimmunodeficient patient. However, these are more common in immunodeficient subjects such as organ recipients, those with congenital immunodeficiency syndromes, AIDS, or systemic immunodeficiency states (56). In a nonimmunocompromised patient, these lesions tend to be large, solitary, and seen in deep parietal lobe (57). Typically, primary lymphomas are non-Hodgkin's type, and involve deep gray matter, periventricular regions, and corpus callosum (Fig.8) (58). Subependymal spread or ventricular encasement is seen in up to 38% of cases (59). Multiplicity is common but the edema is less than expected. Calcification and hemorrhage are rare. Enhancement is intense, although steroid therapy may render the lesions nonenhancing (60). Lesions in immunocompromised patients may show rim-enhancing lesions, which reveal more necrosis on histology (62). On T2 images, the lymphoma appears slightly hyperintense, but usually shows some hypointense areas (59). On histology, typically the lesions show dense concentric lymphoma cells without necrosis, which may be responsible for the hypointensity on T2 images (61). There was no histopathologic difference between the hyperintense and hypointense lesions seen on T2-weighted images (62). On the other hand, in a recent series of 23 patients, 53% lesions were iso- or hypointense on T2 images and there was a positive correlation between necrosis and hyperintensity on T2-weighted images (60). Sarcoid may behave in a similar manner and may be difficult to differentiate.

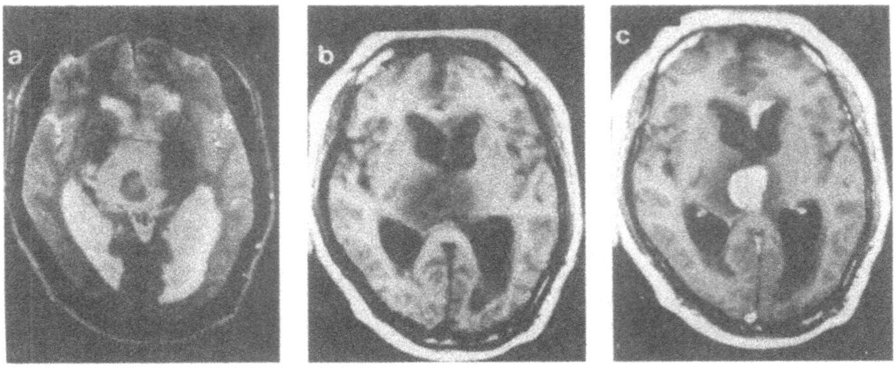

Figure 8. Lymphoma. T2-weighted axial image (a) showing a large lesion with central hypointense area in the right thalamic region with perifocal edema. Another small similar-appearing lesion noted in genu of corpus callosum, anteromedial to left frontal horn of lateral ventricle. These lesions are hypointense on T1 image (b) and show intense enhancement following contrast (c).

Colloid cysts of the third ventricle also show hypointensity on T2 images except for a slightly hyperintense periphery. Histologic examination shows viscous colloidal material in these patients, which may in some cases be surrounded by an amorphous material and cellular debris (63).

3.9. Infections

Certain intracranial infective lesions appear hypointense on T2-weighted images. These include the following.

3.9.1. Neurocysticercosis

This is caused by ingestion of ova of *Taenia solium*, when humans become intermediate host. Brain is one of the most common sites of infection. The lesions seen on MR vary in signal intensity, surrounding edema and enhancement depending on the stage of evolution (Fig.9). The diagnosis of a neurocysticercus cyst depends on the visualization of the scolex, which can be seen in the innocuous or early degenerating stage. Later stages such as granular nodular or nodular calcified stages may be similar to other granulomatous conditions, where either the periphery or the entire lesion may appear hypointense on T2-weighted images (64,65). In such cases the diagnosis may depend on demonstration of the typical cyst with a scolex elsewhere in the brain or in other locations in the body such as orbit, subcutaneous tissue, and skeletal muscle. A positive serology may also help. A recent study using MT sequences has shown significant difference in the MT ratio (MTR) calculated from the hypointense lesions of the tuberculosis and neurocysticercus cysts (66).

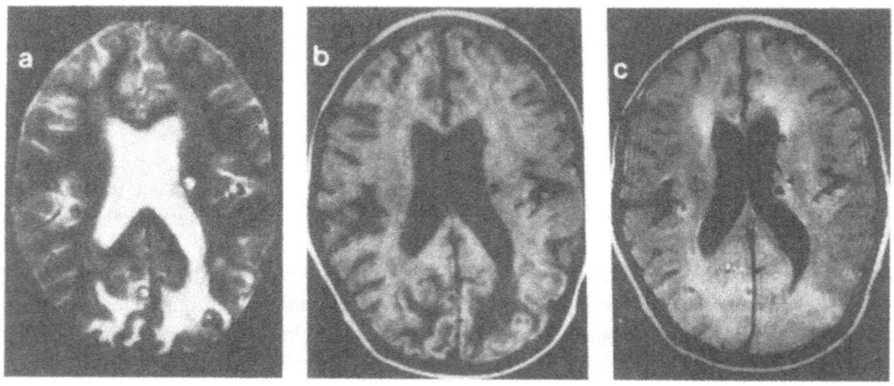

Figure 9. Neurocysticercosis in various stages. T2- (a) and T1-weighted (b) axial images at the level of lateral ventricles show multiple hypointense lesions bilaterally with some showing perifocal edema. In addition, live innocuous cysts are seen in left paraventricular region with scolex. MT T1-weighted image (c) shows poor visibility of the T2 hypointense lesion with MTR between 35 and 40%.

3.9.2. *Tuberculosis*

CNS tuberculosis is frequently seen in developing countries and is now encountered in developed countries due to an increase in AIDS patients. The involvement may be meningeal, parenchymal, or both. The parenchymal granulomas usually appear hypointense on T2-weighted images (Fig.10), at least in part, probably due to the presence of lipid (67). This has been corroborated by the analysis of the localized MR spectrum obtained from the lesion (68). Histologically, a layered capsule has been described as having a hypointense rim outside a hyperintense rim with the center revealing decreased or mixed intensities on T2-weighted images (69). In another study, most of the granulomatous lesions were described as having slight heterogeneous isointensity or hypointensity with small foci of marked hypointensity. In this study the relationship of histologic layers could not be discriminated on the basis of T2 images but corresponded well on T1 images (70). A recent study has also shown difference in the MTR of the tuberculoma compared to similar-appearing lesions of the neurocysticercus cysts (66). Other indicators of the etiology include presence of tubercular lesions in the lung and positive serology.

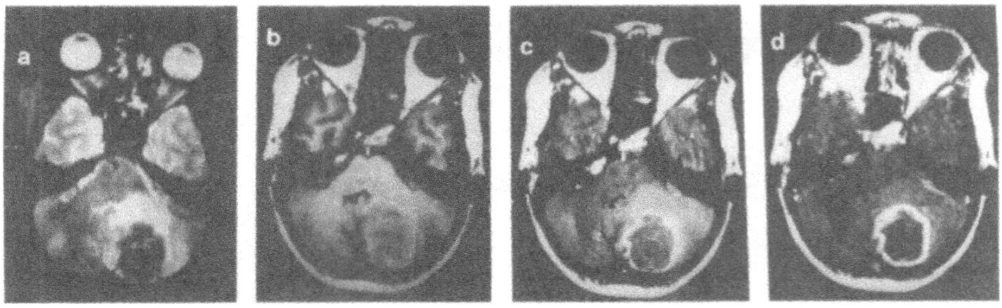

Figure 10. Cerebellar tuberculoma. T2-weighted axial image (a) shows a hypointense area with perifocal edema in the left cerebellar hemisphere. T1-weighted image (b) shows a thin hyperintense rim around the lesion. MT T1-weighted image (c) shows a well-defined hyperintensity rim with strands within the rim that is not well seen in T2- and T1-weighted images. MT ratios from the hyperintense rim and hypointense center were 18 and 20%, respectively. Postcontrast MT T1-weighted image (d) shows thick rim enhancement. Histology confirmed the diagnosis of tuberculoma.

3.9.3. *Toxoplasma Infection*

Toxoplasma encephalitis, a protozoal infection caused by *Toxoplasma gondii,* is more commonly observed in patients with HIV infection (71,72). Lesions are usually multiple, seen more frequently in basal ganglia and corticomedullary junctions, and are surrounded by edema (73). On T2-weighted images, the lesions vary in signal intensity, some being hypointense (Fig.11) (74, 75). Pathologically, it is suggested that T2-weighted hyperintensity correlates with necrotizing encephalitis and T2-weighted isointensity with organizing abscesses. Also, T2-weighted MR appearance may show a transition from hyperintensity to isointensity as a function of a positive response to antibiotic treatment in patients on medical therapy (76). Hemorrhage is uncommon. Postcontrast studies show ring or nodular enhancement (77). Lesions respond well to treatment and healed lesions may calcify. Differentiation from primary CNS lymphoma in the immunocompromised patient is important but may be difficult on imaging as well as spectroscopy as both conditions can show a similar pattern (78). Another study, however, reported some difference in MRS of toxoplasma, lymphoma, and progressive multifocal leukoencephalopathy enabling a correct diagnosis in a majority of cases (79). Increased levels of lactate and lipids were seen in toxoplasma lesions with reduction or absence of normal brain metabolites (Cho, Cr, NAA) whereas Cho was markedly increased in lymphoma with mild to moderate increase in lactate and lipids. A recent study used perfusion MRI to differentiate these two conditions and found reduced regional cerebral blood volume (rCBV) in toxoplasmosis and an increase in rCBV in lymphoma (80). Toxoplasma infection may occur *in utero* and cause hydrocephalus due to aqueductal stenosis and calcification in the basal ganglia and cortex of the affected infant (81).

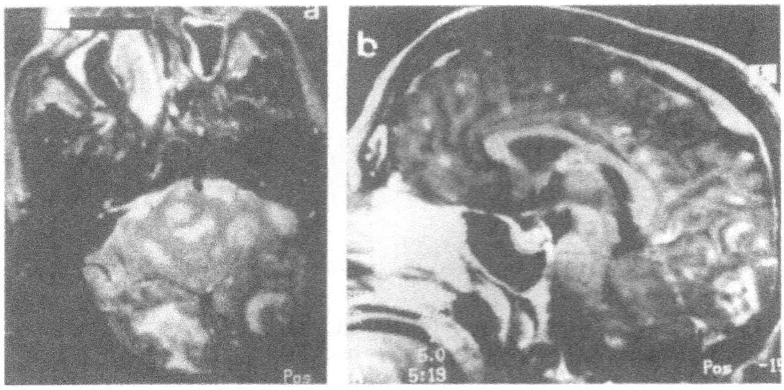

Figure 11. Toxoplasmosis. T2 axial image (a) showing multiple lesions with perifocal edema. Lesions in the occipital region are predominantly hypointense. Postcontrast T1-weighted sagittal image (b) shows multiple enhancing lesions of varying sizes.

3.9.4. Fungal Infections

Some fungal infections involving CNS may appear iso- to hypointense on T2-weighted images. These include *Histoplasma, Candida,* and *Aspergillus* (72,82,83). *Aspergillus* and *Candida* are usually seen infecting immunodeficient people while *Histoplasma* can be seen in immunocompetent subjects as well (84).

The spread of *Histoplasma* is usually hematogenous and causes meningeal involvement more commonly compared to *Aspergillus* and *Candida*, which involve parenchyma more often and cause infarction or cerebritis.

Aspergillus involvement of CNS may be through direct extension from the nasal cavity or paranasal sinuses, but more commonly, it is hematogenous. The former route may lead to involvement of cavernous sinus and thus to arterial thrombosis and infarction. Rarely this spread may reach subarachnoid space and cause meningitis (85). Hematogenous spread also leads to arterial thrombosis and infarction, which may lead to an abscess formation, usually in the anterior or middle cerebral arterial territories (86). Unusual low signal on T2 images (Fig.12) was attributed to increased concentration of iron, manganese, magnesium as well as increased calcium content as compared to bacterial sinusitis (87).

MR imaging of a thalamic histoplasma has also shown the lesion to be T2 hypointense (88). The reason could be similar to that suggested above for *Aspergillus* (87) or it may be due to the presence of paramagnetic free radicals in macrophages in the walls of abscess capsules as described below for chronic abscess.

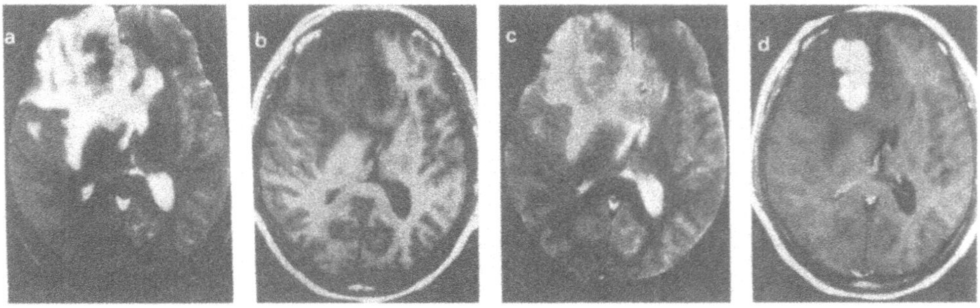

Figure 12. CNS aspergillosis. A large mixed-intensity lesion with predominantly central hypointensity seen in the right frontal region (a) with perifocal edema and mass effect on T2-weighted image. T1-weighted image (b) shows no hemorrhage and GRE image (c) does not reveal any T2 * effect. Postcontrast T1-weighted image (d) shows intense enhancement of the lesion.

Candida may cause abscess formation or meningeal involvement (72,82). Dense fibrous connective tissue with infiltration of Langhans giant cells, lymphocytes, and fibroblasts may be responsible for the observed T2 hypointensity (89).

Paracoccidiodomycosis is a systemic mycosis, endemic in South and Central America, which affects the CNS in almost 10% of cases. On MR, multiple round or lobulated lesions predominantly hypointense in T2-weighted images are described, distributed in the supratentorial as well as infratentorial compartments (90).

3.9.5. Chronic Abscess

Mature pyogenic abscesses have a rim that is hypointense on T2 images. This signal is attributed to collagen, paramagnetic free radicals within phagocytosing macrophages, or probably microhemorrhage (82, 91). This rim may show thinning on its medial aspect, which is thought to be due to differences in the perfusion of the white and the gray matter. The rim also shows thinning as a response to the therapy (82). In our experience, a few abscesses on treatment have changed into a complete hypointense lesion on serial MR (Fig. 13).

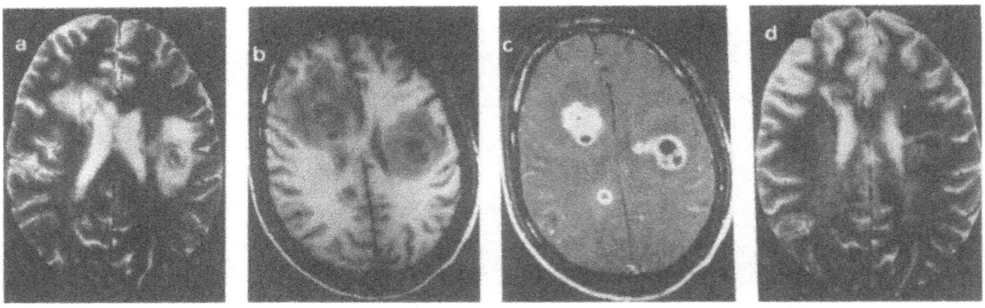

Figure 13. Chronic abscess. T2-weighted image (a) shows multiple bilateral lesions with perifocal edema. Lesion on the left side shows a subtle hypointense rim. Pre- (b) and postcontrast (c) T1-weighted images show multiple rounded confluent hypointense lesions bilaterally with ring enhancement. Blood culture showed *Staphylococcus aureus*. Follow-up after 6 months of antibiotics reveals that most of the lesions have disappeared. The lesion in the left periventricular region shows hypointensity on T2-weighted image (d).

3.9.6. Miscellaneous

American trypanosomiasis (Chagas' disease) is a zoonosis caused by *Trypanosoma cruzi* and it may induce an uncommon form of localized encephalitis termed "chagoma," found in few immunocompromised patients. The imaging pattern of brain chagoma is similar to that of cerebral toxoplasmosis (92). Hence, it may be considered in the differential diagnosis of an intracerebral mass lesion in immunocompromised patients in endemic areas.

4. APPROACH TO HYPOINTENSE T2 LESION

The main purpose of most MR examinations is to demonstrate the abnormality and then to tailor the rest of the examination so as to get as close to the probable diagnosis as possible.

T2-weighted images demonstrate abnormal regions in the brain quite consistently; hence, the initial sequence is usually a T2-weighted sequence, either an SE or an FSE sequence. FSE images are almost comparable to SE images and can be obtained in less than half the time (93). There is an increase in signal-to-noise ratio (SNR) and contrast-to-noise ratio (CNR) in FSE images except in gray–white matter contrast (94). However, FSE has certain shortcomings such as increase in motion artifacts, and may miss small hypointense lesions (94). Interestingly, newer sequences like half-Fourier acquisition single-shot turbo spin-echo (HASTE) and segmented-HASTE (s-HASTE) have even lower pick-up of T2 hypointense lesions compared to FSE (95). The initial T2-weighted images are examined for any abnormality. The abnormal region is then analyzed on the basis of its location, size, shape, signal characteristics, multiplicity, and surrounding edema. Here we will consider only hypointense lesions.

A GRE sequence is usually the next sequence once the lesion shows hypointensity on T2-weighted SE or FSE images and the presence or absence of a T2* effect, i.e., an increase in the hypointensity will determine the path to be followed.

1. If the T2* effect is seen, the differential diagnosis is limited to hemorrhage/hemorrhage-containing lesions, calcification, occult cerebrovascular malformations, and degenerative conditions.

   A T1 image from an SE sequence will usually help in identifying certain stages of blood where the lesion appears hyperintense. Calcification on the other hand will be hypointense in most instances. Occasionally, a calcification will appear hyperintense on T1 images, and thus may appear similar to a hematoma (8). Under such circumstances, calcification may be differentiated from small hemorrhagic lesions by the acquisition of GRE phase images along with magnitude images. A positive phase is indicative of calcification whereas hemorrhage will typically reveal a negative phase (12). Alternatively, a method of identifying calcification would be a plain CT scan.

   To decide whether the lesion is a benign hematoma or a hemorrhagic infarct, one looks for typical findings such as time-related evolution of signal changes, lack of heterogenesity, and lack of significant edema. If one or more of these features are absent, then a contrast study may help by enhancing the abnormal tissue, which can be a neoplasm.

   An occult cerebrovascular malformation usually shows characteristic changes, e.g., a cavernous angioma, has a hypointense rim around a unilocular or multilocular hyperintense lesion, usually without significant mass effect and shows contrast enhancement. T2 hypointensity in degenerative conditions usually occurs in deep gray matter regions such as globus pallidus or putamen.

2. If there is no T2* effect, the differential diagnosis usually rests between a granuloma and a tumor. The common granulomas that appear T2 hypointense include tuberculoma, healing neurocysticercus cysts, and fungal granuloma. Presence of a

scolex is characteristic of a neurocysticercus cyst. A set of pre- and postcontrast T1 studies is usually performed next. Granulomas will usually show small lesions with ring or nodular enhancement whereas a tumor usually has variable enhancement. Metastases may behave like a granuloma. The choice of the next sequence lies between an MT sequence and MR spectroscopy (MRS) or both.

MRS will usually help to differentiate a tumor from tuberculoma. Tumors usually show an increase in choline, variable decrease in NAA and creatine along with presence of lipids or lactate or both. On the other hand, different lipids usually dominate the spectrum in a tuberculoma (68). If MRS suggests a neoplasm, examination of the previous images along with postcontrast study will help. An enhancing lesion in periventricular or a deep gray matter location will rule in favor of a lymphoma. Small, multiple, rounded lesions with edema in excess of their size and ring or nodular enhancement will favor a diagnosis of metastases. Inhomogeneous enhancement in an irregular mass lesion will point toward a glioma. Perfusion imaging can be helpful in cases of tumors where it shows increased in perfusion even in the absence of any enhancement (96).

In granulomas, MT sequence has been used with advantage to differentiate between tuberculoma and cysticercus granuloma (66). A tuberculoma will have lower MT ratios from the hypointense lesions compared to similar-looking cysticercus granuloma. In addition, abnormal meningeal thickening may be visible on noncontrast or on contrast - enhanced MT images indicating tuberculosis as the probable diagnosis.

In immunocompromised patients, typical features of either an infection or a tumor may not be seen. In such instances, combination of MR techniques such as MT sequences and MRS may have to be supplemented with other modalities like single-photon-emmision computer tomography (SPECT) and with other laboratory investigations like ELISA for specific agent antibody/antigen, PCR for demonstration of agent-specific DNA, and smear and culture of different offending microorganisms. Where time is not crucial, response to a specific therapy may be taken as a definitive clue to confirm or exclude a provisional diagnosis. As a last measure, biopsy may have to be performed to arrive at a definitive diagnosis.

We have attempted to evolve an algorithm for the differential diagnosis of T2 hypointense lesions. This needs to be seen in the light of the geographical distribution, regional prevalence of the disease in the patient population, and should be supported by the relevant laboratory investigation to make this algorithm a successful guide for differentiation of T2 hypointense lesions.

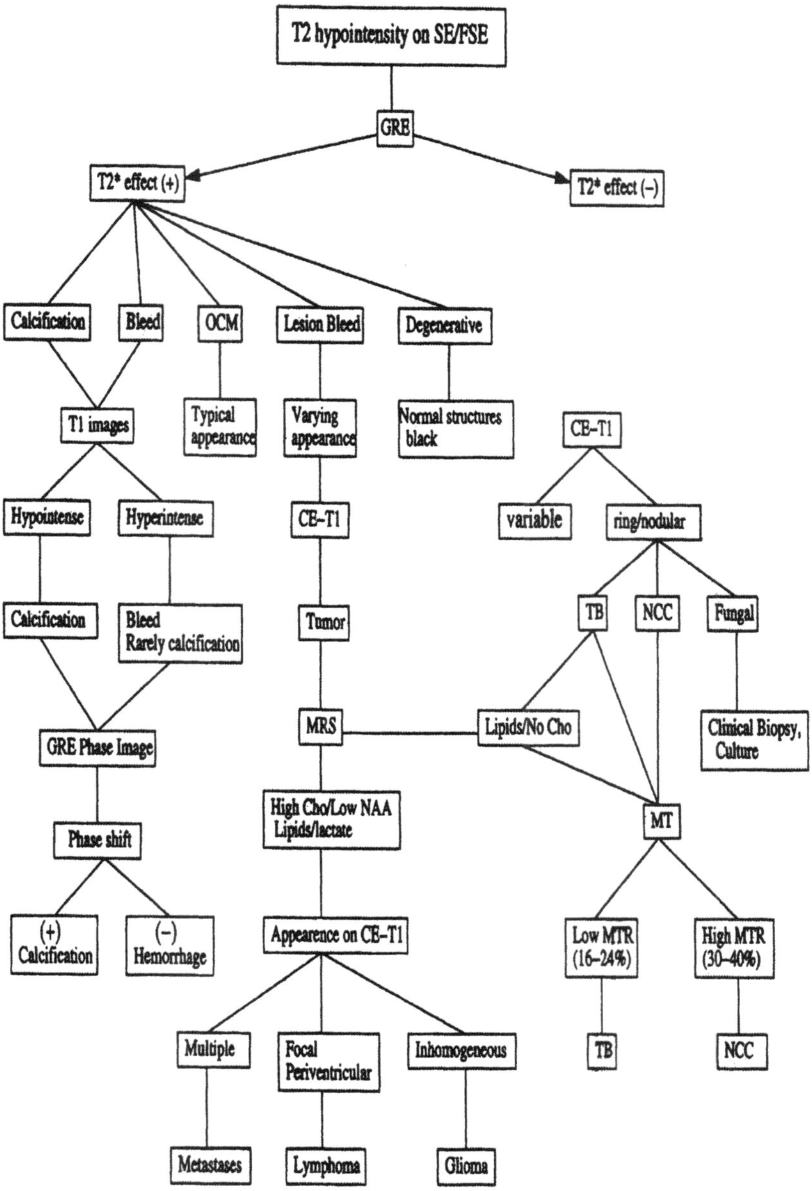

Algorithm for a T2 hypointense lesion

Key for abbreviations and symbols used in the algorithm:

(-)	=	Negative
(+)	=	Positive
SE	=	Spin-echo images
FSE	=	Fast spin-echo images
GRE	=	Gradient echo images
OCM	=	Occult vascular malformation
CE-T1	=	Contrast-enhanced T1 images
TB	=	Tuberculosis
NCC	=	Neurocysticercosis
MRS	=	Magnetic resonance spectroscopy
Cho	=	Choline
NAA	=	*N*-Acetylaspartate
MT	=	Magnetization transfer sequence
MTR	=	Magnetization transfer ratio

REFERENCES

1. Miller DH, Grossman RI, Reingold SC, McFarland HF. The role of magnetic resonance techniques in understanding and managing multiple sclerosis. Brain 1998; 121: 3–24.
2. Brenner RE, Munro PM, Williams SC, Bell JD, Barker GJ, Hawkins CP, Landon DN, McDonald WI. The proton NMR spectrum in acute EAE: The significance of the change in Cho/Cr ratio. Magn Reson Med 1993; 29: 737–745.
3. Jezzard P. Advances in perfusion imaging. Radiology 1998; 208: 296–299
4. Strich G, Hagen PL, Gerber KH, Slutsky RA. Tissue distribution and magnetic resonance spin lattice relaxation effects of gadolinium-DTPA. Radiology 1985; 154: 723–726.
5. LeBihan D, Breton E, Lallemand D, Grenier P, Cabanis EA, Laval-Jeantet M. MR imaging of intravoxel incoherent motions: Applications to diffusion and perfusion in neurologic disorders. Radiology 1986; 161: 401–407.
6. Williams DS, Detre JA, Leigh JS, Koretsky AP. Magnetic resonance imaging of perfusion using spin inversion of arterial water. Proc Natl Acad Sci USA 1992; 89: 212–216.
7. Warach S, Dashe JF, Edelman RR. Clinical outcome in ischemic stroke predicted by early diffusion-weighted and perfusion magnetic resonance imaging: A preliminary analysis. J Cereb Blood Flow Metab 1996; 16: 53–59.
8. Henkelman RM, Watts JF, Kucharczyk W. High signal intensity in MR images of calcified brain tissue. Radiology 1991; 179: 199–206.
9. Thajeb P. Progressive late delayed post irradiation encephalopathy with Kluver-Bucy syndrome. Serial MRI and clinico-pathological studies. Clin Neurol Neurosurg 1995; 97: 264–268.

10. Atlas SW, Grossman RI, Hackney DB, et al. Calcified intracranial lesions: Detection with gradient-echo-acquisition rapid MR imaging. Am J Roentgenol 1988; 150: 1383–1389.
11. Taveras JM. *Neuroradiology*. Baltimore: Williams & Wilkins, 1996.
12. Yamada N, Imakita S, Sakuma RT, Takamiya M. Intracranial calcification on gradient-echo phase image: Depiction of diamagnetic susceptibility. Radiology 1996; 198: 171–178.
13. Lu X, Chen J, Wang L, Liang Y, Gui O. Correlative studies of MR findings with neuropathology in Shy–Drager syndrome and striatonigral degeneration. Chin Med J (Engl) 1997; 110: 628–631.
14. Kuhn MJ, Mikulis DJ, Ayoub DM, Kosofsky BE, Davis KR, Taveras JM. Wallerian degeneration after cerebral infarction: Evaluation with sequential MR imaging. Radiology 1989; 172: 179–182.
15. Ambrosetto P, Nonni R, Bacei A, Gobbi G. Late onset familial Hallovarden–Spatz disease: MR findings in two sisters. Am J Neuroradiol 1992; 13: 394–396.
16. Di Chiro G, Brook RA, Girton ME, Caporale T, Wright DC, Dwyer AJ, Horne MK. Sequential MR studies of intracerebral hematomas in monkeys. Am J Neuroradiol 1986; 7: 193–199.
17. Nose T, Enomoto T, Hyodo A, Narushima K, Yamada T, Sugimoto K, Maki Y, Shirai S, Honma K. Intracerebral hematoma developing during MR examination. J Comput Assist Tomogr 1987; 11: 184–187.
18. Atlas SW, Thulborn KR. MR detection of hyperacute parenchymal hemorrhage of the brain. Am J Neuroradiol 1998; 19: 1471–1477.
19. Zyed A, Hayman LA, Bryan RN. MR imaging of intracerebral blood: Diversity in the temporal pattern at 0.5 and1.0 T. Am J Neuroradiol 1991; 12: 469–474.
20. Campbell BG, Zimmerman RD. Emergency magnetic resonance of the brain. Top Magn Reson Imaging 1998; 9: 208–227.
21. Gomori JM, Grossman RI, Goldberg HI, Zimmerman RA, Bilaniuk LT. Intracranial hematomas: Imaging by high-field MR. Radiology 1985; 157: 87–93.
22. Zuerrer M, Martin E, Boltshauser E. MR imaging of intracranial hemorrhage in neonates and infants at 2.35 tesla. Neuroradiology 1991; 33: 223–229.
23. Thulborn KR, Atlas SW. Intracranial hemorrhage. In *Magnetic resonance imaging of the brain and spine*, SW Atlas, ed. Philadelphia: Lippincott–Raven, 1996; 265–314.
24. Mittl RL Jr, Gomori JM, Schnall MD, Holland GA, Grossman RI, Atlas SW. Magnetization transfer effects in MR imaging of in vivo intracranial hemorrhage. Am J Neuroradiol 1993; 14: 881–891.
25. Clark RA, Watanabe AT, Bradley WG Jr, Roberts JD. Acute hematomas: Affects of deoxygenation, hematocrit, and fibrin-clot formation and retraction on T2 shortening. Radiology 1990; 175:201–206.
26. Melhem ER, Patel RT, Whitehead RE, Bhatia RG, Rockwell DT, Jara H. MR imaging of hemorrhagic brain lesions: A comparison of dual-echo gradient–and spin-echo and fast spin-echo techniques. Am J Roentgenol 1998; 171:797–802.
27. Hayman LA, Ford JJ, Taber KH, Saleem A, Round ME, Bryan RN. T2 effect of hemoglobin concentration: Assessment with in vitro MR spectroscopy. Radiology 1988; 168:489–491.
28. Ebisu T, Tanaka C, Umeda M, et al. Hemorrhagic and non-hemorrhagic stroke: Diagnosis with diffusion-weighted and T2-weighted echo-planar MR imaging. Radiology 1997; 203: 823–828.
29. Enzmann DR, Britt RH, Lyons BE, Buxton JL, Wilson DA. Natural history of experimental intracerebral hemorrhage: Sonography, computed tomography and neuropathology. Am J Neuroradiol 1981; 2:517–526.
30. Gomori JM, Grossman RI, Hackney DB, Goldberg HI, Zimmerman RA, Bilaniuk LT. Variable appearances of subacute intracranial hematomas on high-field spin-echo MR. Am J Roentgenol 1988; 150:171–178.

31. Munro HN, Linder MC. Ferritin: Structure, biosynthesis and role in iron metabolism. Physiol Rev 1978; 58: 317–396.
32. Scott M. Spontaneous intracerebral hematoma caused by cerebral neoplasm: Report of eight verified cases. J Neurosurg 1975; 42: 338–342.
33. Russell DS, Rubinstein LJ. *Pathology of tumours of the nervous system*. Baltimore: Williams & Wilkins, 1989.
34. Mandybur TI. Intracranial hemorrhage caused by metastatic tumours. Neurology 1977; 27: 650–655.
35. Atlas SW, Grossman RI, Gomori JM, et al. Hemorrhagic intracerebral malignant neoplasm: Spin echo MR imaging. Radiology 1987; 164:71–77.
36. Destian S, Sze G, Krol G, Zimmerman RD, Deck MD. MR imaging of hemorrhagic intracranial neoplasms. Am J Roentgenol 1989; 152: 137–144.
37. Tomlinson FH, Houser OW, Scheithauer BW, Sundt TM Jr, Okazaki H, Parisi JE. Angiographically occult vascular malformations: A correlative study of features on magnetic resonance imaging and histological examination. Neurosurgery 1994; 34: 792–799.
38. Gomori JM, Grossman RI, Goldberg HI, Hackney DB, Zimmerman RA, Bilaniuk LT. Occult cerebral vascular malformations: High field MR imaging. Radiology 1986; 158: 707–713.
39. Rigamonti D, Spetzler RF. The assoc.ation of venous and cavernous malformations. Report of four cases and discussion of the pathophysiological, diagnostic and therapeutic implications. Acta Neurochir 1988; 92:100–105.
40. Lee RR, Becher MW, Benson ML, Rigamonti D. Brain capillary telangiectasia: MR imaging appearance and clinicohistopathologic findings. Radiology 1997; 205:797–805.
41. Barr RM, Dillon WP, Wilson CB. Slow-flow vascular malformations of the pons: Capillary telangiectasia? Am J Neuroradiol 1996; 17:71–78.
42. Chan S, Kartha K, Yoon SS, Desmond DW, Hilal SK. Multifocal hypointense cerebral lesions on gradient-echo MR are associated with chronic hypertension. Am J Neuroradiol 1996; 17:1821–1827.
43. Gaensler EH, Dillon WP, Edwards MS, Larson DA, Rosenau W, Wilson CB. Radiation-induced telangiectasia in the brain simulates cryptic vascular malformations at MR imaging. Radiology 1994; 193:629–636.
44. Atlas SW, Braffman BH, LoBrutto R, Elder DE, Herlyn D. Human malignant melanomas with varying degrees of melanin content in nude mice: MR imaging, histopathology, and electron paramagnetic resonance. J Comput Assist Tomogr 1990; 14: 547–554.
45. Isiklar I, Leeds NE, Fuller GN, Kumar AJ. Intracranial metastatic melanoma: Correlation between MR imaging characteristics and melanin content. Am J Roentgenol 1995; 165:1503–1512.
46. Premkumar A, Marincola F, Taubenberger J, Chow C, Venzon D, Schwartzentruber D. Metastatic melanoma: Correlation of MRI characteristics and histopathology. J Magn Reson Imaging 1996; 6: 190–194.
47. Atlas SW, Grossman RI, Gomori JM,et al. MR imaging of intracranial metastatic melanoma. J Comput Assist Tomogr 1987; 11: 577–582.
48. Ginaldi S, Wallace S, Shalen P, Luna M, Handel S. Cranial computed tomography of malignant melanoma. Am J Neuroradiol 1980; 1: 531–533.
49. Atlas SW, Lavi E. Intra-axial brain tumors. In *Magnetic resonance imaging of the brain and spine*, SW Atlas, ed. Philadelphia: Lippincott–Raven, 1996; 315–422.

50. Suzuki M, Takashima T, Kadoya M, Ueda T, Arakawa F, Ueda F, Yamashita J, Yamashima T, Nonomura A. Signal intensity of brain metastases onT2-weighted images: Specificity for metastases from colonic cancers. Neurochirurgica 1993; 36: 151–155.
51. Carrier DA, Mawad ME, Kirkpatrick JB, Schmid MF. Metastatic adenocarcinoma to the brain: MR with pathologic correlation. Am J Neuroradiol 1994; 15: 155–159.
52. Sijens PE, van Dijk P, Oudkerk M. Correlation between choline level and Gd-DTPA enhancement in patients with brain metastases of mammary carcinoma. Magn Reson Med 1994; 32:549–555.
53. Poptani H, Gupta RK, Roy R, Pandey R, Jain VK, Chhabra DK. Characterization of intracranial mass lesions with in vivo proton MR spectroscopy. Am J Neuroradiol 1995; 16:1593–1603.
54. Sijens PE, Levendag PC, Vecht CJ, van Dijk P, Oudkerk M. 1H MR spectroscopy detection of lipids and lactate in metastatic brain tumors. NMR Biomed 1996; 9:65–71.
55. Elster AD, Challa VR, Gilbert TH, Richardson DN, Contento JC. Meningiomas: MR and histopathologic features. Radiology 1989; 170:857–862.
56. Hochberg FH, Miller DC. Primary central nervous system lymphoma. J Neurosurg 1988; 68: 835–853.
57. Schwaighofer BW, Hesselink JR, Press GA, Wolf RL, Healy ME, Berthoty DP. Primary intracranial CNS lymphoma: MR manifestations. Am J Neuroradiol 1989; 10:725–729.
58. Jack CR Jr, O'Neill BP, Banks PM, Reese DF. Central nervous system lymphoma: Histologic type and CT appearance. Radiology 1988; 168: 211–215.
59. Dina TS. Primary central nervous system lymphoma versus toxoplasmosis in AIDS. Radiology 1991; 179:823–828.
60. Johnson BA, Fram EK, Johnson PC, Jacobowitz R. The variable MR appearance of primary lymphoma of the central nervous system: Comparison with histopathologic features. Am J Neuroradiol 1997; 18:563–572.
61. Koeller KK, Smirniotopoulos JG, Jones RV. Primary central nervous system lymphoma: Radiologic–pathologic correlation. Radiographics 1997; 17:1497–1526.
62. Ueda F, Takashima T, Suzuki M, Kadoya M, Yamashita J, Kida T. MR imaging of primary intracranial malignant lymphoma. Radiat Med 1995; 13:51-57.
63. Wilms G, Marchal G, Van Hecke P, Plets C, Lammens M, Goffin J, Baert AL. Colloid cysts of the third ventricle: MR findings. J Comput Assist Tomogr 1990; 14:527–531.
64. Lotz J, Hewlett R, Alheit B, Bowen R. Neurocysticercosis: Correlative pathomorphology and MR imaging. Neuroradiology 1988; 30:35–41.
65. Chang K-H, Lee JH, Han MH, Han MC. The role of contrast enhanced MR imaging in the diagnosis of neurocysticercosis. Am J Neuroradiol 1991; 12:509–512.
66. Gupta RK, Kathuria, Pradhan S. Magnetization transfer MR imaging in CNS tuberculosis. Am J Neuroradiol 1999; 20: 867–875.
67. Jinkins JR, Gupta R, Chang KH, Carbajal JR. MR imaging of the central nervous system tuberculosis. Radiol Clin North Am 1995; 33:771–786.
68. Gupta RK, Roy R, Dev R, et al. Finger printing in patients with intracranial tuberculomas by using in vivo, ex vivo, and in vitro magnetic resonance spectroscopy. Magn Reson Med 1996; 36: 829–833.
69. Brismar J, Hugosson C, Larsson SG, Lundstedt C, Nyman R. Imaging of tuberculosis. III. Tuberculosis as a mimicker of brain tumor. Acta Radiol 1996; 37: 496–505.
70. Kim TK, Chang KH, Kim CJ, Goo JM, Kook MC, Han MH. Intracranial tuberculoma: Comparison of MR with pathologic findings. Am J Neuroradiol 1995; 16: 1903–1908.

71. Post MJD, Chan JC, Hensley GT, Hoffman TA, Moskowitz LB, Lippman S. Toxoplasma encephalitis in Haitian adults with acquired immunodeficiency syndrome: A clinical–pathologic–CT correlation. Am J Roentgenol 1983; 140: 861–868.
72. Levy RM, Bredesen DE, Rosenblum ML. Neurological manifestations of the acquired immunodeficiency syndrome (AIDS): Experience at UCSF and review of literature. J Neurosurg 1985; 62:475–495.
73. Porter SB, Sande MA. Toxoplasma in the central nervous system in the acquired immunodeficiency syndrome. N Engl J Med 1993; 2:1643–1648.
74. Sze G, Brandt-Zawadski MN, Norman D, Newton HT. The neuroradiology of AIDS. Semin Roentgenol 1987; 22:42–53.
75. Post MJD, Sheldon JJ, Hensley GT, Solia K, Tobias JA, Chan JC, Quencer RM, Moskowitz LB. Central venous system disease in acquired immunodeficiency syndrome: Prospective correlation using CT, MR imaging and pathologic studies. Radiology 1986; 158:141–148.
76. Brightbill TC, Post MJ, Hensley GT, Ruiz A. MR of toxoplasma encephalitis: Signal characteristics on T2-weighted images and pathologic correlation. J Comput Assist Tomogr 1996; 20:417–422.
77. Post MJD, Kursunoglu SJ, Hensley GT, Chan JC, Moskowitz LB, Hoffman TA. Cranial CT in acquired immunodeficiency syndrome: Spectrum of diseases and optional contrast enhancement technique. Am J Neuroradiol 1985; 6: 743–754.
78. Chinn RJS, Wilkinson ID, Hall-Craggs MA, Paley MN, Miller RF, Kendall BE, Newman SP, Harrison MJG. Toxoplasmosis and primary central nervous system lymphoma in HIV infection: Diagnosis with MR spectroscopy. Radiology 1995; 197: 649–654.
79. Chang L, Miller BL, McBride D, Cornford M, Oropilla G, Buchthal S, Chiang F, Aronow H, Beck CK, Ernst T. Brain lesions in patients with AIDS: H-1 MR spectroscopy. Radiology 1995; 197:525–531.
80. Ernst TM, Chang L, Witt MD, Aronow HA, Cornford ME, Walot I, Goldberg MA. Cerebral toxoplasmosis and lymphoma in AIDS: Perfusion MR imaging experience in 13 patients. Radiology 1998; 208:663–669.
81. Becker LE. Infections of the developing brain. Am J Neuroradiol 1992; 13:537–549.
82. Sze G, Zimmerman RD. The magnetic resonance imaging of infections and inflammatory diseases. Radiol Clin North Am 1988; 26:839–859.
83. Shapiro MD, Som PM. MRI of the paranasal sinuses and nasal cavity. Radiol Clin North Am 1988; 27:447–475.
84. Lyons RW, Andriole VT. Fungal infections of the CNS. Neurol Clin 1986; 159–170.
85. Kurita H, Shiokawa Y, Fumya K, Segawa H, Sano K. Parasellar aspergillus granuloma extending from the sphenoid sinus: Report of two cases. Surg Neurol 1995; 44: 489–494.
86. Kawakami N, Nishizaki T, Sugiyama S, Ito H. Aspergillus brain abscess in a patient with normal immunity—case report. Neurol Med Chir (Tokyo) 1994; 34:237–40.
87. Zinreich SJ, Kennedy DW, Malat J, Curtin HD, Epstein JI, Huff LC, Kumar AJ, Johns ME, Rosenbaum AE. Fungal sinusitis: Diagnosis with CT and MR imaging. Radiology 1988; 169:439–444.
88. Dion FM, Venger BH, Landon G, Handel SF. Thalamic histoplasma: CT and MR imaging. J Comput Assist Tomogr 1987; 11:193–195.
89. Zheng L, Okabe S, Hino K, Kohno T, Kamata K. Intracranial fungal granuloma with CSF space dissemination: A case report. No Shinkei Geka 1996; 24: 389–392.

90. Magalhaes AC, Bacheschi LA, Caramelli P, Lo LS, de Menezes Neto JR, Shikanai-Yasuda MA, Magalhaes A. Paracoccidioidomycosis of the central nervous system: Study of 5 cases by magnetic resonance. Rev Hosp Clin Fac Med Sao Paulo 1993; 48: 94–97 (in Portugese; English abstract).
91. Haimes AB, Zimmerman RD, Morgello S, Weingarten K, Becker RD, Jennis R, Deck MD. MR imaging of brain abscesses. Am J Roentgenol 1989; 152: 1073–1085.
92. Di Lorenzo GA, Pagano MA, Taratuto AL, Garau ML, Meli FJ, Pomsztein MD. Chagasic granulomatous encephalitis in immunosuppressed patients. Computed tomography and magnetic resonance imaging findings. J Neuroimaging 1996; 6: 94–97.
93. Tien RD, Felsberg GJ, MacFall J. Practical choices of fast spin echo pulse sequence parameters: Clinically useful proton density and T2-weighted contrasts. Neuroradiology 1992; 35:38–41.
94. Incesu L, Gunes M, Akan H, Selcuk MB. MRI of the intracerebral lesions at 0.5 tesla: Comparison of fast spin echo and conventional spin-echo sequences. Comput Med Imaging Graph 1996; 20:105–113.
95. Sugahara T, Korogi Y, Hirai T, Hamatake S, Ikushima I, Shigematu Y, Takahashi M. Comparison of HASTE and segmented-HASTE sequences with a T2-weighted fast spin-echo sequence in the screening evaluation of the brain. Am J Roentgenol 1997; 169:1401–1410.
96. Knopp EA, Cha S, Johnson G, Mazumdar A, Golfoni JG, Zagzag D, Miller DC, Kelly PJ, Kricheff II. Glial neoplasms: Dynamic contrast enhanced T2*-weighted MR imaging. Radiology 1999; 211:791–798.

INDEX

GPSR Compliance
The European Union's (EU) General Product Safety Regulation (GPSR) is a set of rules that requires consumer products to be safe and our obligations to ensure this.

If you have any concerns about our products, you can contact us on

ProductSafety@springernature.com

In case Publisher is established outside the EU, the EU authorized representative is:

Springer Nature Customer Service Center GmbH
Europaplatz 3
69115 Heidelberg, Germany

www.ingramcontent.com/pod-product-compliance
Ingram Content Group UK Ltd.
Pitfield, Milton Keynes, MK11 3LW, UK
UKHW051828230926
475607UK00012B/118
* 9 7 8 0 3 0 6 4 6 5 5 1 2 *